S0-AWE-981

The Words of Mathematics

An Etymological Dictionary of Mathematical Terms Used in English

© 1994 by

The Mathematical Association of America (Incorporated)

Library of Congress Catalog Card Number 93-80612

ISBN 0-88385-511-9

Printed in the United States of America

Current printing (last digit):

10 9 8 7 6 5 4 3 2 1

The Words of Mathematics

An Etymological Dictionary of Mathematical Terms Used in English

A reference book describing the origins
of over 1500 mathematical terms used
in English, including a glossary that
explains the historical and linguistic
terms used in the book.

Steven Schwartzman

The Mathematical Association of America

QA
5
.S375
1994

SPECTRUM SERIES

The Spectrum Series of the Mathematical Association of America was so named to reflect its purpose: to publish a broad range of books including biographies, accessible expositions of old or new mathematical ideas, reprints and revisions of excellent out-of-print books, popular works, and other monographs of high interest that will appeal to a broad range of readers, including students and teachers of mathematics, mathematical amateurs, and researchers.

Committee on Publications
JAMES W. DANIEL, *Chairman*

Spectrum Editorial Board
ROGER HORN, *Chairman*

BART BRADEN	RICHARD GUY
UNDERWOOD DUDLEY	JEANNE LADUKE
HUGH M. EDGAR	LESTER H. LANGE
BONNIE GOLD	MARY PARKER

Complex Numbers and Geometry, by Liang-shin Hahn
Cryptology, by Albrecht Beutelspacher
From Zero to Infinity, by Constance Reid
I Want to be a Mathematician, by Paul R. Halmos
Journey into Geometries, by Marta Sved
The Last Problem, by E. T. Bell (revised and updated by Underwood Dudley)
Lure of the Integers, by Joe Roberts
Mathematical Carnival, by Martin Gardner
Mathematical Circus, by Martin Gardner
Mathematical Cranks, by Underwood Dudley
Mathematical Magic Show, by Martin Gardner
Mathematics: Queen and Servant of Science, by E. T. Bell
Memorabilia Mathematica, by R. E. Moritz
Numerical Methods that Work, by Forman Acton
Out of the Mouths of Mathematicians, by Rosemary Schmalz
Polyominoes, by George Martin
The Search for E. T. Bell, also known as John Taine, by Constance Reid
Shaping Space, edited by Marjorie Senechal and George Fleck
Student Research Projects in Calculus, by Marcus Cohen, Edward D. Gaughan, Arthur
    Knoebel, Douglas S. Kurtz, and David Pengelley
The Words of Mathematics, by Steven Schwartzman

Mathematical Association of America
1529 Eighteenth Street, NW
Washington, DC 20036

800-331-1MAA   FAX 202-265-2384

√59

This book is dedicated to the memory and the continuing influence of my brother-in-law, Alain Levy, who loved science in both the familiar and the etymological senses of the word, and who was well described as a prince.

Contents

Introduction

In recent years I've become very fond of asking my students why the White House is called the White House. When the question first comes up the students are usually dumbfounded. They wonder if I could really be asking them that question in a mathematics class—or any other class, for that matter—and they try to figure out what I might be driving at. More interestingly, though, they seldom know how to answer the question even though the answer is trivial. The White House is called the White House for two reasons: because it's white and because it's a house.

I started asking my trivial and seemingly irrelevant question because I noticed that most students are not good at using mathematical terminology. Many of them haven't realized that technical terms aren't just arbitrary syllables designed to make their lives more difficult. The point I try to make with my White House analogy is that most mathematical terms actually *describe* the things they refer to. The difficulty is that the descriptions are usually in Latin or Greek rather than English, and few students nowadays have been exposed to those ancient languages.

The study of the origins of words is known as etymology: this book is an etymological guide to the most common mathematical terms that occur in the elementary, secondary, and college curricula. Armed with this guide, students may find mathematics a little more understandable. Their non-technical English vocabulary should also improve because the same roots found in technical terms occur in many other words as well, some of which will be pointed out in this book. At a time when many students' English skills are very weak, it is important to stress English even in classes like mathematics and science that no longer focus on language as much as they once did.

Historically speaking, there used to be less of a separation between mathematics and language than there is now. As a teenager the great German mathematician Carl Friedrich Gauss was still having trouble deciding whether to devote his life to mathematics or linguistics. His discovery of a method for constructing a regular heptadecagon (a seventeen-sided polygon) tilted the balance in favor of mathematics, though his knowledge of languages continued to serve him well as a mathematician. His most famous book, *Disquisitiones Arithmeticæ*, was written in Latin, as were many other mathematicians' books. Even as recently as the first decades of the 20th century virtually all college students were required to study Latin, and usually Greek as well; at the end of the school year the valedictory address was typically given in Latin. Times have changed. Because knowledge of Latin and Greek is no longer common, teachers can use the explanations in this guide to supplement their explanations of technical terms; they can, for example, ask students to find other English words that seem to be related to a given technical term.

At some point teachers should also point out that if "traditional" mathematics had been discovered in America, mathematical terminology would be primarily in English. Compare modern physics, much of which has been created in America; one of the "flavors" of quark is *charmed*, which was

1

previously a non-technical English word. Similarly, the relatively recent branch of mathematics that deals with knots classifies knots as *wild* or *tame*, and in game theory a *duel* may be *noisy* or *silent*. Recent number theory deals with *weird, untouchable, hailstone* and *lucky* numbers. The English word *software* is recognized by many computer users all over the world, even though they may not understand the components *soft* and *ware* or the implied contrast with *hardware*.

Although English may seem quite unrelated to Latin and Greek (except for words that English borrowed directly from those languages), all three languages are actually descended from a common ancestor that we usually call Indo-European. Surprisingly, almost all the languages of Europe, plus Persian and many of the languages of India, belong to that same family. That is why the original language, whose speakers even in ancient times spread out as far east as India and as far west as western Europe, is called Indo-European. Some modern Indo-European languages are Russian, German, Gaelic, Spanish, Yiddish, Catalan (spoken in Spain), Serbo-Croatian, Romansch (spoken in Switzerland), Hindi, Pashto (spoken in Afghanistan), Polish, Afrikaans (spoken in South Africa), Welsh, Romany (the language of the Gypsies), Lithuanian, and Urdu (spoken in Pakistan). In fact, roughly half the people alive in the world today speak an Indo-European language.

With the passage of time the various Indo-European languages have diverged phonetically and semantically from the original source to the point where almost none of them remain mutually intelligible; the ones that are mutually intelligible, like Swedish and Norwegian, have diverged from each other only relatively recently. Even in rather different languages, however, the relationship between certain words is still recognizable even after thousands of years. Compare Russian *mat'* with Spanish *madre* and English *mother*. Not only do they sound similar, but their meanings are identical. None of the three words was borrowed from any of the others; all three are directly descended from a common ancestral word. Such related descendants are called cognates, from Latin *co-* "together" and *gnatus* "born." Cognates may be likened to sisters or cousins, who also come from a common ancestor but don't come from each other. Although much of our mathematical vocabulary in English has been borrowed from Latin and Greek, many of those technical borrowings also have native English cognates, some of which will be pointed out in the following pages.

This guide can hardly go into Indo-European linguistics in much depth. Still, it might be helpful to look at a few groups of recognizable cognates. By comparing the members of such groups, linguists have set up a series of sound correspondences among the Indo-European languages. Consonants are generally more stable than vowels, and among the consonants some are more stable than others. Furthermore, a given sound often developed differently in different phonetic environments and under different types of stress.

The table below shows a few examples of typical Indo-European consonant correspondences. Of course the given correspondences apply only to native words, not to words that were borrowed

Original root	Original Meaning	Native English	Greek root	Borrowing from Greek	Latin root	Borrowing from Latin
ped-	foot	foot	pod-	podiatrist	ped-	pedestal
ten-	stretch	thin	tono-	tone	ten-	tenuous
kerd-	heart	heart	kard-	cardiac	cord-	cordial
bheu-	be, exist	be	phu-	phylum	fu-	future
dhe-	put, set	do	the-	thesis	fac-	fact
wed-	water	water	hudor	hydrant	und-	undulate
gno-	know	know	gnosko-	agnostic	gnoro-	ignorant

directly from a foreign language in relatively recent times. An English word borrowed from Latin or Greek will have the Latin or Greek consonants rather than the corresponding native English ones. For example, *paternal* is clearly borrowed from Latin because it retains the *p* and *t* of Latin *pater* "father," whereas, *fatherly*, with its *f* and *th*, is native English.

Just as someone listening to languages as dissimilar as Icelandic and Hindi would be surprised to learn that they are actually related, so a user of this dictionary will almost certainly be struck from time to time by the unexpected relationships among words that initially seem to have nothing in common. For example, although the terms *parallel* and *alternate* occur together in the study of the alternate interior angles that result when parallel lines are cut by a transversal, who would recognize that both words share a root that means "other"? The evidence for that relationship can be found in cognates such as *alias, alien, alibi,* and *alter ego*.

From another point of view, however, an unexpected linguistic correspondence like the one between *parallel* and *alternate*, or the even less apparent ones between *ecological* and *sandwich* or *obtuse* and *piercing* might *not* surprise someone who has delved into the world of mathematics, where the finding of formerly unknown connections is one of the greatest thrills a researcher can have. After all, who would expect to see any relationship between the trigonometric functions and the natural exponential function? The trigonometric functions arise from relationships among sides of right triangles or chords of a circle, while exponential functions arise from taking powers of a strange number whose value is roughly 2.718; and yet one of the great discoveries of the 18th century was that $e^{ix} = \cos x + i \sin x$, an equation that inextricably links trigonometry to algebra. I hope that readers or even casual consulters of this book will be fascinated by some of the surprising linguistic connections that lie in wait in the paragraphs and pictures that follow.

The origins of English mathematical terms

The mathematical words that we use in English come from many sources and have assumed their current forms as a result of various processes, either alone or in combination. Some English words used in mathematics are of native origin. Many of those are descended directly from Indo-European. A native word like *five* is what we might call monomorphemic: it contains a single morpheme, or unit of meaning, as did its Indo-European predecessor of the same meaning, *penk^w e-*. The changes from the ancient form to the modern one have been strictly phonetic, but there have been enough of them to make *five* look quite different from *penk^w e-*.

Although a native English word like *twelve* contains only one syllable, it incorporates two morphemes, or units of meaning. The first part of the word is related to *two*, and the second part to an Indo-European root meaning "leave." The concept underlying *twelve* is "the number that leaves 2 behind when 10, the base in which we do our calculating, is subtracted from it." That's a lot of meaning to be packed into a single syllable. Here again sound changes have taken an assumed original *twa-leof*, whose components were still "transparent" to speakers of Old English, and turned it into what linguists call an "opaque" form. The original components are no longer obvious, although modern speakers might still recognize that *twelve* is related to *two* because of the initial *tw-*.

Sometimes English words that are not exclusively or specifically mathematical find their way into the mathematics classroom. Examples are *down, greater, left, or, slope, length, top, fit, least,* and *smooth.* Because of their frequent occurrence and/or technical definition in the world of mathematics, such words have been included in this dictionary. Some of them have uninteresting origins. Others such as *down* have surprising histories: etymologicaly speaking, *down* has to do with hills. Also part of the English vocabulary of mathematics are certain nonmathematical words that few people would suspect are used with a mathematical meaning. Examples are *curious, kissing, happy, pigeonhole, sociable, sandwich, weird,* and *hailstone.*

Intermingled with the native English vocabulary are many mathematical words that have been taken directly from classical Greek and Latin. Terms that have been borrowed with no change include *vertex, maximum, radius, directrix, modus ponens, criterion, axis, reductio ad absurdum, lituus,* and *latus rectum.* Nouns in this group often retain their Latin or Greek plurals, so we speak of *vertices, maxima, radii, directrices, criteria, axes, litui,* and *latera recta.* Occasionally borrowed nouns lose their original plurals and acquire anglicized plurals. With a word like *index* we vacillate and say either *indices* or *indexes* for the plural. With *formula* and *parabola* we almost always say *formulas* and *parabolas*; only in writing, and even then only rarely, might we find *formulæ* and *parabolæ.*

In addition to outright borrowings, we use many terms that are only slightly altered from words that the ancient Greeks and Romans used. Our word *intercept* is almost identical to Latin *interceptus* "taken between, interrupted." *Oblique* is little changed in form or meaning from Latin *obliquus* "bent aside, twisted awry." Sometimes an existing Greek or Latin word is only slightly modified in appearance but used in a new, often metaphorical way. The name of the two-looped curve now known as a *lemniscate* is easily recognizable in Greek *lemniskos* "a ribbon."

Probably most often of all English has followed other European languages in using borrowed Greek and Latin roots as elements of new compounds that name things or concepts unknown in the ancient world. For hundreds of years mathematicians in France, the Netherlands, Germany, Italy, England, and other countries have invented such compounds. Many people felt that the classical languages were somehow "nobler" than modern languages. Inventors of new terms also chose Greek and Latin roots because words based on them had a good chance of being understood and accepted by mathematicians who normally spoke different languages. For that reason an American mathematician with an elementary knowledge of French or Spanish can usually understand a mathematics book published in Paris or Mexico City.

By resorting to classical roots mathematicians have come up with some pretty fancy terms. From Latin we have words like *escribed, superadditive, sexagesimal, interquartile, semimean, contragredient, commensurable,* and *equitangential.* From Greek we have words like *leptokurtic, brachistochrone, homoscedastic, isomorphism, paradromic,* and a tongue-twister like *rhombicosidodecahedron.*

Sometimes mathematicians use both a common English word (whether native or borrowed) and a fancy Greek- or Latin-derived term of similar meaning. Thus we have the *kissing* number of equal circles that can touch a given circle, as well as the *osculating* circle drawn to a curve. We speak of *untouchable* numbers as well as *tangents* and *tac-nodes.* Statistics deals with *scatter* diagrams and *homoscedastic* distributions. A curve may happen to *fall* as it approaches its *asymptote*, and two congruent figures can be made to *coincide*; all three italicized words embody the notion of falling.

In geometry, curves have frequently been named after objects that they resemble, even if the resemblance is sometimes rather fanciful. The words for those objects are usually taken from a classical language. Examples of curves (and the objects they purportedly resemble) are the *clothoid* (spinning wheel), *cochlioid* and *limaçon* (snail), *nephroid* (kidney), *bicorn* (two horns), *cardioid* (heart), and *lituus* (augur's staff). One word that doesn't resort to a foreign root is native English *kite*, named after the flying toy, which was named for its resemblance to the tail of the bird of the same name. (The name of the bird is apparently due to the sound it makes, not to its shape.)

In general, when two or more roots are put together to form a compound they come from the same language. The word *asymptote*, for instance, comes from three Greek elements: *a-* "not," *sum-* "together," and *pte-* "to fall." The commonly found suffix *-oid* "looking like," is of Greek origin, and it regularly attaches to Greek rather than Latin roots, yielding compounds like *epicycloid, trapezoid,* and *trochoid.* We have Latin-Latin *unicursal* alongside Greek-Greek *monodromy*; both words involve the notion of running.

On rare occasions Greek and Latin elements are mixed in a single word: that often happens when a foreign element becomes so much a part of English that it takes on a life of its own and loses its "foreignness." Examples are the Greek prefixes *hyper-* and *pseudo-* and the Greek suffix *-oid*. We have *hyperspace* rather than the etymologically "pure" *superspace*. We have *pseudoprime*, based on Latin-derived *prime*. In *ovoid* and *toroid* the Greek-derived suffix *-oid* has been attached to Latin roots, as it has been in a nontechnical term like *humanoid*. Similarly, Latin-Greek *concyclic* is patterned after all-Latin *collinear* and *coplanar*. In a case like Latin-Greek *supersphere*, languages had to be mixed because the Greek-Greek *hypersphere* was already in use with a different meaning. Greek-Latin *hexadecimal* takes the place of all-Latin *sexadecimal*; *hexadecimal* is frequently abbreviated *hex*, but *sexadecimal* can hardly be shortened to *sex*. We have the linguistically pure Latin-Latin *uniform*, but instead of *unidigit* we have the Greek-Latin hybrid *monodigit*.

Looking farther back than either Greek or Latin, sometimes the same set of Indo-European roots appears in different mathematical terms. Etymologically speaking, *profit* and *perfect* are the same word. So are *cost* and *consistent* and *constant*; *consecutive* and *consequent*; *compute* and *count*; *common* and *commute*; *hypotenuse* and *subtend*; *null* and *none*; *sufficient* and *subfactorial*; *complete* and *complement*; *contain* and *continuous*; *quatrefoil* and *quadrifolium*.

Because mathematicians enjoy playing with numbers and patterns, it isn't surprising that they also often enjoy playing with words. Some new vocabulary has been generated by starting with an existing term and either altering a part of it or adding to it. The term *factor* led to *factorial*, which in turn led to *primorial* and *oddorial*. Joining the two-dimensional *polygon* and the three-dimensional *polyhedron* is the higher-dimensioned *polytope*. Joining the two-dimensional *rectangle* and the three-dimensional *rectangular solid* is the higher-dimensioned *orthotope*.

During the Renaissance, Latin *mille* "a thousand," was augmented to *million*. Later on, the *m-* of *million*, though it means nothing by itself, was replaced by Latin number roots; the result was the sequence *billion*, *trillion*, *quadrillion*, etc., and even the non-mathematical (because non-specific) *zillion*. More recently the *do-* of *domino*—which once again is meaningless by itself—was replaced by Greek number roots; the result is the series *triomino*, *tetromino*, *pentomino*, and other *polyominoes*. Similarly *diamond* has as its progeny *triamond*, *tetriamond*, and succeeding *polyiamonds*.

On rare occasions mathematicians have broken with analogy by choosing not to continue with a sequence. The existence of the two-dimensional *quadrant* and the three-dimensional *octant* might seem to imply a four-dimensional *hexadecimant*, but that word and the ones for even higher dimensions are too unwieldy. Instead, the word *orthant* is used for the analogical terms in four or more dimensions.

The English mathematician Charles Lutwidge Dodgson (1832–1898), better known under the pseudonym Lewis Carroll, invented the term *portmanteau word*. The French word *portmanteau* means literally a "coat carrier." It is a piece of luggage consisting of two halves that fold together to make a single piece. "Folding" two words into one, the playful mathematician came up with such terms as *chortle* (*chuckle* + *snort*) and *slythy* (*slimy* + *lithe*). The mathematical (and plain English) word *diminish* is a portmanteau word composed of the related but now obsolete verbs *diminue* and *minish*. Some other portmanteau words are *alphametic* (*alphabet* + *arithmetic*), *tranjugate* (*transpose* + *conjugate*), *bit* (*binary* + *digit*), *cryptarithm* (*cryptogram* + *arithmetic*), and *osculinflection* (*osculation* + *inflection*). The polyominoes and polyiamonds mentioned two paragraphs ago are at least "demi-portmanteaux," as are *repunit* (*repeated* + *unit*), *digraph* (*directed* + *graph*), *steradian* (*stereo* + *radian*), *symmedian* (*symmetric* + *median*), *relagraph* (*relation* + *graph*), and *repdigit* (*repeated* + *digit*). The cleverest such word may be *reptile* (*repeated* + *tile*), whose apparent relationship to alligators and lizards lends it extra charm.

One infrequent but appropriate technique used in inventing new terms is the reversal of the letters of an existing term. Whereas the *slope* of a straight line is defined as change in *y* divided by change in *x*, the *epols* is defined by dividing in the opposite order. In calculus, *atled* is the word *delta* spelled backwards. The *atled* operator is symbolized by an equilateral triangle with a vertex pointing down; that symbol is the reverse of the more familiar *delta* symbol, an equilateral triangle with a vertex pointing up.

The *atled* example shows that mathematicians sometimes name things for the symbol used to represent them rather than for any inherent property of the thing being named. The English mathematician William Rowan Hamilton referred to the same operator as the nabla operator because the ∇ symbol reminded him of an ancient stringed instrument known as a nabla. In the realm of vectors, the dot product and cross product are named for the "·" and the "×" that represent them.

Beyond the ways in which mathematical words are created and put together, it is also interesting to look at semantics, the part of linguistics that has to do with meaning. As expected, many of the words of mathematics deal with numbers, quantities, measurements, shapes, and operations. Probably less expected is the large number of mathematical terms that have been drawn from biology. The natural worlds of people and animals have contributed terms like *arm, braces, cardioid, cornicular, cochleoid, conchoid, face, family, foot, hermit, hippopede, keratoid, kissing, leg, limaçon, matrix, narcissistic, nephroid, nested, osculation, oval, pearl, pigeonhole, pons asinorum, ramphoid, serpentine, tail, tera-*, and *umbilic*. The botanical world has contributed terms like *arboricity, branch, cactus, canonical, cissoid, épi, folium, forest, hull, kernel, leaf, petal, planted, quatrefoil, radical, root, rose, stem*, and *tree*. The biological origins of some of those terms are obvious, but others will almost certainly need to be looked up in the main part of this dictionary if you want to see the connection to the natural world. Understanding how some of those biological terms are used mathematically will probably also require a trip to the corresponding dictionary entries.

How this book is organized

Entries are arranged alphabetically in a single list. Immediately following each entry is its grammatical part of speech. If a noun has an irregular plural, it is also noted, as in "*directrix,* plural *directrices.*" Different forms of a word are usually grouped together, since the forms share a common linguistic explanation. An example is the entry containing *distribute, distribution*, and *distributive*. Occasionally a related form of a word is listed apart from the main entry; that usually happens if there is something to be said about that form that doesn't apply to the group as a whole. An example is *infinitesimal*, which is clearly from the same source as *infinite*, but which appears separately in order to accomodate a discussion of the suffix *-esimal*.

After each word and its part of speech comes its source or sources. Most of the words in this book come from Greek or Latin. Occasionally a reference is made to Vulgar Latin, the form of Latin that was spoken in the latter days of the Roman Empire and that eventually developed into the Romance languages. Sometimes words are traced back to Late or Medieval Latin; during the Middle Ages and the Renaissance people coined Latin words that never existed during the classical period but which were needed to deal with expanding knowledge and recent discoveries. Some of the words in this dictionary, of course, are native English. Still other words are of common Germanic stock; they come from the Indo-European subgroup that includes Scandinavian, Dutch, and English, as well as the specific language called German.

In most cases, the Indo-European root of a word is given, assuming that there is one and that it is known. I have generally used the form given in Calvert Watkins' *The American Heritage Dictionary of Indo-European Roots*, a book I highly recommend to those people who want to know more about Indo-European. (Many of the entries from that book appear in a supplement to the 3rd edition of the *American Heritage Dictionary of the English Language*.) The Indo-European forms listed there are reconstructions of the original roots based on recent information. Those forms are subject to change as more research is done, though some connections may never be discovered because, the farther back in time we go, the less evidence we have to draw conclusions from. In any case, this guide focuses on the connections among words rather than on the phonetic details of the original language. If a connection isn't certain, a qualifier like *plausible* or *possible* or *may* is used.

Just as English has similar-sounding or even identically spelled words like *left* (the past of *leave*) and *left* (the opposite of *right*) that are historically unrelated, so, presumably, did Indo-European. For example, the word *cycle* evolved from an Indo-European root $k^w el$- "to revolve," while *tele-* in *telescope* evolved from the different Indo-European root $k^w el$- "far." Although the *American Heritage Dictionary* distinguishes between those roots by designating them $k^w el$-^{1} and $k^w el$-^{2}, I haven't done so in this book; nor have I distinguished long vowels from short vowels, because that phonetic distinction won't interest most users of this book.

As stated in the introduction, the main purpose of this dictionary is to make students and teachers aware of the origins of mathematical terms, and a secondary purpose is to make them aware of the origins of and relationships among the words of English in general. To that end, after the source of a mathematical term is given, related non-mathematical words also often appear. For example, the entry for *paradox* mentions *orthodox* and *dogma*. On the other hand, since this *is* a book about mathematical terms, linguistic information is often followed by mathematical information. That information may be the name of the mathematician who first used a term, as in Leibniz for *abscissa*; the symbol used to represent a word, as in ! for *factorial*; an equation corresponding to a curve, as in $r = a + b\sin\theta$ for the *limaçon*; a graph of a curve; a chart showing the general distributive property; etc.

Many entries end with one or more numbers enclosed in brackets. The bracketed numbers refer to the corresponding numbered items in the book's appendix, where etymologically related words are grouped together. For example, the dictionary's discussion of *trisectrix* ends with the numbers [235, 185, 236]. Entry [235] in the appendix lists mathematical terms derived from the Indo-European root *trei-* "three." Entry [185] lists mathematical terms derived from the Indo-European root *sek-* "to cut." Entry [236] in the appendix lists mathematical terms like *directrix* and *separatrix* that end with the Latin suffix *-trix* "a female thing or person that performs the indicated function."

The etymological relationship among items grouped together in the appendix may be fairly obvious, as with *fractal*, *fractile*, and *fraction*, or *bisect*, *intersect* and *section*. It may also be obscure, as with *obtuse*, *piercing*, and *type*, or with *eleven*, *inch*, and *union*. Sound changes affecting these groups have been so drastic over time and distance that only a specialist can recognize the original relationship. I hope that you will be skeptical enough and curious enough about some of the seemingly unrelated words grouped together in the appendix to read the individual entries and find out how a common root developed into such different words. To my way of thinking, following up these cross-references often proves to be the most interesting part of the book.

Of the known Indo-European roots, many don't occur in this dictionary at all. Some, like *bhel-* "to cry out," have given rise to just a single English word used in mathemathics (in this case *bell*), and therefore do not appear in the appendix of cross-referenced Indo-European roots. Still others have given rise to many mathematical terms. The two most prominent among them are [103] *kom-* and [165] *per-*, whose progeny permeates English in general and mathematics in particular. (In fact the words *prominent*, *progeny*, and *permeate* all contain the root *per-*.)

When I started this book, I included technical terms like *asymptote*, *spinode*, and *modus ponens* that were obvious candidates for an explanation. The terms were drawn from elementary algebra, geometry, analysis, trigonometry, abstract algebra, topology, logic, number theory, and statistics. I also decided to include mathematical elementary-school vocabulary like *denominator* and *minuend* as well as common numerical units like *year*, *mile*, and *ounce*. As the project grew, I added some of the most common terms from recreational mathematics. I also included a few of the most common terms from physics, especially words like *mass* and *work* that occur in most calculus textbooks. I later broadened the scope of the dictionary to include normal English words that are used in mathematics; examples include *large, carry, center,* and *saddle,* some of which turned out to have interesting origins and to be connected to some of the more technical terms already in the dictionary. For example, *saddle* is etymologically related to *polyhedron, residue,* and *nested.* I also added entries for common number words like *five* and *half,* as well as for "hidden" number words like *any, dime, noon,* and *September.*

Although mathematicians are fond of studying infinity, every dictionary must be finite. There are probably words that I have overlooked. One category that I have systematically omitted is proper nouns derived from names of people. So, although the Fibonacci numbers are often discussed in mathematics, *Fibonacci* doesn't appear as an entry in this dictionary. Only in rare cases like *abelian* and *algorithm,* where a person's name has begun to function as part of a noncapitalized word, have I included the word. Compounds have usually been left out if the basic word has an entry of its own and the meaning of the compound is obvious: *consistent* appears in this dictionary, but *inconsistent* does not. Although computers and calculators are increasingly pervasive in mathematics, I ultimately decided to include only a few computer terms that happen to be derived from number words. Absent are words like *compiler, mouse, processor,* and *modem.*

End of the beginning

There have been mathematical dictionaries before. Although they defined mathematical terms, they rarely explained where any of the terms came from; when they did explain them, the explanations were usually very brief. There have also been etymological dictionaries before. Although they have included the most common mathematical terms found in the general vocabulary of English, they haven't elaborated much on those words; what's more, general etymological dictionaries have omitted many technical mathematical terms. This is the first time that a mathematical dictionary and an etymological dictionary have been so closely and extensively intertwined. (The word *intertwined,* by the way, is a compound of *two,* because two things are being twisted [another cognate] together.)

Whether you are an elementary, secondary, or college teacher, a student, a mathematician, or a lover of words, there is at least some material in this book that will be appropriate for your level of mathematics and language. No matter how much mathematics you have been exposed to, there will almost certainly be some words here that you've never run across before. Until I lived with this project for three years, I had no idea how many mathematical words there are, nor how much energy has gone into their creation. Browse through the entries and I think you'll be amazed at the word wealth you'll find. In any case, let me close with a comment that Abraham Lincoln is supposed to have made about another book: "People who like this sort of thing will find this is the sort of thing they like."

Steven Schwartzman
Austin, Texas
April 1993

Acknowledgments

Many people have contributed to this book, the most obvious being the authors listed in the bibliography. Thanks are also due to the many people who created my Macintosh® computer and the software I used on it to write this book; if I had had to rely on the tools available to me just a decade ago, namely pens, paper, and a typewriter, I don't think I could have completed a project of this magnitude.

I would also like to recognize three people who helped me personally. The first is my father, Jacob Schwartzman, who instilled in me a lifelong love of learning and made me a perpetual reader. The second is my wife, Evangeline Diaz Schwartzman, who kept the house running (and me fed) during the three years that I worked on this project. The third is Helen-Jo Hewitt, who, though not a mathophile, proofread a near-final draft of the book and made many helpful suggestions.

An Invitation

If you are aware of any mathematical terms that don't appear in this dictionary, please send them (and their definitions, if not apparent) to the author so that they can be included in a future edition of the book. At the time of printing, the address is: Steven Schwartzman, Box 4351, Austin, Texas 78765. If at some future time that address is no longer valid, you can write to Steven Schwartzman in care of the Mathematical Association of America.

Explanation of terms and symbols

[ ]: the number in brackets at the end of many dictionary entries corresponds to the similarly numbered item in the appendix, where mathematical terms that come from the same root are listed together in alphabetical order. Not every dictionary entry contains complete etymological or historical information, so looking in the appendix may lead you to a related term that contains extra information.

adjective: a word which describes a noun. Examples are *good*, *difficult*, *happy*, and *pusillanimous*. The word *adjective* is borrowed from Latin and means literally "thrown against," because an adjective is most commonly in direct contact with the noun it modifies.

adverb: a word that modifies a verb, an adjective, or another adverb. The first part of that definition is apparent in the word *adverb* itself, which means "to or for a verb." The most common adverbial ending in English is *-ly*, as in *quickly*, *clearly*, and *decisively*. See more under the dictionary entry for *like*.

agental suffix: when attached to a verb, an agental suffix indicates the person who performs an action. In English the most common agental suffix is *-er*, as in *baker* "a person who bakes." The Latin spelling *-or* is also used, as in *sailor*.

assimilation: a process by which a sound is changed so that it more closely resembles another sound, most often one that immediately follows it. For example, the final *n* of the Latin negative prefix *in*-regularly assimilated to a following *l*, *m*, or *r*, as in our borrowed words *illogical*, *immature*, and *irrational*.

augmentative: a form that in its most literal usage indicates large size. By extension, an augmentative may convey a sense of clumsiness, awkwardness, or inferior quality. English doesn't really have augmentatives, but the Romance languages do. The one that occurs in this dictionary is *-on*, as in *salon* or *saloon*, originally "a large room," and *carton* or *cartoon*, originally "thick cardboard."

Catalan: a Romance language spoken primarily in northeastern Spain, the Balearic Islands in the Mediterranean, and to a lesser extent extreme southeastern France. Because of its position between Spanish and French, Catalan shares some features of each of those languages. The focal point of Catalan culture is the city of Barcelona. The dictator Franco largely banned the public use of Catalan after the Spanish Civil War ended in 1939, but he couldn't prevent people from speaking it privately. Catalan has enjoyed a great public revival since Franco's death in 1975.

causative: a form of a verb that indicates that something is being made to happen. English used to have a well-developed causative system, but only a few causatives survive in the modern language. Examples are *raise* "cause to rise," *set* "cause to sit," *lay* "cause to lie," and *fell* "cause to fall."

Celtic: one of the subfamilies that developed from Indo-European. The Celtic family includes modern Gaelic, Welsh, and Breton, as well as the extinct Gaulish spoken in ancient Gaul. The Celtic family once covered much of Europe, but the surviving Celtic languages are restricted to the British Isles, Ireland, and Brittany.

Classical Latin: the type of Latin written when the Roman Empire was at its height; it is often associated with Roman literature. Many people don't realize that Latin, like any living language, kept developing; Classical Latin was one stage in the development of the language from Old Latin to the modern Romance languages.

cognate: from Latin *co-* "together" and *(g)natus* "born." The term *cognate* refers to two or more words that are independently descended from a common source; for example, Latin *frater* and English *brother* are cognates because both words can be traced back to a common Indo-European ancestor.

comparative: When you compare two things, you match them up to check for equality. Typically one of the things will be wider, heavier, taller, etc. As seen in those examples, the English comparative form of an adjective typically ends in *-er*. In Latin, the comparative often ended in *-ior*.

conjunction: a word which joins (*junct-*) together (*con-*) two parts of a sentence. Some common English conjunctions are *and*, *but*, *as*, *because*, and *or*.

demonstrative: a type of pronoun or adjective that singles out something or someone. English demonstratives include *this*, *that*, *these*, and *those*.

diminutive: a form that in its most literal usage indicates small size. By extension, a diminutive may convey a sense of endearment. By even further extension, a diminutive expresses the notion "derived from," just as children come from parents. Latin diminutives typically had *-ul-* in the ending, as in *calculus* "a little stone." In French, a common derivative ending is *-ette*, as in *rosette* "a small rose." The "derived from" meaning is seen in *roulette*, a curve that is traced out as one thing rolls on another thing. English doesn't often use diminutives of common nouns, but we have many diminutives of proper nouns such as *Johnny* from *John* and *Rosie* from *Rose*.

ə: an unstressed vowel that linguists often call *schwa*. It has the sound of the *a* in the English words *sofa* and *machine*.

Etruscan: an ancient, non-Indo-European language once spoken in that part of the Italian peninsula now known as Tuscany. The Etruscan civilization preceded and influenced the Roman civilization.

extension: sometimes a basic Indo-European root is combined with a second element to create a compound that takes on a life of its own. Such a compound is called an extension. An example is *plek-* "to plait," which is an extension of *pel-* "to fold."

feminine: English nouns like *queen* and *lioness* are feminine because they refer to female people or animals. A noun like *line* in English is neuter because a line is a thing. In Latin and Greek, however, even inanimate objects could be masculine or feminine. Latin *linea* "line" was feminine, for example.

Frankish: the language spoken by the Franks, a Germanic tribe that lived near the Rhine River in the first centuries A.D. The Franks conquered much of the region that is now named for them, France.

Frisian: a Germanic language spoken in Friesland, in the northern part of the Netherlands. Frisian is the closest Germanic relative to English.

Gaulish: an extinct Indo-European language belonging to the Celtic subfamily. It was spoken in ancient Gaul, which is now known as France.

Germanic: one of the subfamilies that developed from Indo-European. The Germanic subfamily includes English, Dutch, the Scandinavian languages, Afrikaans, Yiddish, and the specific language we call German.

Greek: one of the subfamilies that developed from Indo-European. Mathematical terms of Greek origin generally come from Attic Greek, the type that was spoken in ancient Athens. Our English word *Greek* comes from the Latin word *graecus*, from Greek *graikos*, the name used to describe one of the tribes of Epirus. Epirus is a region in the northwestern part of modern Greece and the southern part of modern Albania.

High German: the dialects of German spoken in the highlands. Geographically speaking, the higher parts of Germany are found in the south. The modern standard German language evolved primarily from High German dialects.

i.e.: an abbreviation of the Latin words *id est* "that is."

Indo-European: a name given to the original language from which Latin, Greek, English, and most of the languages of Europe are descended. The homeland of the original Indo-Europeans is still unknown. Their linguistic descendants originally migrated as far east as India and as far west as western Europe, hence the term Indo-European. Largely because of extensive colonization, especially by the English, French, Spanish, and Portuguese, Indo-European languages are now natively spoken by about half of the earth's inhabitants.

Italic: one of the subfamilies that developed from Indo-European. Languages of this family spoken in ancient times in the Italian peninsula included Oscan, Umbrian, and Latin. Of those, only Latin survived, and it developed into the modern Romance languages.

infinitive: a form of a verb that might be considered the most basic. The infinitive is the form to which various endings may be added. English *walk* is an infinitive; it is the base for forms like *walks*, *walked*, *walking*, and *walker*. An English infinitive is often preceded by *to*. In ancient Greek, an infinitive typically ended in *-ein*. In Latin, almost all infinitives ended in *-are*, *-ere*, or *-ire*.

interjection: a word that is thrown (*ject-*) into (*inter-*) a sentence to express emotion. Some examples are *oh*!, *wow*!, *hooray*!, and *yuk*!.

Late Latin: a variety of Latin used in Europe in the Middle Ages or Renaissance. It included words for things and actions that didn't yet exist during the time when Classical Latin was a living language.

Latin: the language that was spoken in ancient Rome and the Roman Empire. Latin was one of the languages in the Italic subfamily of Indo-European. Long after the demise of the Roman Empire, Latin continued to be spoken as a common language by educated people throughout Europe. Our English word *Latin* is from Latin *latinus* "from Latium," a region in the west-central part of what is now Italy. The capital of Latium was and still is Rome.

Low German: the German dialects spoken in the lowlands. Geographically speaking, that means the Netherlands (literally "lower lands") and parts of Belgium and northern Germany.

masculine: in English, nouns like *king* and *lion* are masculine because they refer to male people or animals. A noun like *radius* in English is neuter because a radius is a thing. In Latin and Greek, however, even inanimate objects could be masculine or feminine. Latin *radius*, for example, was masculine.

metathesis: a change in the order of sounds within a word (from Greek *meta* "beyond" and *thesis* "a placing"). Sometimes the metathesized pronunciation (if not the spelling) becomes standard, as with "comfterble" from *comfortable* and *bird* from the earlier *brid*. In other cases the metathesized pronunciation remains colloquial, as in "perty" from *pretty*. Speakers of the standard language regard some metatheses as distinctly substandard; examples are "larnyx" instead of *larynx* and "aks" instead of *ask*.

Middle English: the stage of the English language from around 1100 to around 1500. A modern English speaker would be able to understand a fair amount of Middle English because a lot of vocabulary derived from French and Latin and still in use today entered the language during the Middle English period.

Old English: the stage of the English language from the 5th century to the beginning of the 12th century. A modern English speaker would be able to understand very little Old English because English has changed so much since the Norman invasion in 1066, which brought huge changes in grammar and vocabulary.

Old Norse: a Scandinavian, and hence North Germanic, language that was spoken about a thousand years ago; it was the predecessor of modern Icelandic, Norwegian, and Faeroese. The word *Norse* means "north." Old English borrowed quite a few words from Old Norse.

passive: a form of a verb whose subject is really the object of the action indicated by the verb. An example is "A large rock was removed by the bulldozer." The passive verb form in that sentence is *was removed*. If the sentence were reworded in active form, it would become "The bulldozer removed a large rock."

past participle: a form of a verb that is used in English compound tenses with the verb *have*. For example, the past participle of English *speak* is *spoken*; examples of compound tenses are *I have spoken, he had spoken, you will have spoken, they should have spoken*. A past participle often functions like an adjective: we talk about the *spoken* language as opposed to the *written* language. Many of our learned borrowings from Latin are based on past participles of verbs rather than on infinitives.

preposition: a function word that indicates the relationship of a noun or pronoun to other words in the sentence. Such relationships most commonly (but by no means always) involve time or space. Common examples of prepositional phrases are *in the city, before a snowstorm, after John*, and *because of his opposition*. Like the etymologically related word *exponent*, a preposition is named for its most common position, which is before (= *pre*-) a noun or pronoun. Nevertheless, prepositions could often be called postpositions in English, because they frequently follow the noun or pronoun to which they are connected. Examples are *What for?* and *Her bad behavior is something that I won't put up with*.

present participle: a form of a verb that is used in English compound tenses with the verb *be*. The participle itself ends in -*ing*, as in *we were running, they are singing*, etc. Like a past participle, a present participle can be used as an adjective: she is *amazing*, that was *confusing*, etc. In Latin, the stem of the present participle typically ended in -*nt*-.

q.v.: an abbreviation of the Latin words *quod vide* "which see." The plural is *qq.v.*, for *quae vide*. For example, in this dictionary's entry on *acnode*, the *ac*- is explained fully. The *(q.v.)* which appears after *node* tells you to look up the separate entry *node* for more information about that part of the compound *acnode*.

Romance: the name for the family of languages that developed out of Latin. The word itself is a variant of *Roman*. Some of the Romance languages are Portuguese, Spanish, Catalan (spoken in northeast Spain and southern France), Occitan or Provençal (spoken in southern France), French, Romansch (a group of languages spoken in Switzerland), Italian, Sardinian, and Romanian. The Romance languages have done quite well in the world: they are still spoken in their European homelands, and some have crossed oceans. Spanish and Portuguese have taken over most of South and Central America. French is spoken as a first or second language in large parts of Africa and also in Canada. French is also the Romance language that has most influenced English.

root: the most basic part of a word. For example, in the English compound *aimlessness*, the root is *aim*, to which the suffixes *-less* and *-ness* have been added.

Scandinavian: a group of languages within the Germanic subfamily of Indo-European. Some of the Scandinavian languages are Norwegian, Swedish, Danish, Icelandic, and Faeroese. Old English borrowed quite a few words from Old Norse, a predecessor of some of the modern Scandinavian languages.

Scythian: an extinct Iranian language once spoken in Scythia, a region in Europe and Asia between the Danube River on the west and China on the east.

Semitic: a family of ancient and modern languages spoken in the Middle East and the northeastern part of Africa. Some Semitic languages are Arabic, Hebrew, ancient Egyptian, ancient Aramaic (the language presumably spoken by Jesus), and Amharic (spoken in Ethiopia).

stem: the form of a word to which endings are added. For example, the stem of English *talks*, *talked*, and *talking* is *talk*.

superlative: from Latin *super* "above" and *latus* "carried." The superlative form of an adjective is the form that is most "carried above" the basic form. In English, if the basic form is *high*, then the comparative is *higher* and the superlative is *highest*. In Latin, the superlative often ended in *-simus*.

Vulgar Latin: the variety of Latin spoken by common people during the centuries before the fall of Rome, as opposed to the literary language that educated Romans used in literature. The modern Romance languages evolved primarily from Vulgar Latin, rather than from literary Latin.

Yiddish: a Germanic language that developed out of Old High German. For centuries it was the common language of Jews throughout Eastern Europe. Millions of Yiddish speakers were killed during World War II, and since Hebrew has become the national language of Israel, Yiddish is now in danger of dying out.

A

a posteriori (adverb, adjective): a Latin phrase. The preposition *ab*, abbreviated to *a* before a consonant, means "from." *Posterior* is a Latin comparative adjective based on *post* "after," perhaps from the Indo-European root *apo-* "away, off" (*off* is a native English derivative of that root). Surprisingly, in light of their phonetic differences, both Latin *a(b)* and *post* may have descended from the same Indo-European root: when you go *off* (*ab*), you move away to a place where you will be *after*wards (*post*). In mathematics or science, when you reason *a posteriori*, you start with what comes "after," namely facts or observations, and try to go back to the general principles that are presumed to underlie those specific facts or observations. Compare *induction*; contrast *a priori* and *deduction*. [14]

a priori (adverb, adjective): a Latin phrase. The preposition *ab*, abbreviated to *a* before a consonant, means "from." The Indo-European root may be *apo-* "away, off." *Prior* is a Latin comparative adjective based on *pro-* "before." The Indo-European root is *per-* "forward, before." When you reason *a priori*, you start with axioms and postulates that "come before," and try to find effects, facts, or relationships that necessarily follow. Compare *deduction*; contrast *a posteriori* and *induction*. [14, 165]

abacus, plural *abaci* (noun), **abacist** (noun): the Latin word *abacus* was taken from Greek *abax* "a slab," probably from the Hebrew *abhaq* "dust." The connection between dust and arithmetic is explained by the fact that many ancient peoples performed mathematical calculations by writing on a slab covered with dust or sand, much as children nowadays sometimes use a finger to write a message like "Wash me!" on the back of a dirty car or truck. The word *abacus* was formerly used to indicate calculating in general as well as specific algorithms for calculating. Fibonacci's influential book, published in 1202, was titled *Liber Abaci*, or *Book of Calculating*, though a more modern translation would be simply *Arithmetic Book*. Europeans who argued against adopting the Hindu-Arabic numerals, and who preferred using Roman numerals and calculating with an abacus or similar device, were known as abacists. We now use *abacus* exclusively as the name of a kind of manual calculating device found primarily in Asia and consisting of beads that slide on parallel wires. The Chinese abacus is called a *suan pan*, literally "a calculation square" or "counting board." The Japanese abacus is called a *soroban*. Both types combine number systems based on five and ten.

abelian (adjective): from the last name of the Norwegian mathematician Niels Henrik Abel (1802–1829). The English suffix *-ian* is commonly used to turn a proper noun into an adjective; other mathematical examples are *Euclidian*, *Gaussian*, and *Wronskian*. Although *abelian* was originally capitalized, it is now frequently used in lower case, so that people are increasingly unaware of the connection with Abel. In abstract algebra an abelian group is one which is commutative.

-able, -ible (suffix): via French from Latin *habilis* "capable of being had," from *habere* "to have." If you are able to do something, you can have that thing. The Indo-European root is *ghabh-* "to give" or "to receive." The English cognate *give* obviously reflects the meaning "give," while Latin *habere* reflects the meaning "receive." Although it may seem paradoxical for the same root to have opposite meanings, such polarization is common in language; compare the confusion over English *bring* and *take*. The *-able* suffix has become so anglicized that we freely apply it to native English roots to create adjectives such as *doable* and *unthinkable*. The noun corresponding to *-able* and *-ible* is *ability*, as seen in *doability* and *unthinkability*. [69]

abridged (adjective): from Old French *abregier*, from Late Latin *abbreviare* "to shorten." The first component is from Latin *ad-* (*q.v.*) "to." The second component is from Latin *brevis* "short," which, via French, was borrowed into English as *brief*. The Indo-European root is *mregh-u-* "short." The word *abridged* is unrelated to native English *bridge* (*q.v.*). Abridged multiplication is a form of multiplication in which digits beyond a certain decimal place are deleted in each term of the partial product. [1, 138]

abscissa (noun): from Latin *linea abscissa* "a cut-off line." *Abscissa* is the past participle of the verb *abscindere* "to break off," from *ab-* "off" and *scindere* "to cut." The Indo-European root is *skei-* "to cut, split," itself an extension of *sek-* "to cut." The slightly different *abscisa* is the past participle of *abscidere* "to cut off," from *abs* "off" and *cædere* "to cut." The Indo-European root is *kaə-id-* "to cut." The German mathematician Gottfried Wilhelm Leibniz (1646–1716) coined the term *abscissa*, but it's not clear which of the two similar Latin verbs he was using. In any case, the *abscissa* or *x*-value of a point in a two-dimensional rectangular coordinate system is the length of the segment that is "cut off" on the x-axis between the origin and the place directly above or beneath the point in question. The concept wasn't actually new: the ancient Greeks had also used an expression involving the notion of a segment being cut off. In botany, an abscission layer is a layer of plant cells where the stem of a fruit or leaf breaks off the branch. [14, 185, 84]

absolute (adjective): from *absolutus*, past participle of the Latin verb *absolvere*, a compound of *ab* "off, away from," and *solvere* "to free, to loosen." The Indo-European root is *leu-* "to loosen, divide, cut apart," as seen in native English *lose*. Something *absolute* is free from restrictions. In mathematics, the absolute value of a number is free of any sign, so to speak. The current symbol for the absolute value of a number n, $|n|$, has been in use only since the middle of the 20th century. Before then, the common term for *absolute value* was *numerical value*, and there was no special symbol to represent it. In mathematics books published as recently as the first third of the 20th century, the expression *absolute term* referred to the constant term of a polynomial because, unlike all the other terms, the constant is "free from" a variable. [14, 116]

absorb (verb), **absorption** (noun): from Latin *absorbere* "to swallow up, devour." The first element is from Latin *ab* "away from, off," from the Indo-European root *apo-* "away," as found in English *of*, *off*, and *ebb*. The second element is from Latin *sorbere* "to suck in, drink down, swallow." The Indo-European root is *srebh-* "to suck, absorb," as seen in *slurp*, borrowed from Dutch. In Boolean algebra the two statements $A \cup (A \cap B) = A$ and $A \cap (A \cup B) = A$ exemplify the Law of Absorption. In each case B has been "swallowed up" or "absorbed," leaving only A as the result. [14]

abstract (adjective): the first element is from Latin *abs* "away from, off," from the Indo-European root *apo-* "away," as seen in English *of*, *off*, and *ebb*. The second element is from Latin *tractus*, past participle of *trahere* "to draw, drag, haul." The Indo-European root is *tragh-* "to draw, drag." An Indo-European variant *dhragh-* is the source of native English *draw* as well as *drag*, borrowed from Old Norse. An abstract principle is one which has been drawn away from any specific examples of that principle. Modern abstract algebra, for example, is far removed from the physical problems that originally gave rise to the

methods and manipulations of elementary algebra. [14, 234]

abundant (adjective): the first element is from Latin *ab* "away from, off," from the Indo-European root *apo-* "away," as seen in English *of*, *off*, and *ebb*. The second element is from Latin *unda* "wave," plus the present participial ending *-nt*. The Indo-European root is *wed-*, whose meaning is seen in the native English cognates *water* and *wet*. A Russian cognate is *vodka*, literally and euphemistically "a little bit of water." If a wave carries things away, it swells or overflows. A number like 12 is said to be abundant because when its proper divisors are added, the total swells past the original number: the number 12 is abundant because $1 + 2 + 3 + 4 + 6 = 16$, which is more than 12. Contrast *deficient* and *perfect*. [14, 242, 146]

accelerate (verb), **acceleration** (noun): from Latin *ad-* "relating to" and *celer* "speedy." The Indo-European root is *kel-* "to drive, set in motion." In mathematics acceleration relates to speed, or more technically velocity; specifically, it is the rate at which the velocity is changing or "speeding up." [1]

accrue (verb): from French *accru(e)*, past participle of *accroître* "to grow, increase"; from Latin *ad* "(on)to" and *crescere* "to grow." The Indo-European root is *ker-* "to grow." A related borrowing from French is *crew*, a group of people who increase your numbers. In finance, when interest accrues, it is added on to your capital, and the total amount of money you have grows bigger. [1, 92]

accumulate (verb), **accumulation** (noun): from Latin *ad* "in addition to" and *cumulare* "to pile up," based on the noun *cumulus* "a pile." The Indo-European root is *keuə-* "to swell; a vault, a hole." In mathematics an accumulation point of a sequence is a point that has an infinite number of terms of the sequence in any neighborhood of it; the terms "pile up" at the accumulation point. [1, 98]

accuracy (noun), **accurate** (adjective): from Latin *accurare* "to apply care to," from *ad* "to" and *cura* "care," of unknown prior origin. When a person puts a lot of care into something mathematical, the result is likely to be accurate. Related borrowings from Latin include *curator* and *curious*. [1]

acnode (noun): the first component is from Latin *acus* "needle," from the Indo-European root *ak-* "sharp." The second component is *node* (*q.v.*). In mathematics an acnode is an isolated double point on a curve; the isolated point looks like the prick of a needle in the graph paper. [4, 142]

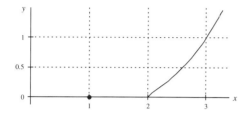

Acnode at $(1,0)$ on the curve $y = (x-1)^2 \sqrt{x-2}$.

acre (noun): a native English word, the original meaning of which was "a piece of tilled or arable land." Only around the year 1000 did the word come to mean a specific amount of land, originally as much as a yoke of oxen could plow in one day. Later the amount was taken to be 32 furrows each a furlong (= furrow long) in length. Nowadays an acre equals 160 square rods or 43,650 square feet. The word has also at times been used as a linear measurement equal to one furlong (220 yards). The Indo-European root *agro-* "field," is recognizable in borrowings from Latin such as *agriculture* and *agrarian*.

actuary (noun), **actuarial** (adjective): from Latin *actuarius* "a clerk, bookkeeper, registrar," from *actus*, past participle of *agere* "to drive, to do, to act, to deal with." The Indo-European root is *ag-* "to drive." A modern actuary is a lot more than a clerk or bookkeeper; an actuary is a statistician who specializes in insurance. [3]

acute (adjective): via Latin *acutus* "pointed, sharp," from *acus* "needle." The Indo-European root is *ak-* "sharp." A native English cognate is *edge*. Related borrowings include: from Latin, *acid*, which is sharp; from Greek, *acrobat*, who works up at the top or pointy part of the tent in a circus; from Old Norse, *egg on* "to goad someone with a pointy stick." In geometry an acute angle is one that looks pointy or sharp, as opposed to an obtuse or blunt angle. Numerically speaking, an acute angle is between 0° and 90°. [4]

acyclic (adjective): the first element is from Greek *a(n)-* "not," from the Indo-European root *ne* "not." The second element is from Greek *kuklos* "a circle," from the Indo-European root *kwel-* "to move around," as found in native English *wheel*. Something cyclic repeats in the same order, while some-

thing acyclic does not. In an infinite continued fraction in which the partial quotients eventually begin repeating, the non-repeating partial quotients are called the acyclic part of the continued fraction. [141, 107]

ad- (prefix): Latin *ad* is from the Indo-European root *ad-* "to, at, near," as seen in native English *at*. A great many compounds borrowed from Latin bear this suffix. It is recognizable in mathematical words like *additive*, *adjoint*, and *admissible*. The *-d-* of the prefix frequently assimilates to the following consonant; *ad-* is present but harder to recognize in mathematical words like *acceleration*, *affine*, *alligation*, *approach*, and *associative*. [1]

add (verb), **addition** (noun), **additive** (adjective): from the Latin verb *addere*, itself a compound of *ad* (q.v.) "to" and *dare* "to give." When you add an amount to something, you "give more to" that something. The Indo-European root is *do-* "to give." Related borrowings from Latin include *donate* and *condone* "to give sanction or approval." A related borrowing from Greek is the *dose* of medicine given to a patient. In arithmetic, the "+" symbol that we use to indicate addition appeared as early as 1489 in an arithmetic book by the Bohemian-born Johann Widman. The symbol may have been abstracted from the *t* of the Latin word *et*, meaning "and." While the only modern synonyms of *to add* are *to sum (up)* and *to total (up)*, Renaissance terms for *addition* included *aggregation* and *collection*. In algebra, the additive inverse of a number n is the number which must be added to n to produce the identity element 0; the additive inverse of n is $-n$. [1, 47]

addend (noun): from Latin *addere* "to add" (q.v.), with the suffix *-nd*, which creates a type of passive causative, so that Latin *numerus addendus* meant "the number to be added." In the statement 3 + 4 = 7, many Renaissance writers would have referred to the 4 alone as an addend, because it and it alone is to be added to the original 3. Due to the commutativity of arithmetic, however, we have now allowed the term *addend* to apply to both the 3 and the 4. In nonmathematical English, an addendum is text that is to be added to a book after the book has already been printed. [1, 47, 139]

adherent (adjective): from Latin *ad* (q.v.) "to" and the present participle of *haerere* "to stick, cling, adhere." The Indo-European root is *ghais-* "to adhere." In mathematics an adherent point of a set of points is any point that is either in the set or is an accumulation point of the set. [1, 70, 146]

adjacent (adjective): from Latin *ad* (*q.v.*) "at" and the verb *jacere* "to be thrown down, to lie." The Indo-European root is *ye-* "to throw." Something adjacent lies close at hand. In a right triangle, the leg adjacent to a given acute angle is the leg that "lies close at hand" to the angle. Related borrowings from Latin include *reject*, *inject*, and *subject*. [1, 257, 146]

adjoin (verb), **adjoint** (adjective, noun), **adjugate** (adjective, noun): via French *adjoindre*, from Latin *adjungere*, a compound of *ad* (*q.v.*) "to" and *jungere* "to join." The Indo-European root of the same meaning is *yeug-*. Related words include native English *yoke*, which joins two animals together, and Sanskrit *yoga*, which joins the mind to its spiritual nature. In mathematics an adjoined field is formed by "joining" a new element to an existing field using addition, subtraction, multiplication or division. An adjoint of a matrix is made by "joining to" the transpose of the matrix the cofactor of each element; the adjoint is also known by the Latinate name *adjugate*. [1, 259]

admissible (noun): from Latin *ad* (*q.v.*) "(in)to" and *missus*, past participle of *mittere* "to let go, to send." The Indo-European root may be *(s)meit(ə)-* "to throw." Related borrowings from Latin include *missile*, *emit*, and *omit*. To admit something is literally to let it go in. For an explanation of the suffix see *-able*. In mathematics an admissible hypothesis is any hypothesis that may possibly be true, and which therefore should be allowed into consideration. [1, 200, 69]

affine (adjective): via French, from Latin *ad* (*q.v.*) "to, at," and *finis* "boundary, border," of unknown prior origin. Although the original meaning of the Latin adjective *affinis* was "neighboring, adjacent," the most common meaning eventually became "related by marriage." That notion of kinship or *affinity* was carried into French, and from there into English. In mathematics, the affine transformations include translation, rotation, stretching (or shrinking), simple elongation, and reflection. These transformations are related because all of them can be expressed via the equations $x' = a_1x + b_1y + c_1$ and $y' = a_2x + b_2y + c_2$. The transformations also maintain a kinship between a set and its image: parallel lines are carried into parallel lines, finite points into finite points, and the line at infinity into itself. [1, 60]

aggregation (noun): from Latin *ad* (*q.v.*) "to" and *grex*, stem *greg-*, "flock, herd, swarm, company, troop, crowd." The Indo-European root is *ger-* "to gather." A related borrowing from Latin is *gregarious*, said of someone who likes to be in other people's company. In mathematics, symbols of aggregation such as parentheses and brackets are now usually known as grouping symbols. *Aggregation* was once a synonym of *addition*. [1, 67]

aleatory (adjective): from Latin *alea* "a game of chance"; the word may originally have meant "a cube or die" because the Romans often used dice in their games. Even in Classical Latin the word *alea* took on the figurative meanings "uncertainty, accident, venture, risk, chance." When Julius Cæsar defied Roman law by leading his army across the Rubicon River, he is supposed to have said "*Alea jacta est*," meaning "The die is cast." In mathematics *aleatory* means "depending on chance, random."

aleph (noun): the first letter of the Hebrew alphabet, ℵ, meaning literally "ox." In general, each letter in the alphabet was named after a word beginning with the sound represented by that letter. Even today, to avoid confusion when speaking on the phone, we use the same kind of scheme when we say things like "B as in Boy." The shape of each letter in the Hebrew alphabet was originally a picture corresponding to the object that the letter was named after, although as the pictures became more and more stylized over time, the resemblance to the original object was often obscured. The concept of an alphabet seems to have evolved from the writing system of ancient Egypt. Around 1500 B.C. the Canaanites were using a form of the Hebrew alphabet, which the Phoenicians later borrowed and modified. The Greeks in turn adapted the Phoenician writing system to their own language, reassigning certain sounds and adding characters for other sounds. In the process, Hebrew *aleph* became Greek *alpha* (*q.v.*). The Etruscans borrowed the alphabet from the Greeks, the Romans borrowed it from the Etruscans, and later on Europeans borrowed it from the Romans. Just as ℵ is the first letter in the Hebrew alphabet, $\aleph_0$ is the first or smallest type of infinity dealt with in mathematics. The German mathematician Georg Cantor (1845–1918) called the symbol $\aleph_0$ "aleph null," since *null* is the German word for "zero." Some Americans say "aleph naught," because *naught* is an English word for "zero." Other Americans say "aleph zero." The symbol $\aleph_\alpha$ is also used in mathematics; pronounced "aleph alpha," it contains Hebrew and Greek versions of the first letter of the alphabet. [7]

algebra (noun): from the title of a work written around 825 by the Arabic mathematician known as al-Khowarizmi, entitled *al-jebr w' al-muqabalah*. In Arabic, *al-* is the definite article "the." The first noun in the title is *jebr* "reunion of broken parts," from the verb *jabara* "to reunite, to consolidate." The second noun is from the verb *qabala*, with meanings that include "to place in front of, to balance, to oppose, to set equal." Together the two nouns describe some of the manipulations so common in algebra: combining like terms, transposing a term to the opposite side of an equation, setting two quantities equal, etc. Because the original Arabic title was so long, and because it was in Arabic, Europeans soon shortened it. The result was *algeber* or something phonetically similar, which then took on the meanings of both nouns and eventually acquired its modern sense. In its earliest English usage in the 14th century, *algeber* still meant "bone-setting," a reflection of the word's original meaning. (When you think about manipulating equations or complicated fractions, you can see why early mathematicians imagined themselves putting broken bones back together.) By the 16th century, the form *algebra* appeared in English with its mathematical meaning. Recorde, the earliest English user of the term in its mathematical sense, spelled it *algeber*, staying closer to the original Arabic form than did mathematicians who used the spelling *algebra*. Nevertheless, the less accurate version prevailed. Although *algebra* referred originally to what we might now call manipulative algebra, the term is also used much more abstractly in modern mathematics, where many different algebras exist. Contrast the etymological meanings of *fraction* and *fractal*.

algorism (noun), **algorithm** (noun), **algorist** (noun): these words come from the now-quite-distorted name of a person, Ja'far Mohammed Ben Musa, who was known as al-Khowarazmi, meaning "the man from Khwarazm." (In a similar way, Leonardo da Vinci was actually Leonard, a man from the town of Vinci). Around the year 825 al-Khowarazmi wrote an arithmetic book explaining how to use the Hindu-Arabic numerals. That book was later translated for Europeans and appeared with the Latin title *Liber Algorismi*, meaning "Book of al-Khowarazmi." As a consequence, the term *algorism* came to refer to the decimal system of numeration. Any use or manipulation of Arabic numerals—especially a pattern

used to add, subtract, multiply, etc.—was known as an *algorism*. Arithmetic itself was sometimes called *algorism*, and in a similar fashion Europeans who advocated the adoption of Hindu-Arabic numerals were known as algorists. Over the centuries the word *algorism* underwent many changes in form. In Old French it became *augorisme*, which then developed into the now obsolete English *augrim*, *agrim*, and *agrum*. The current form *algorithm* exhibits what the *Oxford English Dictionary* calls a "pseudo-etymological perversion": it got confused with the word *arithmetic* (which was one of its meanings, and which has several letters in common with it); the result was the current *algorithm*. Current dictionaries still list the older form *algorism* in the sense of "the decimal or Arabic system of numeration."

aliquot (noun): from the Latin compound *aliquot* "some, several, a few." The first component is from *alius* "other" (where more than two are involved), from the Indo-European root *al-* "beyond." The second component is *quot* "how many, as many as," from the Indo-European root $k^w o$-, the base of many relative and interrogative pronouns. In mathematics an aliquot part of an integer is any of the "so many other" exact divisors of an integer (but not the integer itself); for instance, the aliquot parts of 21 are 1, 3, and 7. Compare *submultiple*. [6, 109]

all (adjective): a native English word with cognates in other Germanic languages. The root is also found in compounds such as *alone* (= all one), *although*, and *always*. In mathematics, the quantifier "for all" is represented by the symbol $\forall$, an upside-down capitalized version of the first letter of *all*. [8]

alligation (noun): from Latin *ad* (*q.v.*) "to" and *ligare* "to bind," from the Indo-European root *leig-* "to bind." From the same source, via French, comes our word *alloy*. In mathematics books that were in use up to the 20th century, the term *alligation* referred to mixture (= binding) problems. An example of a typical alligation problem was: how many pounds of cashews selling for $4.50 per pound should be mixed with a pound of raisins selling for $2.25 per pound to create a mixture selling for $3 per pound? Arithmetic books used to have lots of such problems, but because of their artificiality they are slowly beginning to disappear from the curriculum, as has the term *alligation* itself. [1, 113]

allometric (adjective): the first component is from Greek *allos* "other," from the Indo-European root *al-* "beyond," as seen in native English *else*. The second component is from Greek *metron* "measure, length"; the Indo-European root is *me-* "to measure," or possibly *med-* "to take appropriate measures." In mathematics allometric growth occurs when one feature grows at a rate proportional to a power of another feature. Allometric growth is represented by an equation of the type $y = kx^n$. [6, 126]

almost (adverb): a native English compound of *all* and *most* (qq.v.). The use of *all* intensifies the meaning of *most*. Notice the redundancy in the phrase "almost all of," a redundancy that disappears with the alternate wording "most all of." In mathematics, "almost all" means "all except for a countable number." [8, 125]

alpha (noun): the first letter of the Greek alphabet, written A as a capital and α in lower case. The names of the first two Greek letters, *alpha* and *beta*, form our word *alphabet*. We still say, in a similar way but with the names of the first letters of the English alphabet, that someone knows his *ABC*'s. See more details under *aleph*, the Hebrew letter that was borrowed into Greek as *alpha*. In triangle *ABC*, α is sometimes used to designate the angle at vertex *A*. [7]

alphametic (noun): the first component is the Greek letter *alpha*, used here to indicate letters of the alphabet. The second component is the end of the word *arithmetic*. A cryptarithm (q.v.) is a puzzle in which the digits of an arithmetic calculation are replaced by letters in a one-to-one correspondence. An alphametic is a type of cryptarithm in which the replacement letters happen to spell out real words; in the best alphametics, the words even make meaningful phrases. The term *alphametic* was coined in 1955 by J.A.H. Hunter, although puzzles of that type had been around for decades. [7]

	S	E	N	D
	M	O	R	E
M	O	N	E	Y

The first known
alphametic in English.

alternate (verb, noun), **alternating** (adjective), **alternation** (noun): from Latin *alter* "other (of two)." Related Latin borrowings include *alias, alibi, altruist* and *adulterate* (to alter, in an illicit sense). The Indo-European root *al-* "beyond" is found in native

English *else*, meaning "otherwise." In mathematics, the terms in an alternating series keep switching from one sign to the other. In geometry, when there are two alternate interior angles, one is on one side of the transversal and one is on the other. In logic, an alternation is another name for a disjunction because you can choose one thing or, alternatively, another. [6]

altitude (noun): from Latin *altitudo* "height," from *altus* "high." The Indo-European root is *al-* "to grow, to nourish." It can be found in native English *old* (= fully grown) and in Latin borrowings like *exalt* (= to raise up high), *altimeter*, and even *alimony* (nourishing money) and *alma mater* (nourishing mother). In mathematics, the *altitude* of a figure is its height.

alysoid (noun): from the Greek *alusis* "chain," of unknown prior origin, plus the Greek-derived suffix *-oid* (q.v.) "having the appearance of." In mathematics, the alysoid is the curve formed by a flexible chain with infinitesimally tiny links hanging under its own weight. The alysoid is more commonly known by the Latin-derived name *catenary*, and it has also been called a *chaînette* (qq.v.). [244]

ambiguous (adjective): from Latin *ambiguus* "shifting, doubtful." The first element is from the Indo-European *ambhi* "around," also found in native English *by*. The second element is from Latin *agere* "to do, act, drive," with other subsidiary meanings, as seen in many related borrowings: *act, agent, agitate, navigate, exact*, etc. When a situation is ambiguous, you "drive around" not knowing what to do. In trigonometry, the ambiguous case occurs when the sizes of two sides of a triangle and of an angle opposite one of them are given. The case is called ambiguous because, depending on the relative sizes, there may be no triangle, one triangle, or two distinct triangles. [9]

amicable (adjective): the main component is from Latin *amicus* "friend." The Latin root *am-*, as well as words like *mama*, are assumed to be imitative; babies in cultures around the world make similar sounds. For an explanation of the suffix see *-able*. Someone who is amicable is "capable of being had as a friend." In mathematics, two positive whole numbers are said to be amicable when the sum of the proper divisors of one equals the other, and vice versa. The standard example is the pair 220 and 284. The proper divisors of 220 are 1, 2, 4, 5, 10, 11, 20, 22, 44, 55, and 110, all of which add up to 284. The proper divisors of 284 are 1, 2, 4, 71, and 142, all of which add up to 220. Because of this mutual relationship, the two numbers 220 and 284 are anthropomorphically characterized as being friendly. Amicable numbers are sometimes called *sympathetic* (*q.v.*) numbers. Also see *sociable*. [69]

amortize (verb), **amortization** (noun): from French *amortir*, a compound of *a*, "to," and *mort-*, "death," so that to amortize a debt is to "kill it off." The Indo-European root is *mer-*, "to die." A native English cognate is *murder*, and related borrowings from Latin include *mortal* and *mortuary*. [1]

amount (verb, noun): from Old French *amont*, a compound consisting of *a* "to" and *mont* "mountain." To amount is literally to build up a mountain of something. As a noun, an amount is the quantity contained in that "mountain." The sense of ascending is preserved in the slightly shorter form *to mount* (a horse, for example). The underlying Indo-European root *men-* "to project" is found in the names of certain projecting body parts such as French *menton* "chin," and surprisingly even native English *mouth*, which is surrounded by projecting lips. [1]

amphicheiral (adjective): the first component is from Greek *amphi* "around, about," from the Indo-European root *ambhi* "around." A cognate from Old Norse is *ombudsman*, someone who "is around" to help you when you are in trouble. The second component is from Greek *kheir* "hand," from the Indo-European root *ghesor-* "hand." Related borrowings from Greek include the *chiropractor* who presses with his hands and the *surgeon* who works with his hands. In the theory of knots, a knot that can be mapped into itself by an orientation-reversing homeomorphism is said to be amphicheiral. With some "working around" the knot can be changed from right-handed to left-handed. [9]

amplitude (noun): from Latin *amplus* "large, wide, spacious, roomy," of unknown prior origin. As used in connection with periodic functions, *amplitude* refers not to the width of a curve but to its maximum height. As used with complex numbers, the word refers neither to width nor height, but rather to the angle that the radius makes with the positive horizontal axis. The shift in meaning from width to height to angle—an example of what linguists call semantic drift—is characteristic of words describing space and time, as when we ask "how long?" and expect an answer like "fifteen minutes" rather than "fifteen inches."

anallagmatic (adjective): the first element is from Greek *an-* "not," from the Indo-European root *ne-* "not." The second element is from Greek *allagman-* "something given in exchange," from *allatein* "to exchange." The more basic word is *allos* "other," from the Indo-European root *al-* "beyond," as seen in native English *else*. A curve which remains the same when inverted relative to a fixed point and a constant distance is said to be anallagmatic; in other words, the original curve is "not (ex)changed." For example, if a circle is inverted relative to its center, and if the constant distance is taken to be the radius of the circle, then the circle is its own inverse. T. Moutard (1827–1901) first discussed the concept in 1864. The English mathematician James Joseph Sylvester (1814–1897) used the term *anallagmatic* to refer to checkerboard-type arrays in which the squares are colored with two colors in such a way that whenever any two rows or columns are put next to each other, half the adjacent cells match and half don't; in other words, the half-half matching is never altered. [141, 6]

analogy (noun): from Greek *analogia* "equality of ratios, proportion." The first component is *ana-* "according to," from the Indo-European root *an* "on." The second component is *logos* "proportion, word," from the Indo-European root *leg-* "to collect," with derivative meanings having to do with collecting one's thoughts and putting them into words. Related borrowings from Greek include *dyslexia*, *lexicon*, and *catalogue*. In mathematics, analogy is used to infer new theorems from existing ones; hypotheses based on analogy must still be proved, of course. [10, 111]

analysis, plural *analyses* (noun), **analytic** (adjective), **analyticity** (noun), **analyze** (verb): *analysis* is a Greek compound meaning "an unloosing, an undoing." The first component is from *ana-* "up, on," from the Indo-European root *an* "on." The second component is from *luein* "to set loose, to set free," from the Indo-European root *leu-* "to loosen, to cut up." Native English cognates include *lose*, *loose* and *forlorn*. When you analyze something complicated, you "free up" the thing or resolve it into its components. In analytic geometry, the plane (or space) is "cut up" by the scale of a coordinate system, as opposed to *synthetic* (*q.v.*) geometry. Calculus, being a complicated subject that requires much detailed attention, is often known as *analysis*. Around 1590 the French mathematician François Viète opted for the term *analysis* rather than *algebra*, claiming that *algebra* doesn't mean anything in any European language. He didn't succeed in driving out the word *algebra*, but he did popularize *analysis* to the point where it has stayed with us. [10, 116]

anarboricity (noun): the first element is from Greek *an-* "not," from the Indo-European root *ne* "not." The second element is from Latin-derived *arboricity* (*q.v.*). In graph theory, the arboricity of a graph *G* is defined as the minimum number of line-disjoint acyclic subgraphs whose union is *G*. In contrast, the anarboricity of a graph *G* is defined as the maximum number of line-disjoint nonacyclic subgraphs whose union is *G*. [141]

and (conjunction): a native English word that may have developed from the Indo-European word *en* "in." The semantic development would have been something like this: *x* is "in with" *y*, therefore *x* and *y* are "in there together," therefore we are dealing with *x* and *y*. The relationship between *and* and *in* can be seen in the Venn diagram representing the intersection of two sets *A* and *B*: the elements that are in *A* and *B* are inside the "inner" region where *A* and *B* overlap. Linguists used to believe there is an etymological connection between *and* and *end*, but that idea has been largely discarded now. In logic, *and* is often represented by the symbol ∧, which coincidentally looks like a capital *A* without the crossbar. Not so coincidentally, the symbol for *intersection* (*q.v.*) is a rounded version of the same symbol: ∩. [52]

angle (noun), **angular** (adjective): from Latin *angulus* "corner, angle." Latin *-ulus* is a diminutive ending, so *angulus* meant literally "a little bending." In mathematics an angle measures how much "bending" or turning one line (= one side of the angle) does to get into the position of another line (= the other side of the angle). The Indo-European root is *ang-* or *ank-* "to bend"; it is seen in the related native English word *ankle*, the bend between the leg and the foot. It is also seen in the word *English*, since the ancestors of the English lived in the angular-shaped region in Europe known as Angul. A related borrowing from Greek is *anchor*. [11, 238]

anharmonic (adjective): the first component is from Greek *ana-* "according to," from the Indo-European root *an* "on." The second component is from Greek *harmonia* "joint, agreement, concord," from the Indo-European root *ar-* "to fit together"; a joint is a place where bones fit together. In mathematics if *A*, *B*, *C*, and *D* are four distinct points on a number line, and if their values are x_1, x_2, x_3, x_4, then the anhar-

monic ratio is defined as the quotient of the ratios in which *AB* is divided by *C* and *D*, respectively. That ratio, given algebraically by

$$\frac{(x_3 - x_1)}{(x_3 - x_2)} \div \frac{(x_4 - x_1)}{(x_4 - x_2)},$$

takes on its numerical value "according to how the points fit together." The anharmonic ratio is also known as the cross ratio. [10, 15]

annihilate (verb), **annihilator** (noun): from Latin *ad* (*q.v.*) "to" and *nihil* "nothing"; *nihil* is itself a compound of *ne* "not" and *hilum* "a little thing, a trifle." The verb *annihilate* means literally "to make nothing out of, to turn into nothing." The Latin suffix *-ator* indicates a male person or thing that performs a certain action, so an annihilator is something that annihilates. In mathematics an annihilator of a set *S* is the collection of all real functions of the form $f(x) = mx$, such that $f(x) = 0$ for every *x* in set *S*. [1, 141, 153]

annuity (noun): from Latin *annus* "year," from the Indo-European root *at-* "to go." A year must originally have been conceived as the time it takes for the night sky to go around to a given position. Related borrowings from Latin include *biennial*, *triennial*, *quadrennial*, etc. An annuity is a yearly payment. More specifically, it is a type of investment in which periodic payments are made from the accumulated amount of principal plus interest. [17]

annulus, plural *annuli* (noun), **annular** (adjective): a Latin diminutive of *anus* "ring," from the Indo-European root *ano-*, of the same meaning. In mathematics an *annulus* is "a little ring," i.e., the region in a plane contained between two concentric circles. The associated adjective *annular* is used to describe something that resembles an annulus, in particular a type of eclipse in which the moon seems slightly smaller than the sun and therefore leaves a "ring" of brightness. The non-diminutive Latin *anus* is used in anatomy. [238]

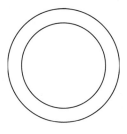

Annulus

anomalous (adjective), **anomaly** (noun): the prefix is from Greek *an-* "not," cognate to native English *un-*, from the Indo-European root *ne* "not." The main part of the word is from *homalos* "even," based on *homos* "same," from the Indo-European root *sem-* "one, one and the same." In mathematics the law of anomalous numbers refers to groups of numbers among which there is no inherent relationship, such as lengths of rivers or areas of countries. In polar coordinates, the anomaly of a point is the polar angle θ formed between the point, the pole, and the polar axis; in general a polar point is "not in the same line as" the polar axis. [141, 188]

ANOVA (noun): an abbreviation for *analysis of variance*. For no discernible reason, most books capitalize all the letters. In statistics ANOVA is a method of comparing two or more means. It may be one-way, two-way, or higher-way, depending on the number of features being simultaneously compared. [10, 116, 250]

answer (verb, noun): from the Old English compound *andswaru*. The first component is from the Indo-European root *ant-* "front, forehead." It is found in native English *un-*, as in *undo*, and in *anti-* "against," borrowed from Latin. The second component is from the Indo-European root *swer-* "to speak, talk," as found with a negative connotation in native English *swear*. To answer is literally "to swear [= talk] back or against." When looking for the answer to a difficult problem, some people are indeed tempted to swear. The word *answer* is the common English equivalent of the fancier Latin-derived *solution* (*q.v.*). [12]

antecedent (noun): from Latin *ante* "before" and *cedere* "to go, to yield," plus the *-nt* ending of the present participle: *antecedent* means literally "going before." The Indo-European root underlying *ante* is *ant-* "front, forehead." Your forehead is the part of your head in *front* of you; the eyes below your forehead see what is *before* you. The Indo-European root underlying *cedere* is *ked-* "to go, yield," as seen in many borrowings from Latin such as *exceed, deceased,* and *secede.* In a mathematical if-then statement, the if-clause, which is usually written before the then-clause, came to be known as the *antecedent.* In spite of the literal meaning of the word, the *antecedent* now always means the if-clause, even if the clauses of the sentence are written in reverse order. For example, in "you will be rich if you win the lottery," the antecedent is "you win the lottery," even though being rich is mentioned before winning the lottery. [12, 88]

anti- (prefix): from Greek *anti-*, from the Indo-European root *ant-* "front, forehead." The semantic connection can be seen in the fact that when you con*front* something, you often fight *against* it. Also from the same root are Latin *ante* "in front of, before," and native English *un-*, as in *undo* and *unblock.* [12]

anticommutative (adjective): a compound of Greek *anti-* "against" and *commutative* (*qq.v.*). An algebraic operation $*$ is said to be commutative if $a * b = b * a$. An operation is said to be anticommutative if $a * b = -b * a$. [12, 103, 130]

antiderivative (noun): a compound of Greek *anti-* "against" and Latin-derived *derivative* (*q.v.*). In calculus an antiderivative is equivalent to an integral because differentiation and integration work "against" each other in the sense of undoing each other. In fact, the native English prefix *un-* is from the same Indo-European root as *anti-*. [12, 32, 180]

antigonal (adjective): the first element is from Greek *anti-* (*q.v.*), literally "against," but used here in the figurative sense of "against the stated order," therefore "opposite." The second element is from Greek *gonia* "angle," from the Indo-European root *genu-* "knee, angle." In one form of geometry, angles are measured directionally; as in trigonometry, the positive direction is counterclockwise. An angle is considered the opposite of another angle if the two angles add up to an integral multiple of π radians (including 0). With respect to the two points A and B, the points X and Y are called antigonal points if $\angle AXB = -\angle AYB$. [12, 65]

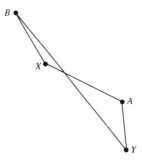

Directional angles *AXB* and *AYB* are antigonal.

antilogarithm (noun): a compound of Greek *anti-* "against" and *logarithm* (*qq.v.*). In algebra if $\log_b x = y$, y is the logarithm of x, and x is said to be the antilogarithm of y. Logarithms and antilogarithms work "against" each other in the sense that they undo each other. In fact, the native English prefix *un-* is from the same Indo-European root as *anti-*. [12, 111, 15]

antimorph (noun): the first element is from Greek *anti-* (*q.v.*), literally "against," but used here in the figurative sense of "against the stated order," therefore "reversed." The second element is from Greek *morph-* "shape, form, beauty, outward appearance," of unknown prior origin. In number theory an integer that can be expressed in the form $x_1^2 - Dy_1^2$ as well as in the form $Dx_2^2 - y_2^2$ is said to be an antimorph. Also, given the binomials $Ax^2 + y^2$ and $x^2 - By^2$, their respective antimorphs are $x^2 + Ay^2$ and $-x^2 + By^2$. Contrast *monomorph* and *polymorph*. [12, 137]

antinomy (noun): the first element is from Greek *anti-* (*q.v.*). literally "against," but used here in the figurative sense of "against the stated order," therefore "reversed." The second element is *nomos* "custom, usage, law," from the Indo-European root *nem-* "assign, allot, take." An antinomy is a paradox or a contradiction. An antinomy goes "against the law" of logic because two seemingly reasonable principles lead to mutually exclusive or irreconcilable conclusions. The word *antinomy* is stressed on the second syllable. [12, 143]

antiparallel (adjective): the first element is from Greek *anti-* (*q.v.*), literally "against," but used here in the figurative sense of "against the stated order," therefore "reversed." The second element is *parallel* (*q.v.*). Suppose two fixed lines are cut by two transversals. If corresponding angles considered in the same order are equal, then the two transversals are parallel. If corresponding angles considered in reverse (= *anti-*) order are equal, then the two transversals are said to be antiparallel. Two vectors are said to be antiparallel if the arrows representing them are parallel but point in opposite directions. [12, 165, 6]

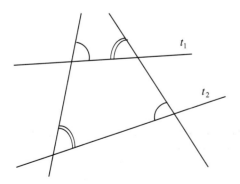

The two transversals are antiparallel.

antipodal (adjective), **antipodes** (plural noun): the first element is from Greek *anti-* (*q.v.*), literally "against," but used here in the figurative sense of "against the stated order," therefore "reversed." The second element is Greek *pous,* stem *pod-,* "foot." The literal meaning of the compound is "[with] feet [planted] on opposite [sides of something]." In mathematics two points that are diametrically opposite each other on the surface of a sphere are said to be antipodal. In geography, Australia and New Zealand were originally referred to by their European colonists as the Antipodes because those two countries are directly opposite Europe on the surface of the earth. The four-syllable words *antipodes* and *antipodal* are stressed on the second syllable. [12, 158]

antiprism (noun): the first element is from Greek *anti-* (*q.v.*), literally "against," but used here in the figurative sense of "against the stated order," therefore "reversed." The second element is *prism* (*q.v.*). A prism is a solid with two parallel, congruent bases: each vertex of the upper base is connected to the corresponding vertex of the lower base directly beneath it. An antiprism goes "against" that arrangement. In an antiprism one base is rotated somewhat relative to the other, so that the upper set of vertices no longer lines up with the lower set. Each vertex of one base is connected to the two closest vertices of the other base. The lateral surface of an antiprism is made up of a zig-zag band of triangular faces. An antiprism is also known as a prismoid. [12, 174]

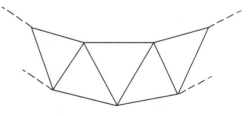

Part of an antiprism

any (adjective): a disguised number word, from the same source as native English *one*: The Indo-European root is *oi-no-* "one." If a set has *any* members, it has at least *one*. When we say something like "For any element of a set . . ." we are really saying "Pick out *one* element of the set . . . ." [148]

apeirogon (noun): from Greek *apeiron* "boundless, endless," plus the suffix *-gon* (*q.v.*) used to indicate a polygon. Greek *apeiron* is itself a compound of *a(n)* "not" and *peira* "experience, attempt, endeavor." The

Indo-European root is *per-* "to try, to risk," as seen in Latin-derived *experience* and *peril*, and native English *fear*. The connection between the meanings of the components "not" and "experience" and the compound's meaning of "boundless" is that something boundless "experiences no [end]." Another explanation is that someone who is without fear can go forward unchecked. In geometry, an apeirogon is a limiting case of a regular polygon. The number of sides in an apeirogon is becoming infinite, so the apeirogon as a whole approaches a circle. A magnified view of a small piece of the apeirogon looks like a straight line. [141, 165, 65]

aperiodic (adjective): the first element is from Greek *an* "not," reduced to *a* before a consonant. The Indo-European root is *ne* "not," as seen in the native English prefix *un-*. The second element of the compound is periodic (*q.v.*). If a function is aperiodic then it is not periodic. The circular trigonometric functions are periodic but the hyperbolic trigonometric functions are aperiodic. [141, 165, 183]

apex, plural *apices* (noun): a Latin word meaning "tip, peak," from the Indo-European root *ap-* "to take, to reach." The apex is the highest (or lowest) point that an object "reaches," the greatest (or least) height that it "takes on." [13]

apothem (noun): the first component is from Greek and Indo-European *apo-* "away from." That root is also seen in the native English cognate *ebb*, the tide which flows away from the shore. The second element is from Greek *thema* "position," from the Indo-European root *dhe-* "to put, to set," as seen in native English *do*. In a regular polygon, an apothem is a line segment that "puts out away from" the center and ends at the midpoint of any side of the poly-

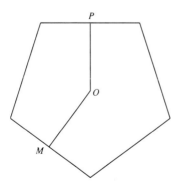

Segments $\overline{OP}$ and $\overline{OM}$ are two of the five apothems of this regular pentagon.

gon. The term *short radius* has occasionally been used in place of *apothem*. By analogy with a two-dimensional apothem, the slant height of a face of a regular pyramid is also known as an apothem. [14, 41]

applied (adjective): via French, from Latin *ad* (*q.v.*) "to" and *plicare* "to fold." The Indo-European root is *plek-* "to plait," an extension of *pel-* "to fold." The basic meaning of *apply* entails putting two things in contact, as when a piece of paper is folded onto itself or a coat of paint is applied to a wall. Applied mathematics is mathematics that is "put in contact" with practical things, as opposed to being studied for its own sake. Contrast *pure*. [1, 160]

appreciate (verb), **appreciation** (noun): from Latin *ad* (*q.v.*) "toward" and *pretium* "price, worth, value." The Indo-European root is *per-* "to distribute, to traffic in, to sell." Related borrowings from Latin via French include *appraise* and *precious*. When the price of something goes up, that thing appreciates in value. Contrast *depreciate*. [1, 165]

approach (verb): from French *approcher*, from Latin *ad* (*q.v.*) "to" and *prope* "near." The Indo-European root is *per-* "forward, through, in front of," with many other extended meanings. When you approach something, it gradually takes on greater importance in front of you until eventually it becomes foremost. Paradoxically, native English *far* comes from the same Indo-European root, since if you keep going forward you reach things that are far away. In mathematics, the notion of a quantity approaching a certain value is usually symbolized by an arrow, as in "$x \rightarrow 3$," i.e., "x approaches 3." In American textbooks of the late 19th century, the same concept was indicated by putting a dot over an equal sign: "$x \doteq 3$" [1, 165]

approximate (verb, adjective): from Latin *ad* (*q.v.*) "to" and *proximus* "nearest," the superlative of *prope* "near." The Indo-European root is *per-* "forward, through, in front of," with many other extended meanings. When you approximate a quantity, you "move forward" and get nearer and nearer to its true value. [1, 165]

arbelos (noun): a Greek word meaning "cobbler's knife." The word is of unknown prior origin. The typical cobbler's knife used in ancient Greece must have looked like the geometric figure known as an arbelos. Imagine a semicircle whose horizontal base is below the arc of the semicircle. Two adjacent, smaller, concave-down semicircular arcs are placed on, and exactly cover, the diameter of the larger semi-circle. The region bounded by all three of the semi-circular arcs is the arbelos. Archimedes calculated the area of the arbelos: $A = \frac{\pi}{4} d_1 d_2$, where d_1 and d_2 are the diameters of the two smaller semicircles. Compare *salinon*.

The arbelos is bounded by
three semicircles with
collinear diameters.

arbitrary (adjective): from Latin *arbitrarius* "un-certain, not fixed," from the noun *arbiter* "a judge." The meanings of the Latin noun and adjective may seem contradictory, but a judge is called in precisely when a situation is uncertain. The origin of Latin *arbiter* is also uncertain. One hypothesis explains it as *ad* "to," plus the Indo-European root $g^w a$- or $g^w em$-, with the same meaning as native English *to come*. If that is the origin, then an arbiter is someone who comes in to settle a dispute. In calculus the constant of integration is an arbitrary constant because, unless otherwise specified, it can have any value. [1, 77]

arborescence (noun), **arboricity** (noun): from Latin *arbor* "tree," of unknown prior origin. (Surprisingly unrelated, at least etymologically, is English *arbor* "a shady place in a garden"; the word is ultimately from Latin *herba*, from which we have also bor-rowed *herb*.) Graph theory deals with mathematical objects called trees and forests. Mathematicians have extended the metaphor with *arborescence*, literally "the state of becoming a tree," and *arboricity*, lit-erally "tree-ness," though the definitions are quite technical. The arboricity of a graph G, for example, is defined as the minimum number of line-disjoint acyclic subgraphs whose union is G.

arc (noun), **arcwise** (adverb): from Latin *arcus* "bow" (the kind that shoots arrows). Because a bow is curved, the word came to be applied to a curved section of anything, particularly a circle. The Indo-European *arku*- meant either the bow or the arrow, as is evident in the native English cognate *arrow*. In geometry, an arc is a (necessarily curved) portion of a circle. In trigonometry, the prefix *arc*- is used to designate the inverse trigonometric functions. For example, in $y = \arcsin x$, y is the angle ($=$ arc, which in a circle is measured by the central angle) whose sine is x. The native English suffix *-wise* means "with regard to," so *arcwise* means "with re-gard to arcs." The Indo-European root *weid-*, which gave rise to *-wise*, meant "to see." It is found in the borrowed words *video*, *vision*, and *view*. A region is said to be arcwise connected if each pair of points can be joined by a curve ($=$ arc) all of whose points are in the region. [244]

are (noun), **area** (noun), **areal** (adjective): *area* is a Latin word with many related meanings: "a vacant piece of ground, a plot of ground for building, the site of a house, a playground, an open court or quad-rangle, a threshing floor." The word is of unknown prior origin. From the piece of ground or the floor itself, the meaning shifted to the size of the floor and eventually to the size of any two-dimensional plot, whether physical or more abstract. In the Interna-tional System of Units, the *are* was chosen as a unit of area equal to $100\, m^2$.

argument (noun): from Latin *arguere*, originally "to make clear." The Indo-European root is *arg*- "to shine; white." A related word is *Argentina*, named for its silver deposits; compare the chemical sym-bol for silver, *Ag*. From the sense of "brightness, clarity," Latin *arguere* meant "to have a dispute or discussion," because each party is trying to make his point of view clear to the other. The word *argu-ment* came to mean not only the dispute itself, but also the proof or evidence used in the dispute, hence the topic or theme that was to be discussed. The mathematical use of *argument* as "the independent variable of a function" came about metaphorically as the "topic" that the function would "deal with." Compare *operand*. In a different sense, when a com-plex number $a + bi$ is represented graphically by a point P in a two-dimensional coordinate system, the angle θ between the positive x-axis and the line connecting the point P to the origin is also known as the *argument* of the complex number in question; that's because θ is the independent variable of the function $e^{i\theta}$ that can also be used to represent the same complex number. The argument of a complex number z is abbreviated arg (z).

arithmetic (noun, adjective): from the Greek *arith-mos* "number," from the Indo-European root *ar*- "to fit together." A related borrowing from Greek is *aris-tocrat*, presumably a person in whom the best qual-ities are fitted together. Arithmetic must once have been conceived of as fitting things together or arranging or counting them. An arithmétic (notice the stress on the third syllable) series is one in which each term is a fixed number apart from adjacent terms, just as the counting numbers of arithmetic are equally spaced. Interestingly enough, the same Indo-European root found in *arithmetic* appears in native English *read*, since when you read you have to fit the sounds together into words. So of the so-called three R's—reading, (w)riting, and (a)rithmetic—two of them are etymologically related. Because *arithmetic* is a foreign word, English speakers have sometimes misconstrued it. In the 14th and 15th centuries it was known in England by the Latin-like name *ars metrik* "the metric art," out of confusion with *metric*. It has similarly been called *arithmetric*. [15]

arm (noun): a native English word, from the Indo-European root *ar-* "to fit together," since a human arm is fitted together out of the radius, ulna, and humerus bones. The English word *arm* in the sense of "weapon" is borrowed from Latin but also comes from the same Indo-European root, since weapons are often fitted together from components. In geometry each of the two sides of an angle is sometimes called an arm by analogy with the way a person's two outstretched arms form an angle. Compare *leg*. [15]

arrange (verb), **arrangement** (noun): from Old French *arangier*, composed of *a* "to" and *reng* "a line, a row." An arrangement was originally a group of objects placed in a row, whether the row was straight or curved. The linear connotation is still common in mathematics in "arrangement" problems involving permutations and combinations. The Old French noun *reng* was borrowed from Germanic, as evidenced by the native English cognate *ring*, originally *hring*. The Indo-European root is *(s)ker-* "to turn," as seen in the meaning of English *ring*. [1, 198]

array (verb, noun): from an assumed Vulgar Latin *arredare* "to put in order." The first element is from Latin *ad* (*q.v.*) "(in)to." The Indo-European root that underlies Latin *redare* is *reidh-* with the same meaning as the English cognate *ride*. The basic meaning of *array* is "to ride into formation." In centuries past, when troops were arrayed they were formed into rectangular arrangements. In mathematics, an array is a rectangular arrangement of numbers, as in a determinant or matrix. All the numbers in a matrix can be imagined riding along like a cavalry in their well-defined rows and columns. [1]

ascending (adjective): from Latin *ad* (*q.v.*) "to" and *scandere* "to climb, leap." The Indo-European root is *skand-* "to climb." Surprisingly related is *scandal*, from Greek *skandalos* "a snare, trap, stumbling block" (something you trip or leap over). In mathematics when the terms of a polynomial are arranged in ascending order, the exponents "climb" higher from left to right. [1, 194]

associative (adjective), **associativity** (noun): from Latin *ad* (*q.v.*) "to" and *socius* "partner, companion"; an associate is someone who is a companion to you. The underlying Indo-European root *sek^w-* meant "to follow." An associate is literally someone you follow around or keep company with. Related borrowings from Latin and French include *suitor*, *ensue* and *sequel*. In mathematics the associative property of addition says that $(a + b) + c = a + (b + c)$. In other words, it doesn't matter how the three quantities "keep company with" one another, the result will still be the same. [1, 186]

assume (verb), **assumption** (noun): from Latin *ad* (*q.v.*) "to" and *sumere* "to take, obtain." When you assume something, you take it to heart or accept it as true. Latin *sumere* is itself a compound. Its first element is *sus-*, more commonly *sub-* "up from under," from Indo-European *upo*, of similar meaning to the native English cognate *up*. The second element is the verb *emere* "to take, receive, buy," from the Indo-European root *em-* "to take, distribute." In mathematics we often make assumptions based on evidence, but then we must prove that our assumptions are true. [1, 240, 51]

asterisk (noun): from Late Latin *asteriscus*, from Greek *asteriskos*, diminutive of *aster* "star." (The flower known as an *aster* is so called because it looks star-like.) The Indo-European root is *ster-*, which is easily recognizable in the native English cognate *star*. Related borrowings from Latin include *stellar* and *constellation*. An asterisk is literally a little star. In mathematics the asterisk (∗) may represent any binary operation. As used in computer notation, the asterisk indicates multiplication. Many people mistakenly pronounce *asterisk* as if it were *asterick*. [212]

astroid (noun): from Greek *aster* "a star," plus the suffix *-oid* (*q.v.*) "having the shape of." The Indo-European root *ster-* is little changed in the native English cognate *star*. Related borrowings include *disaster* (away from the good influence of a star) and, from Persian, the woman's name *Esther* (compare

Latin *Stella*). In mathematics the astroid is a star-like closed curve. Because it is a hypocycloid of four cusps, it has also been called a tetracuspid. Johann Bernoulli first studied the astroid in 1691. [212, 244]

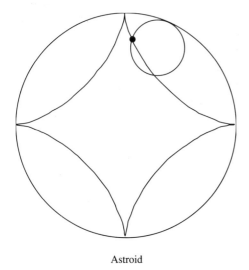

Astroid

asymmetric (adjective), **asymmetry** (noun): the first component is from Greek *an-*, shortened to *a-* before a consonant, meaning "not." The second component is *symmetric* (*q.v.*). In geometry, asymmetry is the lack of symmetry. In algebra, an asymmetric relation $*$ is a relation for which if $x * y$ is true, then $y * x$ is false. For example, the relation "is greater than" is asymmetric; since it is true that $4 > 3$, it is therefore false that $3 > 4$. [141, 106, 126]

asymptote (noun), **asymptotic** (adjective), **asymptotically** (adverb): from three Greek words. The particle *an-*, shortened to *a-* before a consonant, means "not." The native English cognate is *un-*, as seen in *unhappy* and *unloved*. The Greek preposition *sun* or *sum* "together with" is from the Indo-European root *ksun* "with." The Greek verb *piptein* means "to fall." The Indo-European root *pet-* "to rush or fly forward" (and hence to fall) can be seen in Latin-derived *impetuous*, Greek-derived *helicopter*, and native English *feather*. An asymptote is a curve—most often a straight line—that another curve "doesn't fall together with." In other words, the second curve "runs alongside" its asymptote, getting closer to it but never hitting it. Linear asymptotes are classified as horizontal, vertical, or slant (oblique). Two functions f and g are said to be asymptotically equal if

$$\lim_{x \to \infty} \frac{f(x)}{g(x)} = 1.$$

The word *asymptoTe* is not pronounced as if it were *asymptoPe*. People who mispronounce it that way are probably getting confused by the word *isotope*, but no radioactive asymptote has ever been reported. Our word *asymptomatic* is etymologically identical to *asymptotic*: a symptom is literally "a falling together" of a bodily condition and a given disease associated with that condition. Few students of mathematics associate the similarly spelled words *asymptomatic* and *asymptotic*, most likely because the words are used in different disciplines and because different syllables are stressed. [141, 106, 168, 117]

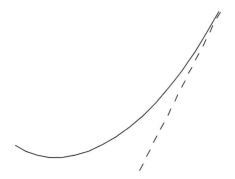

A curve approaching a slant asymptote

atled (noun): the word *delta* spelled backwards. The atled operator,

$$\mathbf{i} \, \frac{\partial}{\partial x} + \mathbf{j} \, \frac{\partial}{\partial y} + \mathbf{k} \, \frac{\partial}{\partial z},$$

gets its name from the fact that it is symbolized by an upside-down (= backwards) Greek capital delta, ∇. See more under *del* and *nabla*. [31]

atomic (noun), **atomism** (noun): the first element is from Greek *an-* "not," shortened to *a-* before a consonant. From the same Indo-European root *ne* "not," comes the English cognate *un-*. The second element is from Greek *temnein* "to cut." The Indo-European root is *tem-* "to cut." An atom is literally something that can't be cut into smaller pieces. (What a misnomer for the thing that physicists now split into all sorts of "subatomic" particles!) In logic, an atomic statement is one which can't be broken down into smaller statements; any further subdivision produces only elements that fall short of being statements. "Eve likes rice," for example, consists of a subject "Eve" and a predicate "likes rice," neither of which is a statement. Atomism is the belief that space or time (or matter, for that matter) can be separated into smallest pieces that can't be further divided. [141, 225]

atto-, abbreviated *a* (numerical prefix): from Danish and Norwegian *atten*, which is cognate to, pronounced like, and means the same as native English *eighteen*. In 1964 *atto-* became part of the set of submultiple prefixes used in the International System of Units. *Atto-* multiplies the unit to which it is attached by 10^{-18}. [149, 34]

attractor (noun): the first component is from Latin *ad* (*q.v.*) "to." *Tractor* is a Latin masculine noun meaning "the one that pulls," from *tractus*, past participle of *trahere* "to drag, to draw, to pull." The Indo-European root is *tragh-* "to draw, to drag." In mathematics an attractor is a region or shape to which points are "pulled" as the result of a certain process. Attractors occur in connection with fractals. [1, 234, 153]

augment (verb): from Latin *augere* "to increase," from the Indo-European root *aug-* of the same meaning. That root appears in English *a nickname*, which used to be *an eke name*, meaning an extra name that a person has. A related borrowing from Latin is *auction*, in which prices increase. In mathematics, an augmented matrix contains one or more extra columns to the right of the original matrix. [18]

aut (conjunction): a Latin word pronounced like English *out* and meaning "or." The Indo-European root is *au-*, which is found in various particles and adverbs including archaic English *eke* "also." In Classical Latin the conjunction *aut* was used when an inherent incompatibility was being referred to: is it day or is it night? Mathematicians sometimes resort to Latin *aut* because English *or* is ambiguous. It can be inclusive: this thing or that thing or both things. It can be exclusive: this thing or that thing but not both things. In logic, to avoid ambiguity, *aut* indicates the exclusive *or*. The symbol $\veebar$ is frequently used to stand for the word *aut*. Contrast *vel*.

automaton, plural *automata* (noun): the first element is from Greek *autos* "of or by itself, independently," of unknown prior origin. The second element is from Greek *matos* "willing." The Indo-European root is *men-* "to think." An automaton is literally a self-willed machine; the most common example is a robot. New York used to be host to self-service restaurants called automats: a customer would insert coins into wall slots and remove food

from glass-fronted compartments. In graph theory, finite automata are certain ordered pairs of functions. [19]

automorphic (adjective), **automorphism** (noun): from Greek *autos* "of or by itself, independently," and *morph-* "shape, form, beauty, outward appearance." The origin of both Greek components is still unknown. An automorphism is an isomorphism (*q.v.*) of a set with itself. In number theory, an automorphic number is one whose square ends with the same sequence of digits as the number itself; both endings have the "same shape." A simple example is 25, whose square also ends in 25. A bigger example is 1,787,109,376, whose square ends with that same sequence of 10 digits. [19, 137]

auxiliary (adjective): from Latin *auxiliaris* "suitable for aid, helping," from the noun *auxilium* "help, aid, assistance." The Latin noun is based on the verb *augere* "to increase, to make grow, exalt" from the Indo-European root *aug-* "to increase, promote, originate." Related Latin borrowings include *author* (an originator of a work) and *augur* (to foretell the future with the hope of bringing about an increase in one's fortune or luck). Native English cognates include the verbs *to eke* (out a living, for example) and *to wax* (as in the moon's waxing and waning). In geometry, an auxiliary line is an extra line added to a diagram to help make certain relationships clearer. An auxiliary equation is an extra equation that helps in the solution of a linear differential equation. [18]

average (noun): ultimately from the Arabic *'awar* "fault, blemish." Upon that root was built the Arabic *'awariyah*, meaning "goods damaged in shipping." The term arose during the many centuries when the Arabs traded extensively around the Mediterranean and elsewhere. The Italians and French borrowed the Arabic word in the sense of "the financial loss incurred when goods are damaged." The meaning then shifted to "that portion of the loss borne by each of the many people who invested in the ship." From there the word came to have its modern sense of "an equal portion." The current *-age* ending of the word is probably due to the influence of *damage*, a word that was a synonym of *average* in its original sense.

avoirdupois (noun): a compound of three French words. *Avoir* "to have" is from Latin *habere*, from the Indo-European root *ghabh-* "to give, to receive." *Du* is a form of *de* (*q.v.*) "of," though in French it developed the meaning "some [of]." *Pois* is an alternate spelling of *poids* "weight," from Latin *pensare* "to weigh," from the Indo-European root *(s)pen-* "to draw, stretch." When a weight is attached to a spring it stretches the spring. *Avoirdupois* means literally "have some weight," which is what can be said of the pound, ounce, and other units of the avoirdupois system. Although *avoirdupois* is a French word, the system is used primarily in English-speaking countries. The word *avoirdupois* has at times been capitalized, leading some English speakers to the false conclusion that, just as there were men named Fahrenheit and Celsius whose names are now applied to temperature scales, there was someone named Avoirdupois whose name is now applied to a system of weights. [69, 32, 206]

axiom (noun), **axiomatic** (adjective): via Latin, from Greek *axioma* "that which is thought fitting; decision; self-evident principle." The Indo-European root is *ag-* "to drive, to lead." A subsidiary Greek meaning, "to weigh," led to *axioma*, literally "something weighty." In mathematical terms, axioms are concepts felt "weighty" or worthy enough that you can base a logical system on them. [3]

axis, plural *axes* (noun), **axial** (adjective): *axis* is a Latin word meaning "axle, pivot." The Indo-European root is *aks-* "axis." The English word *axle* is from the same root, via Old Norse. In mathematics an axis is a line around which a coordinate system "pivots." In calculus a coordinate axis acts as an axle when a curve is rotated about it to create a surface. In a three-dimensional coordinate system, the axial planes are the *yz-*, *xz-*, and *xy*-planes; in other words, they are the planes obtained by letting $x = 0$, $y = 0$, and $z = 0$, respectively. [5]

axonometric (adjective), **axonometry** (noun): the first element is from Greek *axon-* "axis," from the Indo-European root *aks-* "axis." The second element is from Greek *metron* "a measure," from the Indo-European root of the same meaning, *me-*. When a three-dimensional graph is projected onto a flat piece of paper, the positive *x-*, *y-* and *z*-axes are typically drawn radiating from a single point, and then a scale is placed on each axis. The scaling of the axes is known as axonometry, literally "axis measuring." The three types of axonometry are *isometry*, *dimetry* and *trimetry* (*qq.v.*). [5, 126]

azimuth (noun), **azimuthal** (adjective): from Arabic *as-sumut* "the ways, compass bearings," from Latin *semita* "path, lane," of unknown prior origin. In a cylindrical or spherical coordinate system, the angle

θ measured counterclockwise in the horizontal plane is sometimes known as the azimuth. In astronomy, as opposed to mathematics, the azimuth is usually measured clockwise.

B

ball (noun): from Old Norse *böllr*, of the same meaning. The Indo-European root is *bhel-* "to blow, to swell," with extended meanings referring to round objects. English cognates are *bowl* and *boll*. In mathematics an open ball consists of all the points inside a sphere, while a closed ball adds the points on the sphere itself. Technically speaking a sphere is a surface, not a solid, though many people say *sphere* when they mean *ball*. Compare *disc* in two dimensions. [21]

band (noun): from French *bande*, from a presumed Germanic *bendon*. The Indo-European root is *bhendh-* "to bind," which is an English cognate. Any strip of cloth or hide or other material that was used to bind something became known as a band. The meaning was later extended to any object having the shape or appearance of a typical band used for binding. In mathematics a band matrix is a square matrix whose only nonzero elements lie on the main diagonal or either of the two diagonals adjacent to the main diagonal. The nonzero elements form a kind of band that runs diagonally across the matrix. [22]

$$\begin{bmatrix} -4 & 5 & 0 & 0 & 0 \\ 8 & 7 & 0 & 0 & 0 \\ 0 & 0 & -2 & 4 & 0 \\ 0 & 0 & 3 & 5 & 9 \\ 0 & 0 & 0 & 1 & 0 \end{bmatrix}$$

A band matrix.

bar (noun): the Romance languages have a common word *barra*, whose French form *barre* was borrowed into English. The source of the original word is unknown. The earliest English meanings, going back to the 12th century, were "rod" and "barrier." From the shape of a rod comes the subsidiary meaning "stripe," which is the sense used in a bar graph, in which the bars look like stripes arrayed side by side. Also from the shape of a rod comes the name of the horizontal line placed over a letter, as in $\bar{x}$, *x*-bar.

barn (noun): a native English word, from the Indo-European root *bhares-* "barley," so that a barn must originally have been a place where grain is stored. A related borrowing from Latin is *farina*, which comes not from barley but from another grain, wheat. A barn is a unit of area equal to 10^{-24}cm^2. The term appears to have been coined by analogy with the word *area* itself, which is a Latin word that meant "threshing floor." Compare *shed*.

barycentric (adjective): the first component is from Greek *barus* "heavy," from the Indo-European root g^w*erə-* "heavy." A related borrowing from Latin is *gravity*; from Sanskrit comes the spiritual heavyweight known as a *guru*. The second component is from Greek-derived *center* (*q.v.*). The barycentric coordinates of a point in *n*-space may be determined by a linear combination involving a center of mass (= "heaviness"). Barycentric coordinates are also known as areal coordinates. [91]

base (noun), **basis**, plural *bases* (noun): from Greek *basis* "a stepping." The Indo-European root is g^w*a-* or g^w*em-*, cognate to and with the same meaning as native English *come*. The sense of "stepping" is retained in the game *baseball*, in which a runner steps on the bases as he advances. Since you necessarily step with your feet, and your feet support your whole body, *basis* came to mean "support, foundation." Also, since your feet are located at the lowest end of your body, something *base* came to mean something positioned down low, and also, in a moral sense, low down. In music the sense has shifted to sound, so that a bass is the stringed instrument that produces the lowest notes. In a mathematical expression like b^n, *b* is said to be the base because it is lower down than the exponent *n*. In a logarithmic expression like $\log_b n$, once again the base *b* is written down low. [77]

because (conjunction): short for *because of*, which in older English was *by cause of*, meaning "by reason of." Native English *by* is from the Indo-European root *ambhi-* "around." The word *cause* is via Old French from Latin *causa* "cause, reason, purpose," of unknown prior origin. In logic the statement "*p* because *q*" is defined as $(p \wedge q) \wedge (q \rightarrow p)$. In other words, it isn't enough that both *p* and *q* be true; *q* must also imply (= "cause") *p*. [9]

bell (noun): a native English word, from the Indo-European root *bhel-* "to cry out, yell," as found also in English *bellow*. A related borrowing from German is *poltergeist*, a spirit that makes noise. The famous

bell curve of statistics is given by the equation

$$y = \frac{1}{\sigma \sqrt{2\pi}} e^{-x^2/(2\sigma^2)},$$

where σ represents the standard deviation. The bell curve is named for its resemblance to the cross-sectional shape of a typical bell; it has nothing to do with yelling or the sound that a bell makes.

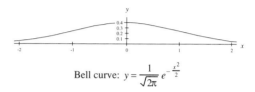

Bell curve: $y = \frac{1}{\sqrt{2\pi}} e^{-\frac{x^2}{2}}$

belong (verb): a native English compound. The first component is *be-*, used as an intensifier. It is from the same Indo-European root *ambhi* "around," found in the English preposition *by*. The second component is *long*, from the Indo-European root *del-*, of the same meaning. The semantic development is complicated. Here are two plausible explanations. 1) Although *long* originally referred to physical dimensions, the meaning was later extended to time. We say, for example "How long did she wait?" As time passes (= grows long), we sometimes miss people who are absent, so *long* came to mean "yearn." As time passes, we also long to accomplish things, to play a part in activities we enjoy and to be a part of groups we admire, so *belong* took on the meaning "be part of." 2) According to the *Oxford English Dictionary*, the original meaning of *belong* was apparently "as long as" or "running alongside of." Subsidiary developments would have been "parallel to," "going along with," "accompanying as a property or attribute," and finally "being a part of." In mathematics an element *x* of a set *A* is said to belong to the set. In symbols, we write $x \in A$. The $\in$ symbol is the first letter of the Greek verb form *estis* "is," because an element that belongs to a set is in the set. Compare *member*. [9, 35]

bend (verb, noun): a native English word, from the Indo-European root *bhendh-* "to bind." If you bind something, you necessarily bend the cord or bond (another cognate) that you use to do the binding. In calculus a bend point on a curve is one at which the tangent line is horizontal and at which there is no point of inflection. In other words, a bend point is a relative maximum or minimum. Compare *turning*. Contrast *stationary*. [22]

beta (noun): the second letter in the Greek alphabet, written B as a capital and β in lower case. The origin of the Greek letter is Hebrew *beth*, ⊐, which meant "house." The original is recognizable in *Bethlehem*, literally "house of bread." Greek β is frequently used in mathematics and physics. In triangle *ABC*, β may represent the angle at vertex *B*. It has also recently gained currency in the computer world: the beta version of computer software comes relatively late in the development of the software, as opposed to the earlier and buggier alpha version. See more information under *aleph* and *alpha*.

between (preposition): a native English compound. The first element is *be-* "at, around," from the Indo-European root *ambhi* "around." A related borrowing from Greek is *amphitheater*. The second component is from the root found in English *two*, *twain*, and *twin*. The Indo-European root is *dwo-* "two." The literal meaning of *between* is "at the place separating two points." [9, 48]

bi-: a Latin prefix that developed from the older form *dui-*, which more closely resembles the underlying Indo-European root *dwo-* "two." The native English cognate is *two*. The prefix *bi-* appears not only in the mathematical entries that follow, but also in many nonmathematical words that English has constructed using Latin and Greek elements. Examples are *bicycle*, *bigamy*, *bilabial*, *biped*, and *biscuit* (= twice cooked). See more under *two*. [48]

bias (noun), **biased** (adjective): from French *biais* "oblique," as can still be seen by the use of the word in sewing, where it means "a cut across the grain of a fabric." From the meaning "oblique" comes the mathematical use of the word to describe a statistical sample which is "slanted" away from the true average. The French word was borrowed from Old Provençal; the original source isn't known. One explanation traces the word to Latin *bifac-* "looking two ways, " from *bi-* (*q.v.*) "two" and *facies* "face." [48, 41]

biconditional (adjective, noun): from Latin *bi-* "two" and *condition* (*qq.v.*). In logic, a biconditional statement such as $a \Leftrightarrow b$ is one that expresses two "conditions" or implications: $a \Rightarrow b$ and $b \Rightarrow a$. A biconditional statement is also known as an equivalence. [48, 103, 33]

bicontinuous (adjective): from Latin *bi-* "two" and *continuous* (*qq.v.*). A mathematical function is said to be bicontinuous if it is continuous and if its inverse

is also continuous. In other words, two continuous functions are involved. [48, 103, 226]

bicorn (noun): from Latin *bi-* (*q.v.*) "two" and *corn-* "horn, " from the Indo-European root *ker-* "horn, head." The native English cognate is *horn*. In mathematics the bicorn is a closed curve with two "horns," i.e., corners, on it. Parametric equations for the bicorn are

$$\left\{ \begin{array}{l} x = a \sin t \\ y = \dfrac{a \cos^2 t\, (2 + \cos t)}{3 + \sin^2 t} \end{array} \right\},$$

for $-\pi \leqslant t \leqslant \pi$. [48, 93]

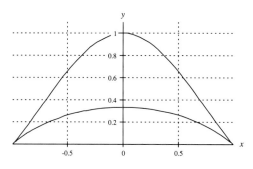

Bicorn: $x = \sin t, \quad y = \dfrac{\cos^2 t (2 + \cos t)}{3 + \sin^2 t}$

bifolium (noun): the first component is from Latin *bi-* (*q.v.*) "two." The second component is Latin *folium* "leaf"; the Indo-European root is *bhel-* "to thrive, to bloom." A related borrowing from French is *flower*. The bifolium is a curve defined by the polar equation $r = 4a \sin \theta \cos^2 \theta$. Its name refers to the curve's two loops or "leaves." Compare *folium* and *trifolium*. [48, 21]

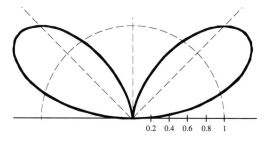

Bifolium: $r = 4 \sin \theta \cos^2 \theta$

bifurcate (verb), **bifurcation** (noun): the first component is from Latin *bi-* (*q.v.*) "two." The second component is from Latin *furca* "a two-pronged fork," of unknown prior origin. Latin *furca* is the source of our borrowed word *fork*. Because the *furca* had two prongs, the word *bifurcate* "to split into two branches" is redundant. Bifurcation points occur in the study of differential equations. [48]

big (adjective): of unknown origin, possibly from Scandinavia. In the 13th century, *big* meant "strong, stout." By the 16th century the word had taken on the meanings "advanced in pregnancy" and "of great bulk." The modern sense dates back only to the 16th century.

bijective (adjective): from Latin *bi-* (*q.v.*) "two" and *ject-*, past participial stem of *jacere* "to throw." The Indo-European root is *ye-* "to throw." Related borrowings from Latin include *inject*, *reject*, and *subject*. In mathematics, a mapping of points from one set onto another set is called a surjective mapping; if the mapping back to the original set is also surjective, then the transformation is said to be *bijective*. [48, 257]

bilinear (adjective): from Latin *bi-* (*q.v.*) "two" and *linear* (*q.v.*). The complex transformation defined by

$$w = \frac{az + b}{cz + d}$$

is called bilinear. It takes its name from the fact that it is the quotient of two linear expressions. [48, 119]

billion (numeral): patterned after *million* by replacing the *mi-* with *bi-* (*q.v.*) "two" in the 16th century. That replacement was rather whimsical, and was based more on orthography than etymology, since the *mi-* of *million* had no meaning in its own right. As indicated by the prefix *bi-*, the new number *billion* had two groups of six zeros, rather than the one group of six zeros in a million; a billion was therefore equal to the second power of a million, or 10^{12}. Because a million was 10^6 and a billion was 10^{12}, the intermediate step of 10^9 got skipped. In the United States and France, a billion was consequently redefined to be 10^9 to fill the gap. In countries where a billion is 10^{12}, the variant *milliard* is sometimes used as the otherwise missing word for 10^9. Only near the end of the 17th century was the word *billion* adopted into English. [48, 72, 151]

bimodal (adjective): from Latin *bi-* (*q.v.*) "two" and *modus* "standard, measure." In statistics a bimodal distribution has two values that appear more prominently than others. The curve representing the distribution has two relative maxima. [48, 128]

A bimodal distribution.

binary (adjective): from Late Latin *binarius*, from Classical Latin *bini* "two by two, two apiece," from *bi-* (*q.v.*) "two." A binary system of numerical notation uses base two. The binary system underlies virtually all computer calculations. Because of its frequent use in the computer world, *binary* is sometimes shortened to *bin*. The Greek-derived word *dyadic* has sometimes been used as a synonym of the Latin-derived *binary*, but *binary* is now used almost exclusively. Compare *denary*, *octonary*, *quinary*, *senary*, *ternary*, and *vicenary*. [48]

binomial (noun): the prefix is from Greek *bi-* (*q.v.*) "two." One explanation for the second part of the compound involves Greek *nomos*, which meant many things: usage, custom, law, division, portion, part. In that case, a trinomial is a mathematical expression consisting of three parts. A second and more likely correct explanation involves Latin *nomen*, cognate to English *name*, so that a binomial is an expression involving two names, i.e., terms. According to this explanation, the Medieval Latin *binomial* was a copy of the Greek phrase *ek duo onomaton* "of two names," found in Euclid. In biology a binomial is a two-part name consisting of a genus and a species. [48, 143, 145]

binormal (adjective): from Greek *bi-* "two," plus *normal* (*qq.v.*). Suppose you have a space curve defined by a vector-valued function. At a certain point on the curve, the first derivative can be used to find a unit tangent vector. There are infinitely many vectors normal to that tangent vector, but the principal unit normal vector is found by differentiating the unit tangent vector. A binormal vector is one which is normal to the other two (= *bi-*) vectors, the tangent vector and the principal normal vector. [48, 75]

bipolar (adjective): from Greek *bi-* "two," plus *polar* (*qq.v.*). In a system of bipolar coordinates, distances are measured from each of two fixed points (= poles). [48, 107]

biquadrate (noun), **biquadratic** (adjective): from Latin *bi-* "two " and *quadratic* (*qq.v.*) "of second degree." The power of a biquadrate is twice that of something quadratic; in other words, a biquadrate is a fourth power. Likewise, an equation of the fourth degree is called biquadratic; it is also known as a quartic equation. [48, 108]

birectangular (adjective): from Latin *bi-* "two" and *rectangular* (*qq.v.*) "involving a right angle." In spherical trigonometry, a birectangular triangle contains two right angles. [48, 179, 11, 238]

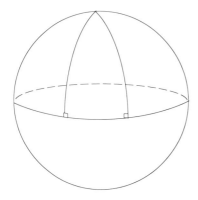

A birectangular
spherical triangle

bisect (verb), **bisection** (noun), **bisector** (noun): from Latin *bi-* (*q.v.*) "two" and *sectus*, past participle of the verb *secui* "to cut." The Indo-European root is *sek-* "to cut," as seen in the native English noun *saw*. When you bisect a line segment you cut it into two (equal) parts. The addition of the Latin agental suffix *-or* creates *bisector* "something (usually a line) which bisects." [48, 185, 153]

bisymmetric (adjective): a hybrid of Latin *bi-* "two" and Greek-derived *symmetric* (*qq.v.*). Something which is bisymmetric has two axes of symmetry. Every ellipse is symmetric with respect to its major and minor axes, and is therefore bisymmetric. A square matrix is said to be bisymmetric if its elements are symmetric to each of the two main diagonals. [48, 106, 126]

$$\begin{bmatrix} 3 & 2 & 0 & 2 & 3 \\ 2 & 9 & 4 & 9 & 2 \\ 0 & 4 & 8 & 4 & 0 \\ 2 & 9 & 4 & 9 & 2 \\ 3 & 2 & 0 & 2 & 3 \end{bmatrix}$$

A bisymmetric
matrix.

bit (noun): an abbreviated form of *binary digit*, the most basic unit of information in the world of computers: a bit answers a yes/no question with 1 for "yes " and 0 for "no." Before the creation of this term, the word *bit* already existed in English with the meaning "small amount"; that is an appropriate definition for a computer bit, given its small numerical value of 0 or 1. Etymologically speaking, English *bit* originally meant "a small amount bitten off." The Indo-European root is *bheid-* "to split." Other native English cognates include *beetle* (an insect with jaws that take a large bite) and *bitter* (= biting). Related borrowings from Latin include *fissure* and *fission*. In the American monetary system, a bit used to be a common unit equivalent to 12.5¢, or one-eighth of a dollar. [48, 33]

biunique (adjective): the first component is from Latin *bi-* (*q.v.*) "two." The second component is *unique*, of French origin, from Latin *unus* "one (*q.v.*)." In mathematics a mapping between two sets is said to be biunique when both the function representing the mapping and its inverse are one-to-one. [48, 148]

borrow (verb): a native English word, from the Indo-European root *bhergh-* "to hide, to protect," as seen also in native English *bury*. According to the *Oxford English Dictionary*, the original meaning of *borrow* involved the security given for the safety of the object taken: to borrow something was to pledge to protect it. In arithmetic, borrowing often takes place as part of the standard subtraction algorithm. As my second-grade teacher used to tell us, "When you borrow, you have to pay back." That's why we were taught that after borrowing we should add 1 to the digit next left in the subtrahend, whereas most students today are told to subtract 1 from the digit next left in the minuend.

bottom (noun): a native English word, from the Indo-European root *bhudh-* "bottom." The same root is found in the Latin borrowings *profound* (deep) and *fundamental* (forming the base of something). Although the lower part of a fraction is "officially" called the denominator (*q.v.*), many English speakers call it the bottom, which is a much simpler word. In a complex fraction, combinations of the words *top* and *bottom* behave like positive and negative numbers under multiplication. For example, the bottom of the bottom of a complex fraction is equivalent to the top of a simple fraction. Compare the picture under *isomorphism*.

bound (noun, verb), **boundary** (noun): from Old French *bunde*, *budne*, presumably from Gaulish *bodina*, of the same meaning. The original source is unknown. The word *bound* was used in English as early as the 13th century with the meaning "landmark," and in the 14th century with the current meaning. The modern form *boundary* didn't appear until the 17th century, perhaps by analogy with the adjective *limitary*.

bow (noun): a native English word, from the Indo-European root *bheug-* "to bend, to curve." From the same source comes native English *to bow (down)*. A related borrowing from Yiddish is *bagel*. In mathematics the curve whose polar equation is $r = a(1 - \tan^2 \theta)$ is called the bow because of its two loops and the long "strings" leading to it. Compare *lemniscate*.

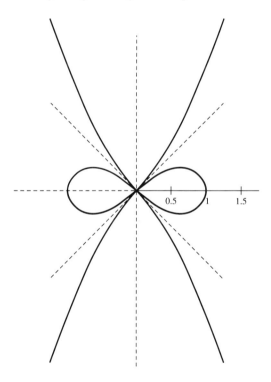

Bow curve: $r = 1 - \tan^2 \theta$

box (noun): from Late Latin *buxis*, a variant of *pyxis* "a box made of boxwood." The Latin word was borrowed from Greek *puxis*, from *puxos* "box tree." The box tree, *buxus sempervirens*, produces a hard, yellowish wood that people in ancient times obviously found suitable for making boxes. In solid geometry, a rectangular parallelepiped, or cuboid, is sometimes called a box. In logic, a box is one name for the symbol used to mean "it is possible that"; the symbol is a square whose diagonals are horizontally and vertically oriented. In analysis, an open or closed box in *n* dimensions is an extension of the concept of an open or closed interval in one dimension.

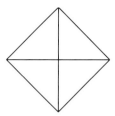

The box symbol used in logic.

brace (noun): an Old French word meaning "arms" (body parts, not weapons). A pair of braces, used initially in carpentry or construction work, and later as the grouping symbol { } in mathematics, must have looked to people like a pair of arms em-*brace*-ing something or someone. The Old French word is from Latin *bracchium* "arm," in turn from Greek *brachion* "shorter," from the Indo-European root *mregh-u-* "short." Originally the word applied to the upper arm of the human body (as opposed to the longer forearm), but eventually came to refer to the arm as a whole. Compare *bracelet*, which is jewelry worn on the arm. [138]

brachistochrone or **brachystochrone** (noun): from the superlative of Greek *brakhus* "short" and *khronos* "time." Although Greek *khronos* is of unknown prior origin, *brakhus* is from the Indo-European *mregh-u-* "short." Surprisingly related is native English *merry*, originally "pleasant, enjoyable," though it isn't clear why something short should be enjoyable. More obviously related is *brief*, which came into English via French from the Latin cognate *brevis*, from which we also have *brevity*. In mathematics, the brachistochrone is the path down which a weighted particle will fall from one point to another in the shortest time: it turns out to be a cycloid. Compare *isochrone* and *tautochrone*. [138, 99]

bracket (noun, verb): from French *braguette*, meaning, of all things "codpiece." A codpiece was a little pouch in the crotch of the tight breeches that men used to wear in Europe hundreds of years ago; the codpiece protected some delicate body parts.

Braguette was the diminutive of *brague* "breeches" (the English word also comes from the same source). A pair of brackets, used at first in architecture or shipbuilding, and later symbolically as [ ] in mathematics, apparently reminded people of a codpiece or perhaps of a pair of breeches. Many American students mistakenly believe that brackets may not be placed inside parentheses or inside other brackets. In fact there is no preferential order among brackets, braces, and parentheses, nor is there any limit to the number of times the same grouping symbol may be repeated. [57]

braid (noun): a native English word, from the Indo-European root *bherək-* "to shine, to glitter." When something glitters, the light makes brief, rapid movements, so in Germanic the root took on the meaning "to move quickly." One common activity that involved quick movements was weaving. In non-mathematical English, a braid is made from strands of hair woven together. In mathematics a braid is formed from a set of parallel strands that interweave in certain ways. Contrast the mathematical definition of *knot*.

branch (noun): from French *branche*, of the same meaning. Related Romance languages have a similar word meaning "claw, paw, hand." Despite being so widespread, the word is of uncertain origin. It may be related to Celtic *bracc* "arm." A paw or hand "branches out" into several claws, making each claw one "branch" or part of the whole. The same notion applies to branches of trees. By extension, in mathematics each separate piece of a curve (for example a hyperbola or the trigonometric tangent function) is known as a branch. [138]

breadth (noun): a native English noun corresponding to the adjective *broad*. The word is of common

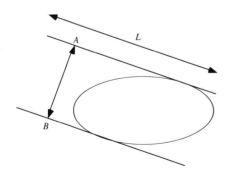

AB is the breadth of the oval in the direction of line *L*.

Germanic origin. See more on the *-th* suffix under *length*. In mathematics the breadth of an oval in a given direction is the distance between two parallel lines that point in the given direction and that are tangent to opposite sides of the oval.

bridge (noun): a native English word, perhaps from the Indo-European root *bhru-* "eyebrow," as seen in *brow* itself. By analogy with a physical bridge, a bridge in graph theory is a segment whose removal leaves the remaining pieces of the graph disconnected.

broken (adjective): a native English word, the past participle of *break*, from the Indo-European root *bhreg-* "to break." Another English cognate is *breach*, and a related borrowing from Latin is *fracture*. Also related is the homonym *brake*, which "breaks" the motion of a vehicle. In non-mathematical English, a broken line is the same as a dashed or dotted line: a straight line from which short sections have been removed. In mathematics, however, a broken line is a curve composed of connected straight line segments. The terminology was adopted around 1898 by the German mathematician David Hilbert (1862–1943). In a magic square, a broken diagonal is parallel to a main diagonal but is broken into two non-contiguous parts. [24]

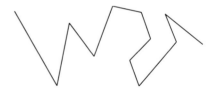

A broken line.

bundle (noun): probably from Middle Dutch *bondel* "a sheaf of papers," from the Indo-European root *bhendh-* "to bind" (which is a native English cognate). Although a bundle may originally have referred to papers, in modern English a bundle may contain objects of any sort. In mathematics the term has been abstracted away from physical objects altogether: a bundle is a collection of all the planes passing through a given point. The physical metaphor is even carried over to the naming of the common point, which is called the *carrier* of the bundle. A mathematical bundle is also known as a *sheaf*. [22]

bushel (noun): from Old French *buissiel* or *boissiel* (modern French *boisseau*), perhaps of Gaulish origin and perhaps related to the word *bin*, which is of

Celtic origin. In England, the bushel has been in use as a unit of dry measure since the 14th century. In the United States 1 bushel = 4 pecks = 32 quarts.

byte (noun): a computer term meaning "8 bits or binary digits." The computer term is spelled with a *y* to distinguish it from the homonym *bite*. In fact, *bite* is etymologically related to the non-computer-related English word *bit*, since when you bite something you chew off a little bit. The first part of the word *byte* is pronounced the same as the prefix *bi-* (*q.v.*) "two, " which is also appropriate because a byte contains 2^3 bits. For more information, see *bit*. [48]

C

cactus, plural *cacti* (noun): a Latin word, borrowed from Greek *kaktos* "cardoon," a prickly, artichoke-like plant. The biologist Linnæus applied the word to a different kind of prickly plant that we now know as cactus. By a second analogy, mathematicians have used the word *cactus* to refer to a type of connected graph in which no line segment lies on more than one cycle (closed path).

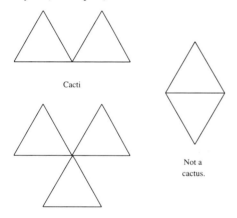

Cacti

Not a cactus.

calculate (verb), **calculus** (noun): a Latin word, based on *calx*, stem *calc-*, "chalk, limestone"; compare the related Latin word *calcium*. In Classical Latin, *-ulus* was a diminutive ending, so a *calculus* was a little stone or pebble. The ancient Romans, not having battery-operated calculators, and in fact not even writing their numbers with place-value notation, used to do simple arithmetic on a pebble board. The verb *calculare*, literally "to use pebbles," came to mean "to do arithmetic," and hence "to calculate." In the 1600's, the most sophisticated form of calculation was the newly invented calculus, which was

named after the humble pebbles used in early calculating. The calculus that your dentist tries to keep you from getting on your teeth looks through the microscope like a bunch of pebbles. In a similar way your doctor wouldn't want you to get a calculus in your kidney or gall bladder. [238]

calendar (noun): from Latin *kalendae* "calends," from the Indo-European root *kelə-* "to shout." The calends were the first day of each Roman month. They received their name from the fact that they were publicly proclaimed by the head priests. Even in Classical Latin the word *kalendæ* took on the meaning "month," and from that sense comes the modern *calendar* "a chart of the months of the year." [90]

cancel (verb): from the Latin plural *cancelli* "crossbars." Latin *cancelli* is a diminutive of *cancer* "a lattice," apparently unrelated to the *cancer* meaning "crab." *Cancer*, in turn, is a variant of *carcer*, also "an enclosure," particularly one used for keeping prisoners, i.e., a jail. Compare the English borrowing *incarcerate*. In mathematics, canceling frequently takes place in arithmetic and algebra. When fractions are being multiplied, you typically cancel numbers by drawing diagonal lines (= *cancelli*) through them. During the 1970's and 1980's the use of the word *cancel* in mathematics education was largely taboo, but the word is quite useful and the prohibition on its use has now been canceled.

canonical (adjective): via Latin *canon*, from the Greek *kanon* "a rod, a rule." The Greek *kanna* "reed," borrowed from the same Semitic original as *kanon*, is the source of our words *cane*, *canal*, *channel*, *canyon*, and *canister* (originally a basket made of reeds). A canon is a set of rules or standards, so *canonical* is a synonym of *standard* or *normal*. For example, in mathematics the canonical form of a square matrix has non-zero elements only in the main diagonal. The canonical form of an equation is the same as the standard form of that equation. [85]

cap (noun): from Late Latin *cappa* "a hood," from Latin *caput* "head," because a hood or cap is a garment that covers the head. The Indo-European root is *kaput*, with the same meaning as the native English cognate *head*. In geometry a spherical cap is that part of the surface of a sphere cut off by an intersecting plane. In the manipulation of sets, the symbol representing intersection, ∩, is sometimes called a cap because of its shape. Contrast *cup* for sound, appearance, and meaning.

capacity (noun): from Latin *capax*, stem *capac-*, "able to hold a lot, wide, spacious, roomy," from *capere* "to take." The Indo-European root is *kap-* "to seize, grasp," as seen in native English *have* and *heavy*, as well as in *capture* and *occupy*, which were borrowed from Latin. The word *capacity* is often used as a synonym of *volume*. [86]

cardinal (adjective), **cardinality** (noun): from Latin *cardo*, stem *cardin-*, "hinge," of unknown prior origin. *Cardinal* virtues were the "pivotal" ones, those on which virtue hinged, so *cardinal* became a synonym for "chief" or "principal." In the Catholic church, a cardinal is a principal official (whose bright red robes account for the name given to the red bird called a cardinal). In mathematics the cardinal or whole numbers (one, two, three, etc.) are the principal ones, as opposed, say, to fractions or irrational numbers. The cardinality of a set is the (necessarily whole) number of elements in the set.

cardioid (noun): the first component is from Greek *kardia*, from the Indo-European root *kerd-*, with the same meaning as the English cognate *heart*. Related borrowings from Latin include *accord* and *cordial*. The second component, *-oid* (*q.v.*), is from Greek *eidos* "form, appearance." In mathematics the *cardioid* is a heart-shaped curve whose polar equation is typically

$$r = a\left(1 \pm \left\{ \begin{matrix} \sin \\ \cos \end{matrix} \right\} \theta \right).$$

[94, 244]

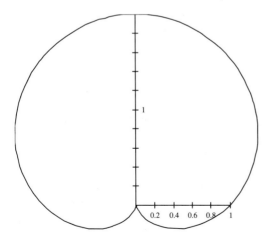

Cardioid: $r = 1 + \sin \theta$

carry (verb), **carrier** (noun): from Old Norman French *carier* "to convey in a wagon," from Old Norman French *car* "a wagon," from Latin *carrus* "a four-wheeled wagon." The Norman French noun is also the source of our word *car*. The Indo-European root is *kers-* "to run." Carts, carriages, chariots and cars are all vehicles used to carry people or freight. In the arithmetic operation of addition, when the total in a column is 10 or more, you have to carry one or more units over to the next column leftward. In geometry, a carrier is the point that all the planes in a sheaf or bundle have in common; that point "carries" the whole bundle. [95]

case (noun): from Latin *casus*, literally "a falling," from the verb *cadere* "to fall." The Indo-European root is *kad-* "to fall." When you say "in that case . . ." you really are saying "when things fall out that way . . ." In mathematics when you prove a theorem by cases you say, in essence, "If things fall out this way, then I'll do such and such; if things fall out this other way, then I'll do something different," etc. Related borrowings from Latin include *decadence* (a falling of moral standards) and *deciduous* (trees whose leaves fall off seasonally; also baby teeth, which fall out). [83]

cast (verb): from Old Norse *kasta*, from *kos* "heap, pile." If you cast away objects, they pile up and form a heap. The word may be related to Latin *gestus* "a pile," as in our borrowed words *digest* (= a collection or "pile" of articles) and *congest*. In mathematics casting out nines is a method of checking base-ten arithmetical calculations. In an arbitrary base b, $(b-1)$'s can be cast out.

catacaustic (adjective): the first component is from Greek *kata-* "downward, reversed," perhaps from the Indo-European root *kata-* "something thrown down." The second component is *caustic* (*q.v.*). Whereas a caustic curve is the envelope of light rays emitted from a radiant source after reflection or refraction in a given curve, a catacaustic curve involves only

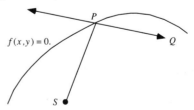

The line through P and Q is one element of the catacaustic of f relative to point S.

reflected rays. When rays are reflected, they reverse direction. [87, 97]

catastrophe (noun): the first component is from Greek *kata-* "down (from)," perhaps from the Indo-European root *kata-* "something thrown down." The second component is from Greek *strophe* "a turning," from the Indo-European root *streb(h)-* "to wind, to turn." A related borrowing from Greek is a *strobe* light. In Greek drama, the catastrophe of a play was the dénouement, at which time there was frequently a sudden reversal of events or situations. As a consequence, *catastrophe* came to mean "a sudden disaster." In mathematics, catastrophe theory creates mathematical models corresponding to disasters such as stock market crashes, wars, metal fatigue, etc. [87, 216]

categorical (adjective): from Greek *kategorein* "to accuse." The first component is from Greek *kata-* "downward, against," perhaps from the Indo-European root *kata-* "something thrown down." The second component is from *agorein* "to speak in public," from *agora* "a gathering place." The Indo-European root is *ger-* as seen in English *cram* and Latin-derived *gregarious*. The original sense of the Greek noun *kategoria* "accusation" was later broadened to "assertion" or "predication," and a collection of things being discussed came to be known as a category. A mathematical theory is said to be categorical when it applies to what is essentially a single type (= category) of situation; as a result, all models of that situation are isomorphic. [87, 67]

catenary (noun): from Latin *catenaria*, from *catena* "chain," of unknown prior origin. The Latin word developed into French *chaîne*, which is the source of English *chain*. In mathematics a *catenary* is the curve that results when a chain or any flexible cord-like object is allowed to hang under its own weight. The German mathematician Gottfried Wilhelm Leibniz (1646–1716) discovered the equation of one type of catenary,

$$y = \frac{a}{2}\left(e^{x/a} + e^{-x/a}\right)$$

—now also known as the hyperbolic cosine of *x*— and gave it its Latin name *catenaria*. The catenary has occasionally been called an *alysoid* or *chaînette* (*qq.v.*). [25]

catenoid (noun): from Latin *catena* "chain," of unknown prior origin, plus the Greek-derived *-oid* (*q.v.*) "having the appearance of." In mathematics, the surface of revolution obtained by rotating a catenary around its axis looks like a catenary, but in three di-

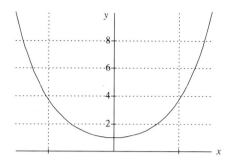

Catenary: $y = \cosh x$.

mensions; because it is generated by and resembles a catenary, it is called a catenoid. [25, 244]

caustic (adjective, noun): from Latin *causticus* "burning," from Greek *kaustikos*, from the verb *kaiein* "to burn." The Indo-European root is *keu-* "to burn." A caustic substance like lye is one which burns the skin. Surprisingly related is the word *ink*, shortened from *encaustic*, a technique in which colored wax is burned into a surface. In mathematics, a caustic curve is the envelope of light rays emitted from a radiant source after reflection (a catacaustic) or refraction (a diacaustic) in a given curve. The *Oxford English Dictionary* explains the semantic connection: "The intensity of the light, and consequently of the heat, is in general greater at a point on this surface than at neighboring points not on it, and at special points may become sufficiently intense to initiate combustion in a body placed there." Caustics were originally studied by the German mathematician Ehrenfried Walter von Tschirnhausen (1651–1708) in 1682. [97]

cavity (adjective): from Latin *cavus* "hollow." The Indo-European root is *keuə-* "to swell; vault, hole." Borrowed from Latin is our word *cave* "a space hollowed out of rock." In dentistry a cavity is a small "hollowing out" of a tooth. In topology a cavity is one of two types of hole that a solid may have. A cavity is entirely surrounded by the solid, like an air bubble trapped inside a solid glass sphere, as opposed to a channel, which passes all the way through the solid. [98]

cell (noun): from Latin *cella* "storeroom, place for depositing grains." The Indo-European root is *kel-* "to cover," so that originally a cell was a covered place for storing things, a meaning preserved in our borrowed word *cellar*. Native English cognates include *helmet, hole, hollow, holster,* and even *hell,* originally "a concealed place." In a magic square, each cell is a "storeroom" for a number. [89]

cent (noun): from Latin *centum* "hundred," from the Indo-European root *dkm-tom-*, with the same meaning as the English cognate *hundred*. The more basic root is *dekm-* "ten": a hundred is ten tens. Related borrowings from Latin are *centenarian* and *centennial*. Just as *decimate* means to select every tenth person in a group for punishment, *centesimate* means to select every hundredth person in a group for punishment. In the American monetary system, a *cent* is a hundredth of a dollar. [34]

center (noun), **central** (adjective): from Greek *kentron* "a sharp point, a peg, a stationary point." The Indo-European root is *kent-* "to prick, to jab," as seen in the Greek-derived *amniocentesis*. In times ancient and modern, people have put a stake in the ground and attached an animal to the stake with a rope. The places that the animal could wander all lie within a circle having the stake as its center. The meaning of the word *center* was later abstracted away from the stake as a pointed object, and the word came to mean the position of the "stake" equidistant from all points on the circle. [91]

centesimal (adjective): from Latin *centesima (pars)* "the hundredth (part)," from *centum* "hundred." The Indo-European root is *dkm-tom-*, with the same meaning as the English cognate *hundred*. The more basic root is *dekm-* "ten": a hundred is ten tens. A centesimal system of numeration is one which is based on 100. The American monetary system is centesimal, as is our frequent use of percents. [34]

centi-, abbreviated *c* (numerical prefix): from Latin *centum* "hundred," from the Indo-European root *dkm-tom-*, with the same meaning as the English cognate *hundred*. The more basic root is *dekm-* "ten": a hundred is ten tens. In the metric system, the prefix *centi-* means "one one-hundredth" of the unit that follows. The prefixes representing the first three negative powers of ten (*deci-* and *milli-* being the other two) were chosen from Latin number words by the Paris Academy of Sciences in 1791. In contrast, the prefixes for the first three positive powers of ten were chosen from Greek number words. The final *-i-* of *centi-* was added at the end of *cent-* to avoid confusion with the French word *cent* "one hundred." Also, for uniformity, all of the numerical prefixes have two syllables and end in a vowel. Newer prefixes representing negative powers of ten all end in *-o* and are no longer necessarily chosen from Latin. [34]

centillion (numeral): patterned after *billion* by replacing the *b-* with Latin *cent-* "a hundred," from the Indo-European root *dkm-tom-*, with the same meaning as the English cognate *hundred*. The more basic root is *dekm-* "ten": a hundred is ten tens. Since in most countries a billion is the second power of a million, a centillion is defined as the hundredth power of a million, or 10^{600}. In the United States, however, a billion is 10^9, and a centillion adds ninety-eight groups of three zeroes, making a centillion equal to 10^{103}. The word is first attested in 1852, when it appeared in its ordinal form *centillionth* in a British magazine. The centillion is the highest named member in the *-illion* sequence. [34, 72, 151]

century (noun): from Latin *centuria* "a group of 100 soldiers," from *centum* "hundred." The Indo-European root is *dkm-tom-*, with the same meaning as the English cognate *hundred*. The more basic root is *dekm-* "ten": a hundred is ten tens. In the Roman military the officer in charge of a *centuria* was logically called a *centurion*. A century eventually came to refer to any group of 100 things, but especially to a period of 100 years. The full phrase used to be "a century of years," but because *century* is seldom used now in any context except that of 100 years, the "of years" has been dropped; few people know that the extra words used to be required. [34]

centroid (noun): from *center* and the Greek-derived *-oid* (*qq.v.*) "looking like." Metaphorically speaking a centroid looks like a center. Actually, it doesn't so much look like a center as behave like one. An irregular shape doesn't really have a center in the same sense that a symmetric shape does; nor do real-world physical objects. Nevertheless, a physical object has a point at which all the mass of the object seems to be concentrated. Since that point acts like a center, it is called a centroid. [91, 244]

chain (noun): from French *chaîne*, from Latin *catena* "a chain," of unknown prior origin. In a chain, one link leads to another, which leads to still another, and so on. In calculus the chain rule for taking derivatives gets its name from the fact that you take the derivative of an "outer" function, then multiply by the derivative of an "inner" function, and then by the derivative of the function inside that one, until you have taken the derivative of the innermost function of all. In surveying, a chain used to be a unit of length; the chain (also called a Gunter's chain) consisted of 100 links whose total length was 66 feet. [25]

chaînette (noun): from French *chaîne* "chain" (*q.v.*), plus the diminutive ending *-ette*. The chaînette is a curve described by a chain made up of very tiny links hanging under its own weight. The chaînette is more commonly called a *catenary*, and has also been called an *alysoid* (*qq.v.*). [25, 57]

chance (noun, adjective): a French word, from a presumed Latin *cadentia* "a falling, a happening," from Latin *cadere* "to fall." The Indo-European root is *kad-* "to fall." Chance events "befall" us at random. From that randomness comes our use of the word *chance* in connection with probability. Related borrowings from Latin include *cadaver* (a body that has fallen down dead), *cascade* (falling water), and *recidivism* (falling back into crime and prison). [83]

change (verb, noun): a French word, from Late Latin *cambiare* "to exchange." The Indo-European root is *(s)kamb-* "to curve, to bend"; when you bend something, you change its shape or location. In calculus the concept of change is often represented by the Latin letter *d* or the Greek letter Δ, for "difference," because after something changes it is different. When speaking of money, change is what you get back when you ex*change* an amount of money for an item that costs less than that amount. The meaning was then further extended to mean "coins," since the change a person gets back when making a purchase often includes coins. [193]

channel (noun): from Old French, from Latin *canalis* "a canal, a reed." The Latin word was borrowed from Greek *kanna* "a reed," which, like a canal, is straight and can be used like a straw to carry water. A related Greek word from the same source is *kanon* "a rod." The words are of Semitic origin. In topology, a channel is a hole that passes completely through a solid, like a hole in a doughnut, as opposed to a cavity (*q.v.*), which is completely surrounded by the solid. [85]

chaos (noun), **chaotic** (adjective): *chaos* is a Greek word that originally meant "empty space." The Indo-European root is *gheu-* "to yawn, gape." A native English cognate is *gum*, because when you yawn, you expose your gums. The scientist J.B. van Helmont (1577–1644), in coining the word *gas*, based it on *chaos*. According to Greek mythology, out of the original great void came the gods, the earth, and all mortal beings. In the Olympian myth, Gaea emerged from Chaos and was the mother of all things. Eventually, because of its association with creation stories, *chaos* came to refer not to the original empty space

but to the confusion of matter that emerged from it. Modern mathematics has created models to account for chaotic happenings in the physical world. As an example, consider M.J. Feigenbaum's transformation of the interval $[-1, 1]$ into itself, via the iteration $x_{n+1} = 1 - \mu x_n^2$, where x is in the stated interval and μ is between 1.401 and 1.75. For general x, successive iterates behave chaotically, (as opposed to a μ-value of 0.5, which produces stable equilibrium).

characteristic (adjective, noun): from Greek *kharax* "a pointed stake." The Indo-European root is *gher-* "to scrape, to scratch." From Greek *kharax* came *kharakhter*, an instrument used to mark or engrave an object. Once an object was marked, it became distinctive, so the character of something came to mean its distinctive nature. The Late Greek suffix *-istikos* converted the noun *character* into the adjective *characteristic*, which, in addition to maintaining its adjectival meaning, later became a noun as well. In linear algebra a square matrix A is said to have the characteristic equation $\det(x\,I - A) = 0$, where I is the identity matrix of the same dimensions as A. Since each square matrix A satisfies its characteristic equation, the equation is closely tied to the intrinsic nature of the matrix. [71]

check (verb, noun): from French *échec*, older *eschac*, and ultimately from the Persian *shah*, meaning "king," as in the Shah of Iran. The Indo-European root may be *tke-* "to gain power over." In the game of chess, when you attack the opponent's king you say "check," in other words "king," as a warning that the king is in danger. (The game of checkers gets its name because it's played on the same board as chess, the game in which you say "check.") From that attack on the king comes the meaning "to stop, to restrain." If you write a check for x dollars on your account, you are telling your bank to "restrain" or hold back x of the dollars in your account and give them to someone else. In mathematics, when you check an equation, you are putting a hold on things until you can verify that the alleged solutions really work. In early 19th century American textbooks, checking was referred to as "proving."

chi (noun): the 22nd letter of the Greek alphabet, written X as a capital and χ in lower case. Its position near the end of the 24-letter alphabet is an indication that χ was a late addition and does not correspond to any of the letters in the earlier Hebrew alphabet. Just as x is often used in algebra to represent a variable, the similar looking χ is used in statistics. In particular,

$$\chi^2 = \sum_{i=1}^{k} x_i^2,$$

where the x_i are independently and normally distributed.

choice (noun), **choose** (verb): from the Old French noun *chois*, from *choisir* "to choose." The predecessor of the French word had originally been borrowed into the Latin spoken in Gaul from a presumed Germanic *kausyan*, of the same meaning. Native English, being a Germanic language, had and still has its own form of the original verb, *to choose*, which developed from the Old English *ceosan*. The past tense included a form *curon*, and the related noun was *cyre* "a choice." That native noun survived into the Middle English period as *kire* or *cüre*, but eventually gave way to the French form *chois*. Such replacements of native words by French equivalents were common because French was considered a more prestigious language than English even by many native speakers of English. English has retained its verb *to choose*; it gave up its own word for *choice* but chose to borrow it back in somewhat different form from French. The Indo-European root is *geus-* "to taste, to choose." The meaning "taste" can be seen in the related borrowing *gusto* in a positive sense, and in *disgust* in a negative sense. The axiom of choice plays an important role in mathematics.

chord (noun), **cord** (noun): from Latin *chorda* "catgut, string," from Greek *khorde*. The cords of a musical instrument were originally made of catgut or similar bodily material. In music, a chord is made when several strings (= cords) are sounded at the same time. In geometry, a chord is a cord-like segment that begins and ends on a given curve. Our word *cord* "a thick string" is just a different spelling of the chord used in geometry and music. A cord is also a measure of volume applied to firewood. It equals 128 cubic feet, and most often the wood is arranged in a stack 8 feet wide, 4 feet high, and 4 feet deep. Presumably a stack of wood was originally measured with a cord, hence the name.

chromatic (adjective): from Greek *khroma* "skin, complexion, color." The Indo-European root is *ghreu-* "to rub, to grind," an extension of *gher-* "to scrape, to scratch." The semantic development is difficult. The grinding of something requires flat surfaces such as those of millstones. The relatively flat surface of the human body is its skin (= Greek *khros*). The extended word *khroma* came to mean "the texture or complexion of the skin," and ultimately its color. In graph theory, a chromatic number is the smallest number of colors that can be used to color a graph on a given surface in such a way that no two adjacent regions have the same color. [71]

cinquefoil (noun): the first component is from French *cinq* "five." For the second component and further explanation see *multifoil*. [164, 21]

cipher or **cypher** (noun, verb): this now somewhat dated word for "zero" is from Old French *cifre*, from Arabic *çifr* "zero." The Arabic word was actually an adjective meaning "empty," which was a translation of the Sanskrit *sunya*. The people of India are credited with discovering the concept of zero, and so their word was the one that was translated into Arabic. The Sanskrit word for "empty" was used as the name for zero because when you write a zero as part of a numeral with place-value notation, you are leaving a place empty. Although when introduced into English in the 14th century *cipher* referred only to the digit zero, by the 16th century its meaning was extended to include all Arabic digits. From the noun came the verb *to cipher* "to use digits to calculate." The noun *cipher* also came to refer to secret writing because in early systems of cryptography letters were replaced with numbers. [26]

circle (noun), **circular** (adjective), **circulant** (noun), **circulating** (adjective): from Latin *circulus*, the diminutive of *circus* "ring, hoop." (We still speak of a three-ring circus.) The Indo-European root is *(s)ker-* "to bend." Although a ring was originally a physical object, the word *circulus* also came to refer to anything physical or abstract that resembles a ring. A circulating decimal is a repeating decimal because, just as a circle closes on itself, the digits in a repeating decimal "go full circle" and start over again. In linear algebra a circulant is a determinant in which the elements of each row after the first are the same as those of the previous row, but with

$$\begin{bmatrix} 3 & 0 & 1 & 9 & 7 \\ 7 & 3 & 0 & 1 & 9 \\ 9 & 7 & 3 & 0 & 1 \\ 1 & 9 & 7 & 3 & 0 \\ 0 & 1 & 9 & 7 & 3 \end{bmatrix}$$

A circulant.

the elements slid one place to the right; what was formerly the last element of a row "circles" around and becomes the first element of the next row. [198, 238, 146]

circuit (noun): the first component is from Latin *circum* "around," from the Indo-European root *(s)ker-* "to bend." The second component is from Latin *itus*, the past participle of *ire* "to go." The Indo-European root is *ei-*, of the same meaning. A circuit is literally a going-around. Simple circuits are frequently used in finite mathematics courses as a way of exemplifying the elementary operations of logic such as negation, conjunction, and disjunction. [198, 50]

circumcenter (noun): from Latin *circum* "around," from the Indo-European root *(s)ker-* "to bend," plus *center* (*q.v.*) With regard to a triangle, the circumcenter is the center of the circumscribed circle. The circumcenter isn't necessarily "centered" inside the triangle: the circumcenter of a right triangle is on the hypotenuse, and the circumcenter of an obtuse triangle is outside the triangle. Contrast *excenter*, *incenter* and *orthocenter*. [198, 91]

circumference (noun): the first element is from Latin *circum* "around," from the Indo-European root *(s)ker-* "to bend." The second element is from Latin *ferre* "to bring, to carry," from the Indo-European root *bher-* "to carry," as seen in the native English cognate *to bear*. Technically speaking, a circle is made up only of the points equidistant from the center, and does not include any of the interior points. Many people, however, use *circle* to refer to all the interior points as well as the boundary points. To avoid confusion, the term *circumference* is sometimes used for the circle itself, the part that is "brought around," as opposed to the interior points. In a different context, the circumference may also mean the distance around the circle. The word *circumference* is a Latin translation of the earlier Greek-based term *periphery*, from *peri* "around" and *pherein* "to carry," from the same Indo-European root *bher* as Latin *ferre*. The generic term *perimeter* could also be used for the circumference of a circle but it rarely is. [198, 23]

circumflex (noun): the first element is from Latin *circum* "around," from the Indo-European root *(s)ker-* "to bend." The second element is from Latin *flexus*, past participle of *flectere* "to bend," of unknown prior origin. A circumflex is a type of accent mark placed over vowels in French, Portuguese, and other languages. It is named for its shape, which

may be described as a bar that begins by rising to the right but then "bends around" in such a way that it ends up descending to the right. In other words, a circumflex looks like an upside down V. In vector calculus, a circumflex placed over a letter, as in $\hat{a}$, indicates that the letter represents a vector. Most mathematics teachers and students call the circumflex a *hat* (*q.v.*). [198, 61]

circumscribe (verb): the first element is from Latin *circum* "around," from the Indo-European root *(s)ker-* "to bend." The second element is from Latin *scribere* "to scratch," hence "to write." The underlying and more physical Indo-European root is *skribh-*, an extension of *sker-* "to cut, separate." The root is found in native English *shear*, *shard*, *short*, and *sharp*, as well as in *scrap* and *scrape*, borrowed from Old Norse. Remember that in ancient times geometric figures were scratched on the ground or onto waxed tablets or other physical objects. In geometry, a circumscribed figure is "drawn around" another one. Contrast *inscribe*. (Don't confuse *circumscribe* with *circumcise*, as a fellow student of mine once did in high school: the class didn't stop laughing for a good five minutes.) [198, 197]

cis (noun): an abbreviation for "$\cos\theta + i \sin\theta$" Although the first letter of cosine is pronounced like a *k*, the abbreviation *cis* is pronounced as if it were "sis," probably to avoid sounding like "kiss." [103, 81, 192]

cissoid (noun): from Greek *kissos* "ivy" and the suffix *-oid* (*q.v.*) "having the shape of." The curve called the cissoid is so named because its cusp resembles that of an ivy leaf. On a larger scale, the curve rising toward its vertical asymptote might remind someone of an ivy plant growing up a wall or fence. A Cartesian equation of the cissoid is $y^2(2a - x) = x^3$. The curve is often referred to as the cissoid of Diocles because it was first studied by Diocles around 200 B.C. in conjunction with the problem of duplicating a cube. [244]

class (noun): from Latin *classis*, originally "a summons for military draft," therefore "an army, a fleet (including the troops in it)," and finally the generic sense of "any group or division of people." The Indo-European root may be *kelə-* "to shout," either as the shouting that takes place to recruit soldiers, or possibly as the shouting of the soldiers themselves. A native English cognate is the *lowing* noise of cattle, and a related borrowing from Latin is *clamor*. Nowadays a class is a group (no longer necessarily human)

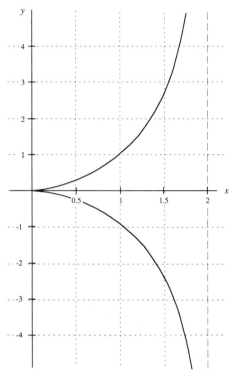

Cissoid of Diocles: $y^2(2-x) = x^3$

with a common property. In mathematics we speak of equivalence classes as well as classes of students clamoring to study equivalence classes. In statistics a class is a subdivision in a frequency distribution. [90]

clear (verb), **clearing** (noun): from French *clair* "of a light color, transparent," from Latin *clarus* "clear, bright, shining." The Indo-European root is *kelə-* "to shout." The original meaning of clear referred to sounds, even as today someone might say "I hear you clearly." The meaning was later extended from sound to sight: in order to hear or see someone clearly, you have to remove any obstacles that are in the way. In mathematics, to clear an equation of fractions means to eliminate the denominators (which are "in your way") by multiplying both sides of the equation by the common denominator of all the fractions. [90]

clinometer (noun): from Greek *klinein*, from the Indo-European root *klei-* "to lean"; plus *meter*, from Greek *metron* "a measuring." A clinometer is a device consisting of an inverted protractor and a string connecting the "vertex" of the protractor to a weight

that pulls the string taught. The clinometer is used to measure angles of inclination. [100, 126]

clique (noun): from Old French *cliquer*, a word of imitative origin meaning "to make noise." The word soon took on the figurative meaning "a group of people gathered together to plot or deceive." It is also used in French to refer to the ensemble of horns and drums in a military band. The English meaning "an exclusive group" isn't quite as negative as the French one. In graph theory, which frequently uses nontechnical English words in a technical way, a clique is defined as a maximal complete subgraph.

clockwise (adjective, adverb): Medieval Latin had a word *clocca* "bell," which presumably imitated the sound a bell makes (compare English *tick-tock* to represent the sound of a clock). During the Renaissance, towns often had towers with bells that rang the hours. The Dutch referred to such bell towers as *klocke*, using a slight variant of the Latin word. The word then came to refer to the instrument in the belltower that kept track of time, namely a clock. The native English suffix *-wise* means "in a certain way," so clockwise means "in the direction of (the turning of) a clock." The Indo-European root *weid-*, which gave rise to *-wise*, meant "to see." It is found in the borrowed words *video*, *vision*, and *view*. In circular trigonometry, motion in a clockwise direction is considered negative. [27, 244]

close (adjective, verb), **closed** (adjective), **closure** (noun): from French *clos(e)*, past participle of *clore* "to close." (Although *close* is a French past participle, English speakers interpreted it as an infinitive, so a *-d* was added to make the new English past participle *closed*.) The French verb is from Latin *claudere* "to shut, to close." The Indo-European root is *kleu-* "a hook, a peg," so that the Latin verb would have meant originally "to lock with a hook or a bolt." That sense is still seen in the musical borrowing *clef*, from Latin *clavis* "a key." If you are closed up in a room, you are in close quarters, so *close* became an adjective meaning "near." In mathematics an interval is said to be closed if its endpoints are "locked into" (= included in) the interval itself. With regard to sets, closure refers to the fact that the result of combining any two elements of the set is also included in the set. Comparing "closeness" in English is ambiguous. Suppose *A* is 3 feet from me and *B* is 6 feet from me: is *A* twice as close to me as *B* is or half as close to me as *B* is? Some English speakers say one thing, some the other. [101]

clothoid (noun): from Greek *kloth-* "to twist by spinning," of unknown prior origin; plus the suffix *-oid* (*q.v.*) "looking like." In Greek mythology, Clotho was one of the three goddesses of fate; she spun the thread of a person's destiny, which was then measured and cut by the other two goddesses. (Although cloth may be spun, the English word *cloth* is apparently unrelated.) In mathematics the clothoid is a type of double spiral that is also known as Cornu's Spiral. It is presumably named for its resemblance to a spinning wheel. The Swiss mathematician Leonhard Euler (1707–1783) studied it in 1781 as part of an investigation of a spring. [244]

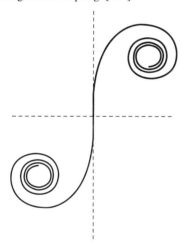

Clothoid

cluster (noun, adjective): a native English word of unknown origin. In mathematics a cluster point is the same as a limit point.

co-, col-, com-, con-, cor- (prefix): from the Old Latin preposition and prefix *com-*, used sometimes to mean "together with," and used at other times as a type of intensifying prefix. The final *-m* of *com-* generally assimilated wholly or partially to the initial consonant of the word to which it was attached. It remained an *-m* when followed by *m-* or *p-*, as in *commit* and *compare*. It became *-l* before *l-* and *-r* before *r-*, as in *collect* and *corrode*. It became *-n* when followed by *c-, d-, f-, g-, j-, n-, s-, t-,* and *v-*; examples are *concrete, condiment, confederate, congregate, conjure, connote, constitute, content,* and *convict*. When the following root began with a vowel, *com-* often took the form *co-*, as in *coalesce* and *coagulate*. In English the form *co-* has taken on a life of its own and can now be attached to any stem, not just

to a stem beginning with a vowel; examples include *co-pilot* and *co-worker*. The Indo-European root is *kom-*. As is common with prepositions, there are many possible meanings: "beside, near, by, with," etc. [103]

coaxial (adjective): from Latin *co-* "together with" and *axis* (*qq.v.*). Two geometric figures are said to be coaxial if they are positioned so that the points of intersection of all corresponding extended sides lie on the same straight line (= axis). [103, 5]

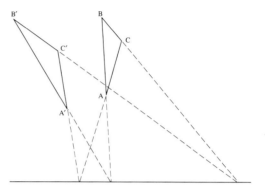

Triangles *ABC* and *A'B'C'* are coaxial.

cobordism (noun): the prefix is from Latin *co-* (*q.v.*) "together with." The root of the word is from French *bord* "side of a ship," therefore the "border" between the ship and the water. The word entered French from Frankish *bord*, cognate to English *board*. The side of a ship was called a bord (as in starboard) because it was made up of boards. The Indo-European root is *bherdh-* "to cut." A related borrowing from Italian is *bordello*, from Old French *bordel*, a little house that must originally have been made of boards. In mathematics cobordism deals with determining whether a compact manifold is the boundary (= border) of another manifold. [103]

cochleoid or **cochlioid** (noun): the stem is from Greek *kokhlias* "a snail," from the Indo-European root *konk(h)o-* "mussel, shellfish." The suffix is Greek-derived *-oid* (*q.v.*) "looking like." In a person's inner ear, the cochlea is a spiral tube that resembles a tiny snail. In mathematics the cochlioid is a spiral whose polar equation is

$$r = \frac{a \sin \theta}{\theta}.$$

Although the curve was studied around 1700, the current name was given to it only in 1884 by Bentham and Falkenburg, who noted the curve's snail-like appearance. Compare *limaçon*. [104, 244]

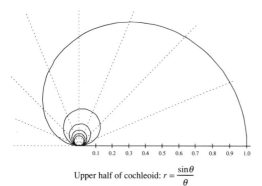

Upper half of cochleoid: $r = \dfrac{\sin\theta}{\theta}$

cochloid: see *conchoid*.

codomain (noun): from Latin *co-* "together with" and *domain* (*qq.v.*). The codomain of a function is commonly called the range; "together with the domain" it describes the set of values taken on by the function. [103, 37]

coefficient (noun): from Latin *co-* (*q.v.*) "together with," plus *efficient-*, the present participial stem of *efficere*, itself a compound of *ex* "out" and *facere* "to do, to make." To *effect* a result is to bring about that result. Similarly, something *efficient* brings about a desired result. A co-efficient (like a co-worker) works together with something else to bring about a final result. In the expression $5x$, the 5 works together with the x to give value to the term as a whole. From examples like this one, *coefficient* came to mean primarily a constant that multiplies a variable expression, though in theory a coefficient could be a variable. The word *coefficient* was coined by the French mathematician François Viète (1540–1603). [103, 49, 41, 146]

cofactor (noun): from Latin *co-* "together with" and *factor* (*qq.v.*). In a determinant, a cofactor is a signed minor that acts as a multiplier (= factor) of (= together with) the appropriate element in the determinant; the value of the determinant is given by the sum of the products of the elements in a row and their respective cofactors. [103, 41, 153]

cofinite (adjective): an abbreviated form of "complement [of a] finite [subset]," from Latin *co-* "together with" and *finite* (*qq.v.*). A set W is a cofinite subset of a set S if $S - W$ is finite. In other words, set W is the complement of finite subset $S - W$. Although

W is called cofinite, which might seem to imply its finiteness, W itself may be finite or infinite. [103, 60]

cogredient (adjective): from Latin *co-* (*q.v.*) "together with" and the present participle of *gradi* "to go." The Indo-European root is *ghredh-* "to walk, to go." A related borrowing from Latin is *ingredient*. Given several sets of variables such that whenever one of the sets is subjected to a transformation, every other set is subjected to the same transformation, the sets of variables are said to be cogredient. Loosely speaking, all the sets undergo the transformation together. [103, 73, 146]

coherently (adverb): the first component is from Latin *co-* (*q.v.*) "together with." The main component is from the present participle of Latin *haerere* "to stick, cling, adhere," from the Indo-European root *ghais-* "to adhere." Related borrowings from Latin include *adhesive* and *hesitate*. When two things cohere, they are "stuck together" in the sense of being alike. In that sense an argument or an essay is said to be coherent if the parts work well together and are arranged logically. In mathematics, given an n-simplex and an $(n - 1)$-simplex derived from it by the deletion of one point, the two structures are said to be coherently oriented when two related mathematical expressions are alike in sign. [103, 70, 146, 117]

coincide (verb), **coincident** (adjective): from Latin *co-* (*q.v.*) "together with," *in* "in, on," and *cadere* "to fall." The Indo-European root is *kad-* "to fall." Related borrowings from Latin include *cadence* and *cadaver* (a body that has fallen down dead). In geometry, when two figures coincide or are coincident, they "fall in together"; in other words, they are congruent. In nonmathematical usage, a coincidence is a set of two or more events that "fall out [the same way] together." [103, 52, 83, 146]

collect (verb): from Latin *collectus*, the past participle of *colligere* "to gather up, assemble, collect." The first element is *co-* (*q.v.*) "together." The second element is from Latin *legere* "to gather, pick," from the Indo-European root *leg-* "to collect." In algebra, to collect like terms is to assemble them (and usually combine them as well). For example, after collecting like terms the expression $4x - 8y - x + 10y$ becomes $3x + 2y$. [103, 111]

collinear (adjective), **collinearity** (noun), **collineation** (noun): from Latin *co-* "together with" and *linea* "line" (*qq.v.*). Distinct points that are "together on the same line" are said to be collinear. Distinct planes that contain a given line might also be called collinear. [103, 119]

collocation (noun): from Latin *co-* "together with" and *locus* "place" (*qq v*). In algebra a collocation polynomial of degree *n* is a polynomial that passes through (*n* + 1) given points. The polynomial is "together in the [same] place with" each of the points. [103, 215]

cologarithm (noun): from Latin *co-* "together with" and *logarithm* (*qq.v.*). The cologarithm of *n*, denoted colog *n*, is defined as $-\log n$, or equivalently $\log \frac{1}{n}$. The cologarithm was used "together with" the logarithm of a number to help carry out computations involving division in the days before hand-held calculators were available. [103, 111, 15]

column (noun): from Latin *columna* "an architectural column." The Indo-European root *kel-* "to be prominent" is also found in native English *hill*. Related borrowings from Latin include *culminate* "to reach the highest part of something," *excel* "to go up in quality," and *colonel* "a prominent officer in the military." Like the columns in front of a building, the columns in a determinant or matrix go up and down, as opposed to the rows, which go from left to right. A column vector, which may have many rows but just one column, looks like an architectural column.

combine (verb), **combination** (noun), **combinatorics** (noun): from Latin *com-* "together" and *bini* "two at a time," itself based on the root *bi-* "two." Originally a combination was a grouping together of two things, but the modern meaning allows for any number of things to be grouped together. Combinatorics is a branch of mathematics that deals with determining the number of ways certain things can be done. [103, 48]

commensurable (adjective): from Latin *co-* (*q.v.*) "together with" and *mensura* "a measuring," from *mensus*, past participle of the verb *metiri* "to measure." The Indo-European root is *me-* "to measure." In mathematics two quantities are commensurable if both can be measured an integral number of times with the same measure. A line segment 4 inches long is commensurable with one 6 inches long because both can be measured exactly with a ruler 2 inches long. A rational number and an irrational number are not commensurable. [103, 126, 69]

common (adjective): via French, from Latin *communis* "shared, common to several or all, common, general." The prefix is from Latin *co-* (*q.v.*) "together with," and the Indo-European root underlying the rest of the word is *mei-* "to change, exchange, go, move." Something held in common is held together with someone else. A native English cognate is *mean* (as in *mean-spirited*), which originally had the sense "common, low-class"; the word is unrelated to the *mean* used in statistics. In geometry, if two polygons have a common side, they share that side. Two circles may have as many as four common tangents; each such line is tangent to both of the circles. In arithmetic a common denominator of two fractions is a denominator that both can share; in other words, both of the original denominators can be converted to the common denominator. [103, 130]

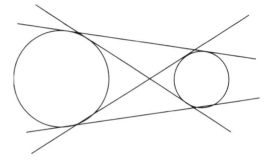

The two circles have four common tangents.

commute (verb), **commutative** (adjective), **commutativity** (noun): from Latin *co-* (*q.v.*) "with" and *mutare* "to move." The Indo-European root is *mei-* "to change, exchange, go, move." When you commute two things, they change place with each other. In common English, if you commute you move to work each day and then go back home again. In arithmetic, an additive instance of the commutative property says that $a + b = b + a$; in other words the *a* and *b* can be commuted, or switched back and forth, without changing the result. [103, 130]

compact (adjective), **compactum**, plural *compacta* (noun): from Latin *co-* (*q.v.*) "together with" and *pactum*, past participle of *pangere* "to fasten, to drive in." When an object is compact, all its component particles have been "driven together." The Indo-European root is *pag-* "to fasten." Related borrowing from French are *peace*, which "fastens" enemies together, and the verbs *to appease* and *to pacify*, From Old Norse comes *fang*, which an animal uses to capture (= fasten) its prey. In mathematics an infinite set of points is said to be compact if every infinite subset of those points has at least one accumulation point in the original set. A compactum is a topological space that is both compact and metrizable. [103, 156]

compare (verb), **comparison** (noun): from Latin *co-* (*q.v.*) "together with" and *par* "equal." When you compare two things, you put them together to see if they are equal. The Indo-European root is *perǝ-* "to grant, to allot," which is possibly the same as *perǝ-* "to produce, procure." In calculus a common way of determining the convergence or divergence of an infinite series is to use a comparison test. One of the most common mistakes students make in the development of an equation is not comparing each new line with the previous line to see if it is equivalent. [103, 167]

compass (noun): from Latin *co-* (*q.v.*) "together with" and *pass-*, the past participial stem of *pandere* "to stretch, to extend." Compare the related borrowing *pace*, which is literally the distance covered when you stretch your legs. In geometry, when you use a compass to trace an arc you are figuratively pacing off a distance from the center of the circle to each of the points on the arc. The use of the same word *compass* to mean "a round dial indicating directions" is believed to have arisen from the resemblance between directional compasses and the circular boxes that drawing compasses used to be stored in. [103, 169]

compatible (adjective): from Latin *co-* (*q.v.*) "together with" and *pati-* "to suffer." The Indo-European root may be *pe(i)-* "to hurt." Related borrowings from Latin include *patient* and *passion*. People who are compatible can "suffer" each other's company. In algebra two equations are said to be compatible (or more commonly consistent) if they have a common solution. In a similar way, equations like $2x - 3y = 8$ and $2x - 3y = 7$ are said to be incompatible (or more commonly inconsistent) because they have no common solution. [103, 69]

complement (noun), **complementary** (adjective): from Latin *co-* (*q.v.*) "together with" and *plere-* "to fill." The Indo-European root is *pelǝ-* "to fill," which is the native English cognate. A complement is an amount that must be added to another quantity to "fill it out." In mathematics the complement of any set with respect to the universal set in question consists of all elements needed to "fill out" that set and make it complete (see next entry). In geometry, the complement of an acute angle is the angle that must

be added to it to make a right angle; in that way the original angle gets "ful*filled*"; the two angles together "fill up" a right angle. See also *cosine, cotangent*, etc. Contrast *supplement*. The identically pronounced but differently spelled English word *compliment* is etymologically the same as *complement*: a verbal compliment and a complimentary item are things that are given to you to please you (= make you feel fulfilled). [103, 163, 146]

complete (verb): from Latin *co-* (*q.v.*) "together with" and *plere-* "to fill." The Indo-European root is *pelǝ-* "to fill," which is the native English cognate. When you complete something, you fill it out. For example, in algebra, when you complete the square of a trinomial you add to the existing terms whatever is necessary to "fill out" the square. [103, 163]

complex (adjective): from Latin *co-* (*q.v.*), which intensifies what follows, and the Indo-European root *plek-* "to plait," an extension of *pel-* "to fold." Something complex is literally "folded many times." Complex numbers are numbers of the form $a + bi$, where a and b are real and i is the square root of -1. Such numbers are more "folded up" or complicated than the real numbers. Contrast *simplex*. [103, 160]

component (noun), **composite** (adjective), **composition** (noun): from Latin *co-* (*q.v.*) "together with" and *ponere* "to put," or its past participle *positus*, so something composite is literally "put together" out of components. Latin *ponere* was contracted from a presumed earlier *posinere*. The second element, *sinere* "to leave," is of unknown prior origin; the first element, *po-* may be from Indo-European *apo-*, with the same meaning as the native English cognate *off*. In mathematics an integer is said to be composite if it can be "put together" (by multiplication) from factors other that itself and 1. The number 15, for instance, is composite because it can be decomposed into 3×5. A composition of two functions involves putting the functions together with one "inside" the other. [103, 14, 146]

compound (verb, noun, adjective): from Old French *compondre* "to compose," from Latin *co-* (*q.v.*) "together" and *ponere* "to put." Etymologically speaking, *compound* is the same word as *compose* (see previous entry). In finance, when an account earns compound interest, each period's interest is put together with the previous amount to accrue even more interest. In algebra a compound expression consists of two or more terms put together, as opposed to a simple expression, which has just one term. [103, 14]

compression (noun): via French, from Latin *co-* (*q.v.*) "together with" and *pressus*, past participle of *primere* "to press." The Indo-European root is *per-* "to strike." When something is compressed, it is literally pressed together. Related words include *pressure*, *express*, and *print* (because a printing press presses the type onto the paper.) Mathematically, a simple compression is a two-dimensional transformation represented by the equations $x' = x$ and $y' = ky$, or $y' = y$ and $x' = kx$, where $0 < k < 1$. Compare *shrink*. Contrast *elongation*. [103, 166]

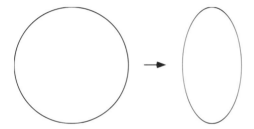

Horizontal compression with a factor of 2.

compute (verb), **computer** (noun): from Latin *co-* (*q.v.*) "together with" and *putare* "to trim, to prune (a tree, for example)." The Indo-European root is *peu-* "to cut, to strike," as seen in Latin-derived *amputate*. Even in Classical Latin the original verb developed the figurative meaning "to put in order mentally, to ponder, to consider, to think." When you *compute*, you put numbers together using mathematical procedures; you have to think about what you're doing to get the correct result. The word *computer* used to mean "a person who computes," and in fact some people used to make a living carrying out arduous computations. Since the middle of the 20th century, however, the primary meaning of *computer* has been "a machine that computes for you." [103, 171]

concatenate (verb), **concatenation** (noun): from Latin *co-* (*q.v.*) "together" and *catena* "chain," of unknown prior origin. In number theory, a concatenation of two integers A and B is the new integer formed by joining the digits of B to those of A while keeping all digits in their original order. The two original integers end up being "chained together." For example, the concatenation of 38 and 257 is 38257. In computer science, strings of characters may be similarly concatenated. [103, 25]

concave (adjective), **concavity** (noun): from Latin *co-* (*q.v.*), which frequently intensified the word it was attached to, and the adjective *cavus* "hollow." The Indo-European root is *keuə-* "to swell; vault, hole." Something concave is "hollowed out," just like a cave. In mathematics a curve is said to be "concave up" if it has its "hollowed out" part facing up. In geometry a figure is said to be concave if any chord passes outside the figure; in other words, part of the boundary of the figure is "caved in." In geometry, a concave polygon is one which is neither convex nor reflex. [103, 98]

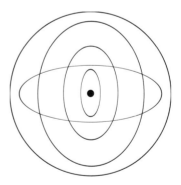

A concave pentagon.

concentric (adjective): from Latin *co-* "together" and Greek-derived *center* (*qq.v.*). If two circles, ellipses, or hyperbolas are concentric, their centers are "together" in the same place. Contrast *confocal*. [103, 91]

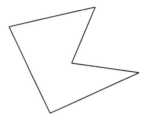

Concentric ellipses.

conchoid (noun): the main element is from Greek *konkhe* "a shell," from the Indo-European root *konk(h)o-* "mussel, shellfish." Related borrowings from Greek include *conch* and *cockle*. The suffix is Greek-derived *-oid* (*q.v.*) "having the shape of." The conchoid is a curve whose polar equation is $r = b + a \sec \theta$. When $b > a$, one branch of the curve has a loop in it, hence the resemblance to a shell. The curve is also known as the conchoid (or *cochloid*, from the same root) of Nicomedes because he studied it around 180 B.C. Compare *limaçon*. [104, 244]

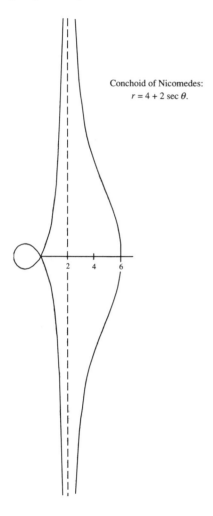

Conchoid of Nicomedes:
$r = 4 + 2 \sec \theta$.

conclude (verb), **conclusion** (noun): from Latin *co-* (*q.v.*), used as an intensifier, and *clausus*, past participle of *claudere* "to close." The Indo-European root is *kleu-* "a hook, a peg," so that the Latin verb would have meant originally "to lock with a hook or a bolt." Related borrowings include *seclude* and *cloister*. In mathematics, the conclusion is the "closing" of a logical argument, the point at which all the evidence is brought together and a final result obtained. [103, 101]

concordantly (adverb): the first component is from Latin *co-* (*q.v.*) "together with." The main component is from Latin *card-* "heart," from the Indo-European root *kerd-*, with the same meaning as the English cognate *heart*. Related borrowings from Latin include *accord* and *cordial*. Metaphorically speaking, when hearts are together there is a state of concord and people agree with one another. Given an *n*-simplex and an (*n* − 1)-simplex derived from it by the deletion of one point, the two structures are said to be concordantly oriented when two related mathematical expressions agree in sign. [103, 94, 146, 117]

concrete (adjective): from Latin *co-* (*q.v.*) "together with" and *cretus*, the past participle of *crescere* "to grow." The Indo-European root is *ker-* "to grow." A related borrowing from Latin is *cereal*, which is produced by plants that have grown mature. The word *concrete* means literally "grown together." The kind of concrete used in construction consists of particles that have "grown together" and become very hard. Something concrete is firm, physical, real. In mathematics concrete numbers refer to specific objects such as animals or stones, as opposed to abstract numbers, which need not have any physical referent. [103, 92]

concurrent (adjective): from Latin *co-* (*q.v.*) "together with" and *current-*, present participial stem of *currere* "to run." A current movie is one that is still running at a local theater. Water that runs in a river creates a current. The native English *horse* may be from the same root, since of all animals the horse is the greatest runner, though the evidence for an etymological connection isn't conclusive. In mathematics, if two or more lines are concurrent in a point, the lines "run together," i.e., intersect, at that point. [103, 95]

concyclic (adjective): from Latin *co-* (*q.v.*) "together" and *cyclic*, the adjective corresponding to Greek *kuklos* "circle." Distinct points which lie "together" on the same circle are said to be concyclic. [103, 107]

condition (noun), **conditional** (adjective, noun): from Latin *condicio*, stem *condicion-*, "an agreement, condition, stipulation"; from *co-* (*q.v.*) "together with" and *dicere* "to say, to tell." The Indo-European root is *deik-* "to show, pronounce solemnly." Related borrowings from Latin include *diction*, *dictation*, and *predict*. A condition is something you must "agree with"; in other words, you must accept the condition before you can proceed. In mathematics if a theorem has a condition, then the theorem applies only when that condition has been satisfied. (Students often make the mistake of trying to apply the conclusion of a theorem even when the requisite conditions haven't been fulfilled.) In differential equations, an initial condition is a stipulation of the values of certain physical quantities at a presumed time-zero. In logic, an if-then statement is known as a conditional statement or simply a conditional; the if-clause is the condition that, if fulfilled, leads to the conclusion stated in the then-clause. [103, 33]

cone (noun), **conic** (adjective, noun), **conical** (adjective): from Greek *konos* "cone." The Indo-European root may be *ko-* "to sharpen," as seen in native English *hone*. A cone would originally have been made by sharpening an object such as a piece of wood until it tapered down to a point at one end. Mathematically speaking, a cone has two parts, known as nappes, that meet at a point called the vertex. (See picture under *nappe*.) In nonmathematical English a cone has only one part, as in an ice cream cone. In mathematics a conic section, or simply a conic, is the intersection of a plane with a cone. [102]

confidence (noun): via French, from Latin *co-* (*q.v.*) "together with," but also used as an intensive prefix, and *fides* "trust, faith." The Indo-European root is *bheidh-* "to persuade, compel, confide." A native English cognate is *abide* (= wait faithfully). English *faith* is borrowed from Old French. In statistics a confidence interval is an interval that is believed, with a certain amount of "faith," to include a given value. [103]

confocal (adjective): from Latin *co-* "together with" and *focus* (*qq.v.*). Two conic sections are confocal if they have their foci "together," in other words, if they have the same foci. Contrast *concentric*. [103, 62]

conformable (adjective): from Latin *co-* "together with," *form*, and the suffix *-able* (*qq.v.*). Two matrices are said to be conformable if the number of columns in the first equals the number of rows in the second, because only then can the two matrices be "put *together*" to form a product. [103, 137, 69]

conformal (adjective): from Latin *co-* (*q.v.*) "together with" and *forma* "form" (*q.v.*). In a conformal mapping, every angle is "together with the form" of its image; in other words, angle size is preserved. [103, 137]

confounding (adjective): via French, from Latin *co-* (*q.v.*) "together with" and *fundere* "to pour,

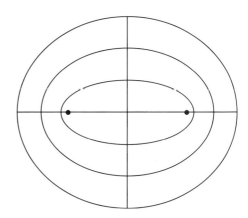

Confocal ellipses with foci indicated.

spill, shed." The Indo-European root is *gheu-* "to pour," as seen in *gust* and *geyser*, borrowed from Old Norse. Latin *fundere* developed the extended sense "to melt," because when things melt you can pour them. The word *confuse* is from the past participle of *confundere*: If you are confused, your ideas have "melted together." In statistics a confounding variable is one which may have "melted together" or become confused with another in the sense that data attributable to one variable may actually be due to the other. A well-known example is a study apparently showing that children with large shoe sizes generally know more than children who wear smaller shoes: the confounding variable is age, because older children—who of course know more—have bigger feet. [103]

congruent (adjective), **congruence** (noun): from Latin *congruent-*, present participial stem of *congruere* "to meet together, agree, correspond," from *co-* (*q.v.*) "together with" and probably the Indo-European root *ghreu-* "to rub, to grind." When two figures are congruent they can be made to "agree" because they have the same size and shape. In 1851 Charles Davies introduced a symbol for congruence that was commonly used in American textbooks for the next fifty years; it looked like a long equal sign, but the central third of the upper part was bent upward into a semiellipse, while the central third of the lower part was bent symmetrically downward into a semiellipse. It was similar to the astrological symbol for Libra, representing balance scales, presumably because congruence is a kind of equality. The "≅" symbol that we now use to indicate congruence is a slight variant of the one introduced by the German mathematician Gottfried

Wilhelm Leibniz (1646–1716); his had only one straight line segment beneath the tilde (*q.v.*) that by itself means "similar." The current symbol combines the meanings "similar" and "equal" (in area). [103, 71, 146]

conicoid (noun): from *conical*, from Greek *konos* "cone" (*q.v.*), plus *-oid* (*q.v.*) "looking like," but used here in the figurative sense of "related to, derived from." In solid geometry, a conicoid is a quadric surface. A conicoid can be obtained by cutting a cone with a plane and then rotating the resulting conic section about one of its axes (and possibly also then stretching the surface in one or two dimensions). [102, 244]

conjecture (noun, verb): from Latin *co-* (*q.v.*) "together with" and *iactus*, past participle of *iacere* "to throw." A conjecture is a set of ideas that you "throw together" as a hypothesis. A conjecture may or may not prove true. Among the many related borrowings from Latin are *subject* (something "thrown under" discussion) and *eject*. [103, 257]

conjugate (adjective): from Latin *co-* (*q.v.*) "together with" and *jugare* "to join, to connect." The Indo-European root is *yeug-* "to join," as seen also in native English *yoke*. In mathematics, given a complex number $a + bi$, the complex conjugate is $a - bi$. The a and the bi are still "yoked together," but with the opposite sign. Similarly the irrational conjugate of $c + \sqrt{d}$ is $c - \sqrt{d}$. The conjugate axis of a hyperbola is the one that "yokes together" the two "nameless" points that would be vertices if only they were on the hyperbola and on the transverse axis. (Although those points have indeed been nameless for centuries, this dictionary refers to them as *pseudovertices*.) In a foreign language class, when you conjugate a verb you list (= join together) the various forms of the verb. In biology, when protozoa conjugate they join to exchange genetic material. [103, 259]

conjunction (noun): from Latin *co-* (*q.v.*) "together with" and *iunct-* past participial stem of *iungere* "to join." The Indo-European root *yeug-* "to join" appears in native English *yoke*, which joins two animals. In general, a conjunction is a joining together of two things. In mathematics, a conjunction of two sets, also known as an intersection, joins them together to produce a new set that contains only elements common to both of the original sets. The conjunction of two statements p and q, symbolized as $p \wedge q$, is true only if both p and q are true. In

grammar, both *and* and *or* (as well as *but, because,* etc.) are categorized as conjunctions because they join clauses together. [103, 259]

connected (adjective), **connectivity** (noun): from Latin *co-* (*q.v.*) "together with" and *nectere* "to bind, tie, join, fasten together, attach." The Indo-European root is *ned-* "to bind, to tie." Native English cognates include *net* and *nettle* (whose fibers can be tied together). In mathematics, a connected set of points is one which can't be separated into two subsets that have no point in common and neither of which contains an accumulation point of the other. The opposite of *connected* is *disconnected*. [103, 142]

conoid (noun): from Greek *konos* "cone" (*q.v.*), plus *-oid* (*q.v.*) "looking like." A conoid is a surface that looks somewhat like a cone. Given a closed curve C, a guide line g not in the plane of C, and a direction plane P not parallel to g, a conoid is the set of lines that meet C and g and are parallel to P. [102, 244]

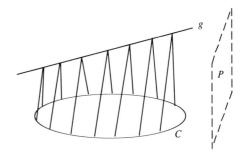

A conoid determined by curve C
and line g with elements
parallel to plane P.

conormal (adjective): from Latin *co-* "together with" and *normal* (*qq.v.*). Two points on a conic

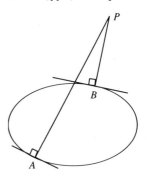

Points A and B on the
ellipse are conormal
with respect to point P.

section are said to be conormal with respect to a point *P* if normals drawn through the two points come together at point *P*. [103, 75]

consecutive (adjective): from Latin *co-* (*q.v.*) "together with" and the past participle of *sequi* "to follow." Related borrowings from Latin and French include *sequel* and *suitor* (a man who follows a woman). If two integers are consecutive, like 3 and 4, one follows right after the other. Of course two consecutive even (or odd) integers like 6 and 8 (or 7 and 9) don't actually follow one right after the other, but one follows the other as closely as possible given that the domain is restricted to even (or odd) integers. [103, 186]

consequent (noun): from Latin *co-* (*q.v.*) "together with" and *sequent-*, present participial stem of *sequi* "to follow." The consequence of an action is whatever follows from that action. In mathematics, the second part of an implication, the then-clause, is called the consequent because it follows logically when the condition in the if-clause is fulfilled. [103, 186, 146]

consistent (adjective): from Latin *co-* (*q.v.*) "together with" and *sistent-*, present participial stem of *sistere* "to place, to put, to cause to stand." The Indo-European root is *sta-*, with the same meaning as the cognate English *stand*. Something consistent is literally "standing together [in one place]"; in colloquial English, we might say it "hangs together." If a system of linear equations is consistent, it has a unique solution. The lines or planes or hyperplanes represented by the equations all meet (= stand together) in a single point. If they do not all meet in a single point the system is said to be inconsistent, where *in-* is a negative Latin prefix. [103, 208, 146]

constant (noun): from the Latin prefix *co-* (*q.v.*), which intensifies the root it is attached to, and *stans*, stem *stant-*, present participle of *stare* "to stand." The Indo-European root is *sta-* "to stand," which is recognizable in the native English cognate *stand*. Something constant literally "stands still." A mathematical constant is a fixed, unchanging number. Strangely enough, the word *constant* as used in mathematics often refers to a variable. For example, in the slope-intercept form of the equation of a straight line, $y = mx + b$, the *m* and *b* are called constants, but they can take on any value. The letters *m* and *b* here are what might be called arbitrary or variable constants, as opposed to constant constants such as *e* and π. We need to coin words to distinguish among the various types of constants. [103, 208, 146]

constellation (noun): from Latin *co-* (*q.v.*) "together with" and *stella* "star," from the Indo-European root of the same meaning, *ster-*. In number theory a star number is a positive integer that fulfills two conditions: the integer has a certain property, and the number of digits in the integer also has the same property. For example, 1681 is a star square because both 1681 and 4 (the number of digits required to write 1681 in base ten) are perfect squares. In nonmathematical English a constellation is a group of stars in a certain pattern. By analogy, in number theory a constellation is a group of star numbers of the same type. For example, all the star squares form one constellation, and all the star primes form another constellation. See more under *star* and *polystar*. [103, 212]

constraint (noun): via French, from Latin *co-* (*q.v.*) "together" and *stringere* "to draw tight, to bind, to tie." The Indo-European root is *streig-* "to stroke, rub, press." Native English cognates include *stroke*, *strike*, and *streak*. Related borrowings from Latin are *strain* and *striation*. If there are constraints on a mathematical problem, you are bound to narrow the scope of possible solutions. [103, 218]

construct (verb), **construction** (noun), **constructible** (adjective), **constructivism** (noun): from Latin *co-* (*q.v.*) "together with" and *struere* "to place in a heap, to pile up," from the Indo-European root *ster-* "to spread." The related Latin *stratus* "stretched, extended," is the origin of our word *street*. In geometry a construction consists of arcs and lines drawn so as to "build up" a certain figure. In classical Greek geometry such a figure was considered constructible if it could be drawn using only a compass and an unmarked straightedge. (For an explanation of the suffix in *constructible* see *-able*.) The fact that arcs and segments are sometimes strewn about in a construction is a reminder that the native English verb *strew* is a cognate. Constructivism is a doctrine which holds that only those mathematical objects which can be constructed in a finite way according to certain principles have meaning. [103, 211, 69]

contact (noun): from Latin *co-* (*q.v.*) "together with" and *tactus*, past participle of *tangere* "to touch." Our borrowed word *tangent* is from the same Latin source, so that in mathematics a point of contact is also known as a point of tangency. The Indo-European root is *tag-* "to touch." Related borrowings from Latin include *contaminate* (bring into contact with), *intact* (with all of its parts still touching), and even *taste* (to touch food with your tongue). [103, 222]

contain (verb): via French *contenir*, from Latin *continere* "to hold together, keep in a mass, hold within, contain." The two components are Latin *co-* (*q.v.*) "together with" and *tenere* "to hold." The Indo-European root is *ten-* "to stretch"; when you stretch something, you hold on to it. In mathematics, if a set contains a certain element, the set "holds" that element "together with" the other elements in the set. [103, 226]

conterminous (adjective): from Latin *co-* (*q.v.*) "together with" and *terminus* "boundary, limit." The Indo-European root is *ter-* "peg, post, boundary." In algebra two repeating decimals are said to be conterminous if their repetends "have a boundary together," i.e., if they end an equal number of places to the right of the decimal point. For example, $0.\overline{234}$ and $0.\overline{519}$ are conterminous. Contrast the use of the etymologically identical *coterminal*. [103, 227]

contingency (noun): from Latin *contingent-*, present participial stem of *contingere* "to take hold of, to seize," and by extension "to be connected with or related to." The components are *co-* (*q.v.*) "together with" and *tangere* "to touch," from the Indo-European root *tag-* "to touch." In nonmathematical English, a contingency is an unlikely event that may "touch" you by happening to you anyway. In statistics a contingency table shows how certain overlapping categories "touch," i.e., how they are related to each other. [103, 222, 146]

	Results of a test for influenza		
	Actually sick	Actually well	Totals
Diagnosed sick	9	91	100
Diagnosed well	7	93	100
Totals	16	184	200

A contingency table.

continuous (adjective), **continuity** (noun), **continuum** (noun): from Latin *co-* (*q.v.*) "together with" and *tenere* "to hold." The Indo-European root is *ten-* "to stretch"; when you stretch something, you hold on to it. Related borrowings include *tenet* (a belief that is held) and *tendon* (which stretches). In mathematics a curve which is continuous at a point literally "holds together" there in the sense that it has no breaks or holes in it, nor does it go shooting off to infinity for the value in question. If a curve isn't

continuous it is said to be discontinuous. A continuum is a set having the same number of points as an interval of the real number line; the interval can be "stretched" in such a way that each point is mapped to a point in the new set. [103, 226]

contour (noun): from Latin *co-* (*q.v.*) "together with" and *tornus* "a lathe." The Latin word was borrowed from Greek *tornos* "a tool for drawing a circle, a lathe." The Indo-European root is *terə-* "to rub, to turn." A native English cognate is *thread*, which is made of twisted fibers. A borrowing from Dutch is *to drill*. Related Latin borrowings include *attrition* and *tribulation*, both of which show the rubbing rather than the turning sense of the Indo-European root. When you take a *tour*, on the other hand, you "go around" to see places of interest. In mathematics contour lines (actually curves) are projections onto a plane of sections of a surface by planes parallel to the plane of projection. Contour lines typically look as if they are "going around together." [103, 228]

contradict (verb), **contradiction** (noun): the first element is from Latin *contra* "against," from the Indo-European root *kom* "with." For the semantic development, compare English "to fight with," meaning "to fight against." The second element is from Latin *dict-*, past participial stem of *dicere* "to say, tell." The Indo-European root is *deik-* "to show, to pronounce solemnly." In mathematics when one statement contradicts another, it "speaks against" the other one. When implications of a given statement lead to an incompatible statement, we say there is a contradiction. [103, 33]

contradictory (adjective, noun): from *contradict* (see previous entry). In logic, the negation of a statement involving the words *all*, *some*, or *none* is called a contradictory. Suppose the original statement is "All people are smart." The contradictory is "Not all people are smart," or "Some people aren't smart." In popular English many speakers mistakenly use the contrary (*q.v.*) rather than the contradictory, and say "All people aren't smart" when they mean "Not all people are smart." [103, 33]

contragredient (adjective): the first element is from Latin *contra* "against," from the Indo-European root *kom* "with." For the semantic development, compare English "to fight with," meaning "to fight against." The second element is from the present participle of Latin *gradi* "to go." The Indo-European root is *ghredh-* "to walk, to go." In mathematics two sets of *n* variables are said to be contragredient if, whenever one is subjected to a non-singular linear transformation, the other undergoes the transformation which has as its matrix the conjugate of the inverse of the matrix of the first. Loosely speaking, the second transformation "goes against" the first. [103, 73, 146]

contrapositive (noun): the first element is from Latin *contra* "against," from the Indo-European root *kom* "with." For the semantic development, compare English "to fight with," meaning "to fight against." The second element is from Latin *positus*, past participle of *ponere* "to put." For a given if-then statement, the contrapositive is the version that is "put against" the original in two senses: the components are both negated, and the original order of the clauses is reversed. Furthermore, a contrapositive is "positive" in the sense that it has the same truth value as the original statement. Suppose the original statement is "If it is raining then my yard is wet." The contrapositive is "If my yard isn't wet then it isn't raining." [103, 14]

contrary (adjective, noun): from Latin *contrarius*, from *contra* "over against, fronting, opposite." *Contra* is from the Indo-European root *kom* "with." For the semantic development, compare English "to fight with," meaning "to fight against." In logic, the contrary of a statement involving the words *all*, *some*, or *none* is not a true negation, but a type of "improper" negation in which only a part of the statement is negated. Suppose the original statement is "All people are smart." The correct negation, or contradictory (*q.v.*), is "Not all people are smart," or its equivalent "Some people aren't smart." The contrary—which many people incorrectly use in place of the contradictory—is "All people aren't smart." [103]

contravariant (adjective): the first element is from Latin *contra* "against," in the sense of "opposite." *Contra* developed from the Indo-European root *kom* "with." For the semantic development, compare English "to fight with," meaning "to fight against." The second element is from the present participle of *variare* "to fleck, diversify, look spotty, alter, change," and finally "to vary." The Indo-European root is *wer-*, which referred to a spot on the skin that represented a change from its normal appearance. In a two-dimensional coordinate system with oblique axes, as the position of a point varies, its contravariant *x*-coordinate is found by drawing a line through the point parallel to the *y*-axis (the one labeled with the contrary letter). To find the *y*-coordinate, a line is drawn parallel to the *x*-axis. The *contra-* in con-

travariant reflects that opposition between *x* and *y*. Contrast *covariant*. [103, 250, 146]

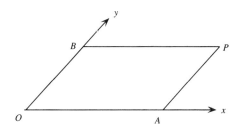

OA is the contravariant *x*-coordinate of point *P*.
OB is the contravariant *y*-coordinate of point *P*.

control (noun, verb): from Norman French *contreroller*, a compound of *contre* and *roller*. The first element is from Latin *contra* "against," in the sense of "opposite." *Contra* developed from the Indo-European root *kom* "with." For the semantic development, compare English "to fight with," meaning "to fight against." *Roller* developed from Latin *rotula*, diminutive of *rota* "a wheel." The Indo-European root is *ret-* "to run, to roll." A related borrowing from Spanish is *rodeo* "a roundup." A control was originally anything that worked against the rolling or turning of a wheel to slow it down. From the extended sense "to keep something in check" comes our modern meaning of the word. In statistics a control group is used to check the results of an experiment and keep from "rolling" to unjustified conclusions. [103, 181]

converge (verb), **convergent** (adjective, noun): from Latin *co-* (*q.v.*) "together with" and *vergere* "to turn, to verge." The Indo-European root is *wer-* "to turn, to bend." Native English cognates include *wrath* and *worry*, both of which twist you up. In mathematics, if an infinite series converges, the running total of the terms eventually "turns" toward a definite value called the limit. In connection with a continued fraction, a convergent is any of the common fractions used to approximate the value of the continued fraction; each successive convergent "turns" closer to the value of the continued fraction than the one before. [103, 251, 146]

converse (noun): from Latin *co-* (*q.v.*) "together with" and *versus*, past participle of *vertere* "to turn." The Indo-European root is *wer-* "to turn, to bend." Native English cognates include *wreath* (which twists around) and *writhe*. In logic, the converse of an if-then statement is the statement that results when you reverse the order of the two clauses. For instance, given the original statement "If it is raining then my lawn gets wet," the converse is "If my lawn gets wet then it is raining." Many people falsely believe that the converse is equivalent to the original statement. Just because my lawn is getting wet doesn't mean that it's raining; my lawn may be getting wet from a sprinkler or a broken water main. [103, 251]

convex (adjective): from Latin *co-* (*q.v.*), which intensifies what follows, and the verb *vehere* "to go, to carry." The Indo-European root is *wegh-* "to go, to transport." Native English cognates include *way* (the path on which you are carried) and *wag* (to carry back and forth). From Dutch comes *wagon*, replacing native English *wain* (which survives in the obsolescent *wainwright* "a builder of wagons"). Related borrowings from Latin include *vehicle* and *vexed* ("all bent out of shape"), and from French *voyage*. Something convex is literally carried or bent forward, as opposed to backward, in which case it would be *concave* (*q.v.*). In geometry, a figure is convex if no chord goes outside the figure. [103, 243]

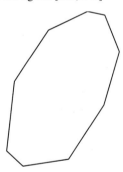

A convex octagon.

convolution (noun): the prefix is from Latin *co-* (*q.v.*) "together with"; the main part of the word is from *volutus*, past participle of *volvere* "to roll." The Indo-European root is *wel-* "to turn, to roll." An English cognate is *walk*, literally to move with a rolling motion. In nonmathematical English, something convoluted is complicated or intricate. We speak of the convolutions of the human brain, for example. In mathematics a convolution of two sequences is a type of "reverse dot product." If the two sequences are a_1, a_2, a_3, a_4 and b_1, b_2, b_3, b_4, then the convolution of the sequences is defined as $a_1b_4 + a_2b_3 + a_3b_2 + a_4b_1$. Figuratively speaking, the sequences have been "rolled together" in a reverse order. [103, 248]

coordinate or **coördinate** (noun): from Latin *co-* (*q.v.*) "together with" and *ordo*, stem *ordin-*, "a straight row." Compare the related word *order*, borrowed from French: when you put numbers in order you put them in a straight row on a number line. The Indo-European root is *ar-* "to fit together," as seen in native English *read*, "to fit words together in order to make sense out of them." In analytic geometry, each coordinate of a point in a rectangular coordinate system corresponds to a distance measured on the ordered scale of an axis; that distance is measured from the origin to the place where the perpendicular dropped from the point intersects the axis. Together those coordinates locate the point in the plane. The notion of order is further involved because the coordinates must be given in a fixed order, with *x* first and *y* second. The term *coordinate* was coined by the German mathematician Gottfried Wilhelm Leibniz (1646–1716). The older spellings *coördinate*, with two dots over the second -*o*-, and *co-ordinate*, with a hyphen, indicate that the two *o*'s are in separate syllables and should not be read like the two *o*'s of *coop* or *poor*. [103, 15]

coplanar (adjective): from Latin *co-* (*q.v.*) "together with" and *plana* "plane." The Indo-European root is *pelǝ-* "flat; to spread." Three or more distinct points are said to be coplanar if they are together on the same plane (but aren't all collinear). [103, 162]

copolar (adjective): from Latin *co-* "together with" and *polar* (*qq.v.*). Two geometric figures are said to be copolar if they are positioned so that lines joining corresponding vertices come together at a point (= pole). Contrast *homothetic*. [103, 107]

coprime (adjective, noun): from Latin *co-* "together with" and *prime* (*qq.v.*). An integer is prime if it has no proper divisor greater than 1. If two or more integers are looked at together (=*co-*) and are found to have no common divisor greater than 1, they are said to be coprime. Note that numbers which are coprime needn't be prime. For example 14 and 15, neither of which is a prime number, are coprime. American books usually use the longer term *relatively prime* instead of the more succinct *coprime*. [103, 165]

co-punctal (adjective): from Latin *co-* (*q.v.*) "together with" and *punctus* "a small spot, a point." The Latin noun is actually the past participle of *pungere* "to prick, to puncture," so a *punctus* was originally the tiny hole that resulted from a pin prick. Latin *punctus* developed into French *point*, in which form it was borrowed into common English. However,

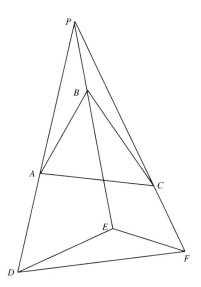

Triangles *ABC* and *DEF* are copolar
relative to point *P*, whether the
figure is construed to be
2-dimensional or 3-dimensional.

root is *ker-* "horn, head." Some of the ancient Greek mathematicians believed that there was an angle between a circle and a tangent line. Such an angle has been called a cornicular angle. When two circles of different radii share a point of tangency and lie on the same side of the tangent line, the arcs of the circles look like the curved horn of an animal. [93, 238]

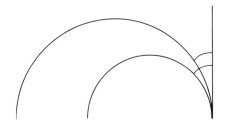

Cornicular angles.

learned borrowings like *punctual* (said of a person who arrives "on the dot") reflect the Latin form. In mathematics three or more planes that have a point in common are said to be co-punctal. [103, 172]

cord: see *chord*.

corner (noun): from Old French *cornere*, ultimately from Latin *cornu*, with the same meaning as the native English cognate *horn*. The Indo-European root is *ker-* "horn, head." Related borrowings from Greek include *carrot* (with its sharp point) and *cranial*. At a corner there is a sharp point, though not necessarily as sharp as that at the tip of a horn. In mathematics a corner is the juncture of two sections of a curve which have limiting tangent lines making a positive (i.e., nonzero) angle. Compare *salient* and contrast *cusp*. [93]

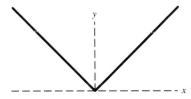

The absolute value function $y = |x|$
has a corner at the origin.

cornicular (adjective): from Latin *corniculus*, diminutive of *cornu* "horn." The Indo-European

corollary (noun): from Latin *corolla*, diminutive of *corona* "garland, wreath." The Indo-European root is *(s)ker-* "curved, bent." A related borrowing from French is *crown*. In mathematics, a corollary is a theorem that follows logically from another, often more important one. Poetically and etymologically speaking, a corollary is a little "garland" that follows the main "bouquet" of a theorem. [198, 238]

correct (adjective, verb): from Latin *correctus*, past participle of *corrigere* "to straighten, to set right." *Corrigere* is a compound of the intensifying prefix *co-* (*q.v.*) and *regere* "to stretch, to keep in a straight line, to guide in the right direction." The Indo-European root is *reg-* "to move in a straight line." When things move forward in a straight line the way they're "supposed to," then everything is all right. A correct answer is a right answer. To correct an error is to "set it right." [103, 179]

correlation (noun): from Latin *co-* (*q.v.*) "together with," *re-* "back, again," and *latus* "carried." The Indo-European root is *telə-* "to lift, support, weigh." It is found in *tolerate*, borrowed from Latin, and *toll* and *Atlas* (who supported the world on his shoulders), borrowed from Greek. In a relation(ship), one thing is figuratively carried back to another. The addition of the prefix *co-* adds a mutual sense to the word, so that a correlation is an interdependence between two sets of numbers. In mathematics a correlation is a type of two-dimensional linear transformation that carries points into lines and vice versa. In statistics a correlation coefficient expresses the degree to which one variable "carries" information about another. [103, 224]

corresponding (adjective): from Latin *co-* (*q.v.*) "together with," plus *re-* "back again, in return," and the verb *spondere* "to pledge, to promise." The Indo-European root is *spend-* "to make an offering." From *sponsum*, past participle of *spondere*, via French, comes our word *spouse*, a person pledged in marriage. Similarly, a *sponsor* is a person who guarantees a pledge. When you respond, you literally pledge or speak in return, that is, after someone else. When two things correspond, they speak together or in the same way. In geometry, corresponding parts of similar triangles are parts that "speak the same way" in their respective triangles. [103, 178]

cosecant (noun): a contraction of *complemental secant*, equivalent to *secant of the complement*, because of the complementary relationship $\csc A = \sec(90° - A)$. [103, 163, 185, 146]

coset (noun): from Latin *co-* "together with" and *set* (*qq.v.*). In group theory, cosets are obtained from a subgroup of a larger group by taking elements of the subgroup and "putting them together with" a set of elements from the larger group. [103, 185]

cosine (noun): a contraction of *complemental sine*, equivalent to *sine of the complement*, because of the complementary relationship $\cos A = \sin(90° - A)$. The term *cosine* was coined by Edmund Gunter (1581–1626). [103, 163, 192]

cost (noun): from Old French *coster*, from Latin *constare*, a compound of *con-* "together with" and *stare* "to stand." The Indo-European root is *sta-* "to stand," which is recognizable in the native English cognate *stand*. The modern meaning of *cost* comes from the notion of checking prices to see how they "stand." From the same Latin verb *constare* comes *constant* (*q.v.*), which is paradoxical, since the price of something rarely stays constant for very long. [103, 208]

cotangent (noun): a contraction of *complemental tangent*, equivalent to *tangent of the complement*, because of the complementary relationship $\cot A = \tan(90° - A)$. The term *cotangent* was coined by Edmund Gunter (1581–1626). [103, 163, 222, 146]

coterminal (adjective): from Latin *co-* (*q.v.*) "together with" and *terminus* "boundary, limit." The Indo-European root is *ter-* "peg, post, boundary." In circular trigonometry two angles are said to be coterminal if their terminal sides end up being together,

that is, in the same position. Contrast the use of the etymologically identical *conterminous*. [103, 227]

count (verb), **countable** (adjective), **counting** (adjective): from French *conter* or *compter*, from Latin *computare* "to sum up, reckon, compute" (*q.v.*). A counter got its name because originally money was counted on it, though nowadays a *counter* can be a flat surface used for any purpose, as the counter in a kitchen or a diner. An accountant computes how much money a business has gained or lost. In mathematics the counting numbers, otherwise known as natural numbers, are that most basic set $\{1, 2, 3, \ldots\}$. A countable set is one whose elements can be put in one-to-one correspondence with the counting numbers. (For an explanation of the suffix in *countable* see *-able*.) [103, 171, 69]

counterclockwise (adjective, adverb): the first element is from French *contre*, from Latin *contra* "against," in the sense of "opposite." *Contra* developed from the Indo-European root *kom* "with." For the semantic development, compare English "to fight with," meaning "to fight against." The rest of the word is the compound *clockwise* (*q.v.*). An object that moves counterclockwise moves "against" or opposite to the direction of the hands of a clock. In British English, the word is *anti-clockwise*. In circular trigonometry, motion in a counterclockwise direction is considered positive. That may be counterintuitive: a word containing the prefix *counter-* is usually presumed to have a negative meaning. [103, 27, 244]

counterexample (noun): the first element is from French *contre*, from Latin *contra* "against," in the sense of "opposite." *Contra* developed from the Indo-European root *kom* "with." For the semantic development, compare English "to fight with," meaning "to fight against." The rest of the word is the compound *example* (*q.v.*). A counterexample is an example which goes against a statement or conjecture. No amount of substantiating examples can prove a mathematical conjecture, but a single counterexample can disprove it. [103, 49, 51]

couple (noun, verb): a French word, from Latin *copula*, originally "a band, rope, thong," and by extension "a tie, a connection." Itself a compound, *copula* is from *co-* (*q.v.*) "together with," an Indo-European root *ap-* "to take, to reach," and a diminutive ending *-ula*. The same Indo-European root can be found in *apex* "the highest point that an object reaches." Two people who are "together" are known as a couple (and, in a sexual sense, those two people may even copulate). More generally, *couple* has come to refer to two similar or identical things that are considered at the same time, for example a couple of apples or a couple of problems. In algebra an expression of the type n^n is sometimes called a coupled exponential because the base and the exponent are identical. Also see *superpower*. [103, 13, 238]

covariant (adjective): from Latin *co-* (*q.v.*) "together with" and the present participle of *variare* "to fleck, diversify, look spotty, alter, change," and finally "to vary." The Indo-European root is *wer-*, which referred to a spot on the skin that represented a change from its normal appearance. In a two-dimensional coordinate system with oblique axes, as the position of a point varies, its covariant x-coordinate is found by dropping a perpendicular from the point to the x-axis. To find the y-coordinate, a perpendicular is dropped to the y-axis. The *co-* in *covariant* reflects that fact that x is "together with" x and y is "together with" y when perpendiculars are dropped. Contrast *contravariant*. [103, 250, 146]

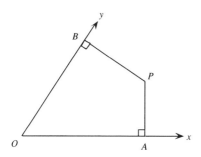

OA is the covariant x-coordinate of point *P*.
OB is the covariant y-coordinate of point *P*.

cover (noun, verb), **covering** (noun): from French *couvrir* "to cover"; from Latin *co-* (*q.v.*), used as an intensifying prefix, and *operire* "to put over, to cover." Latin *operire* is itself a compound of Indo-European *epi-* or *opi-* "near, at, against," and *wer-* "to cover." An English cognate is *weir* "a fence (= covering) placed in a stream to trap fish." In topology, a covering is a family of sets that occupy a region without leaving any holes or gaps in the region. [103, 53]

coversine (noun): short for *coversed sine*, a contraction of *complemental versed sine*. The coversed sine of an angle θ is abbreviated *covers θ* or *cvs θ*. Just as the cosine is the complement of the sine, the coversed sine is the complement of the versed sine. In

other words, since $\text{versin}\,\theta = 1 - \cos\theta$, covers θ is defined as $1 - \sin\theta$. [103, 251, 192].

crescent (noun): from Latin *crescens*, stem *crescent-* "growing," the present participle of *crescere* "to grow." The Indo-European root is *ker-* "to grow." Related borrowings from Latin include *excrescence* (an outgrowth), and *recruit* (a person who causes your ranks to grow). The crescent moon is literally the phase in which the moon is growing. The word *crescent* then came to represent the shape of the moon in that growing (or the symmetric waning) phase. Latin *crescent-* developed into French *croissant*, which is now also a type of pastry named for its crescent shape. In mathematics a crescent is the same as a two-dimensional lune. [92, 146]

crisp (adjective): from Latin *crispus* "curled," from the Indo-European root *sker-* "to turn, bend." A related borrowing from French is the "curled" pancake known as a crepe. How the word *crisp* developed the senses "firm" and "brittle" isn't certain, but with regard to an object like a dried leaf the meaning may have shifted from the curliness of the leaf to the brittleness of the leaf. In logic a crisp set is one whose membership criteria are indisputable. For example, the set of odd integers is a crisp set: there is no doubt about whether a given integer belongs in that set. Contrast *fuzzy*. [198]

criterion, plural *criteria* (noun): from Greek *krites* "a judge, an umpire." A criterion is a standard by which you judge or decide something. The Indo-European root is *krei-* "to sieve, discriminate, distinguish." In number theory, for example, Euler's criterion lets you decide whether a number a is a residue of m of order n. A common mistake in English is to use *criteria* as if it were a singular rather than using the correct singular, *criterion*. [105]

critical (adjective): from the Greek verb *krinein* "to separate, to discern," from the Indo-European root *krei-* "to sieve, discriminate, distinguish." A person who is critical is able to analyze a situation and separate the good from the bad. In calculus, when you find critical points, you analyze the behavior of a curve to find potential maximum and minimum values. You need to be critical, however, and not jump to hasty conclusions: just because the derivative is zero at a critical point doesn't guarantee the point is a relative (local) maximum or minimum. [105]

cross (noun, adjective): via French *croix*, from Latin *crux*, stem *cruc-* "a cross," of unknown prior origin.

In the proportion

$$\frac{a}{b} = \frac{c}{d},$$

the process of cross-multiplying to get $ad = bc$ takes its name from the cross that is formed by connecting a with d, and b with c. The cross sum of a numeral written in place-value notation is the sum of its digits; it is obtained by adding *across* the numeral. For example, the cross sum of 892 is 19. The cross product of two vectors is indicated by the symbol $\times$, hence its name. Furthermore, the cross product is defined formally as a determinant; in evaluating that determinant, the products taken on the downward diagonals "cross over" the products taken on the upward diagonals. The cross product, also known as the vector product or outer product, is distinguished from the dot product, which is indicated by a dot. In an 1812 American arithmetic book, the sign of addition, $+$, was referred to as St. George's cross, and the sign of multiplication, $\times$, as St. Andrew's cross; secular public education does not allow such descriptions now. [28]

cruciform (adjective): from Latin *crux*, stem *cruc-* "a cross," of unknown prior origin, plus *forma* "form (q.v.), figure, shape." The cruciform curve, with rectangular equation $x^2y^2 = a^2x^2 + a^2y^2$, is so named because it has the shape of a cross. [28, 137]

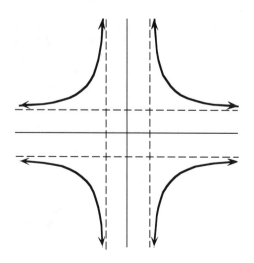

Cruciform curve: $x^2y^2 = x^2 + y^2$

crunode (noun): from Latin *crux* "cross," of unknown prior origin, and *nodus* "a knot," from the Indo-European root *ned-* "to bind, to tie." A crunode

is a place on a curve through which two branches pass with distinct tangents. The curve crosses itself there. Contrast *tacnode* and *osculation*. [28, 142]

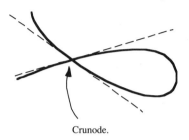

Crunode.

cryptarithm (noun): the first component is from Greek *kruptein* "to hide," from the Indo-European root *krau-* "to conceal, to hide." Related borrowings from Greek include *cryptic* and *cryptogram*. The second component is more than half of the word *arithmetic* (q.v.). A cryptogram is a puzzle in which the letters of a phrase or sentence are replaced by other letters in a one-to-one correspondence. By analogy, a cryptarithm is a puzzle in which the digits of an arithmetic calculation are replaced by letters in a one-to-one correspondence. The term *cryptarithm* was first suggested in 1931 by M. Vatriquant in *Sphinx*, a French magazine of recreational mathematics. Compare *alphametic*. [15]

$$
\begin{array}{ccc}
\text{A} & \text{B} & \text{C} \\
 & \text{D} & \text{E} \\
\hline
 & \text{F} & \text{E} & \text{C} \\
\text{D} & \text{E} & \text{C} \\
\hline
\text{H} & \text{G} & \text{B} & \text{C}
\end{array}
$$

A multiplication
cryptarithm.

cube (noun, verb), **cubed** (adjective), **cubic** (adjective), **cubical** (adjective): from the Greek *kubos* "a six-sided die." The Indo-European root is *keu-* "to bend" and therefore "to be rounded or hollow." A cube isn't rounded, but it "bends" at its edges and may be hollow. Related words include Latin-derived *cup* and Greek-derived *cymbal*, as well as native English *heap* and *hive*. The ancient Greeks interpreted the abstract quantity s^3 as the volume of a cube of side s. That's why something raised to the third power is said to be *cubed*, a word that reflects the three-dimensionality of a cube rather than the number of its edges (12), vertices (8), or faces (6). Compare *squared*. [96]

cuboctahedron (noun): from *cube* and *octahedron* (*qq.v.*). A cuboctahedron is a semiregular polyhedron. Both the name and the figure itself combine elements of the cube and the octahedron. A cube has 6 faces that are squares, and a regular octahedron has 8 faces that are equilateral triangles. The cuboctahedron has 14 faces: 6 squares and 8 equilateral triangles. [96, 149, 184]

cuboid (noun): from *cube* (*q.v.*) plus Greek-derived *-oid* (*q.v.*) "looking like." A cuboid is like a cube in that it is also a hexahedron with all right angles. Unlike a cube, however, a cuboid needn't have equal adjacent sides. In American textbooks, a cuboid is often called a rectangular solid. In nonmathematical English, a cuboid is often called a box. [96, 244]

cumulant (noun): from Latin *cumulant-*, present participial stem of *cumulare* "to pile up," based on the noun *cumulus* "a pile." The Indo-European root is *keuə-* "to swell." In mathematics a cumulant is a coefficient of

$$\frac{x^k}{k!}$$

in a Maclaurin series. As the infinite series advances from term to term, the coefficients keep "accumulating." [98, 146]

cumulative (adjective): from Latin *cumulatus*, past participle of *cumulare* "to pile up." The more basic *cumulus* "a pile" is from the Indo-European root *keuə-* "to swell." In meteorology cumulus clouds are so called because of their "piled up" appearance. In mathematics a cumulative frequency distribution is one in which each ordinate is the sum of frequencies in preceding intervals. Figuratively speaking, the values are accumulating from interval to interval. Compare *ogive*. [98]

cup (noun): from Medieval Latin *cuppa* "a drinking vessel," a variant of Latin *cupa* "a tub, vat." The Indo-European root is *keu-* "to bend; a round or hollow object." In mathematics the symbol used to represent the union of two sets, ∪, is sometimes known as a cup because of its resemblance to the cross-section of a drinking cup. Contrast *cap* for sound, appearance, and meaning. [96]

curious (adjective): from Latin *curiosus* "careful, painstaking, thoughtful, diligent," from Latin *cura* "watchfulness, attention, solicitude, care." As first used in English, *curious* meant "careful, studious, eager to learn." In the 18th century the meaning shifted to "worthy of consideration or study, inter-

esting." From there came the current meaning "interesting because strange or odd." In mathematics, a finite sequence $a_0, a_1, \ldots, a_n$ is said to be curious if a_i is the number of *i*'s in the sequence for each value of *i* from 0 through *n*. For example, the sequence $1, 2, 1, 0$ is curious because it contains 1 zero, 2 ones, 1 two, and 0 threes. Such sequences are mathematically interesting because they are difficult to construct.

curl (verb, noun): a metathesis of Middle Dutch *krull* "curly." Also from Dutch is *cruller*, a "curly" type of pastry. The Indo-European root is *ger-* "curving, crooked." Native English cognates include *cripple*, *creep*, *crutch* and *crank*. In calculus, curl is a measure of circulation per unit area.

curtate (adjective): from Latin *curtatus* "shortened," past participle of *curtare* "to shorten." The Indo-European root is *sker-* "to cut": when a piece is cut off of something, it gets shortened. Native English cognates include *shirt* (a cut garment) and *short*. In mathematics a "regular" cycloid is traced by a point on the circumference of a circle that rolls on a straight line. In a curtate cycloid, the distance from the center of the circle to the tracing point has been shortened; the tracing point of a curtate cycloid is inside of, rather than on, the rolling circle. [197]

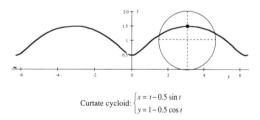

Curtate cycloid: $\begin{cases} x = t - 0.5 \sin t \\ y = 1 - 0.5 \cos t \end{cases}$

curtosis: see *kurtosis*.

curve (noun), **curvature** (noun): from Latin *curvus* "bent, curved." From the same Indo-European root *(s)ker-* "to turn, to bend," comes native English *shrink*. As defined in mathematics, a curve needn't be "curved"; a straight line, for example, is a type of curve. [198]

curvilinear (adjective): from Latin *curvus* "bent, curved" and *linear* (*q.v.*). At one time (and even to some extent now) the term *line* meant curve; a straight line is in fact a special type of curve. The term *curvilinear* was used to indicate that a certain curve was not a straight line. In two-dimensional curvilinear coordinates— as contrasted with rectangular and oblique coordinates—the guide "lines" and "axes" of the system are two sets of parallel curves rather than straight lines. [198, 119]

cusp (noun), **cuspidal** (adjective), **cuspitate** (adjective): from Latin *cuspis*, stem *cuspid-*, originally "the point of a lance," and later "a (whole) lance, a spike, a spear," and ultimately "any pointy thing." The word is of unknown prior origin but is assumed to have been borrowed from another language, as were many Roman military terms. As used in English, a cusp no longer has anything to do with warfare, except perhaps among children, some of whose pointy teeth are called cuspids and bicuspids. In mathematics a cusp is a point on a curve at which two branches of the curve meet and have a common tangent line. In a keratoid cusp, or cusp of the first kind, there is one branch of the curve on each side of the tangent line. In a ramphoid cusp, or cusp of the second kind, both branches of the curve lie on the same side of the tangent line. See pictures under *keratoid* and *ramphoid*. Contrast *corner* and *salient*. [29]

cut (verb, noun), **cutpoint** (noun): *cut* is a native English word with cognates in Scandinavian languages; an earlier source of the word is unknown. There were several Indo-European words with the general meaning "to cut"; compare the Indo-European roots underlying *inscribe*, *shear*, and *section*. Nevertheless, the Nordic peoples apparently felt the need to create still another such word. In mathematics *cut* is often used as a synonym of *meet* or *intersect*, as in "A secant line cuts a circle in two points." In a different sense, a cut is a partitioning of a set into two disjoint subsets. In graph theory, a cutpoint (see *point*) of a graph is a point whose removal, along with the removal of all incident segments, leaves the remaining pieces of the graph disconnected. [172]

cybernetics (noun): from Greek *kubernetes* "steersman," from *kubernan* "to steer." The Greek word was borrowed into Latin as *gubernare* "to steer, direct, rule," which is the basis of our borrowed word *govern*. Cybernetics is the theoretical (and therefore usually mathematical) study of control and communication systems. Cybernetics studies the ways in which such systems are "governed." It deals particularly with the ways in which an artificial system like a computer mimics a natural system like that of the human brain.

cycle (noun), **cyclic** (adjective): from Greek *kuklos* "a circle," from the Indo-European root *kwel-* "to move around," as found in native English *wheel*. (Although *wheel* is a common English word, it is al-

most never used in abstract mathematics.) In graph theory a cycle is a non-trivial closed walk: it "cycles back around" to the starting point. In geometry a cyclic quadrilateral is one which can be inscribed in a "cycle" (= circle). Something cyclic repeats, like a point moving forever around a circle: a cyclic permutation of elements is one which could be obtained if the elements were arranged around a circle. [107]

cyclides (plural noun): the main element is from Greek *kuklos* "a circle" (see previous entry). In its most literal usage the Greek suffix *-ides* was attached to the name of a man's father. For example, the 12th century Jewish philosopher known as Moses Maimonides was literally "Moses, son of Maimon." In a figurative sense, the *-ides* suffix means "descended from, related to, having to do with." In mathematics the cyclides of Dupin comprise the envelope of a family of spheres (= "three-dimensional circles") tangent to three fixed spheres. The envelope is named for the French mathematician François Pierre Charles Dupin (1784–1873). [107]

cycloid (noun), **cycloidal** (adjective): a compound whose main element is from Greek *kuklos* "a circle." Usually the suffix *-oid* (*q.v.*) means "having the appearance of." That is its sense in the adjective *cycloid*, used in zoology to refer to rounded fish scales. In the mathematical noun *cycloid*, however, the *-oid* suffix is used in the sense "having to do with, based on." A cycloid is derived from a circle in the following way: a cycloid is the curve traced by a point on the circumference of a circle that rolls (without slipping) along a straight line. Around 1599 the Italian physicist Galileo Galilei (1564–1642) was apparently the first person to study the cycloid, and he gave it its name. Compare *epicycloid*, *hypocycloid*, and *trochoid*. [107, 244]

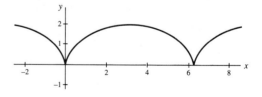

Cycloid: $x = t - \sin t$, $y = 1 - \cos t$

cyclomatic (adjective): the first component is from Greek *kuklos* "a circle," from the Indo-European root *kwel-* "to move around," as found in native English *wheel*. The second component is from Greek *matos* "willing." The Indo-European root is *men-*

"to think." A related borrowing from Greek is *automatic*, literally "self-willing." In graph theory the cyclomatic number of a graph corresponds to the dimension k of the cycle basis $\{\mu^1, \mu^2, \ldots, \mu^k\}$ of independent elementary cycles, a linear combination of which can be used to represent any cycle μ. [107, 134]

cyclosymmetric (adjective): the first element is from Greek *kuklos* "a circle," from the Indo-European root *kwel-* "to move around," as found in native English *wheel*. The second component is from Greek-derived *symmetric* (*q.v.*). A function is said to be cyclosymmetric if it remains unchanged (= "symmetric") under a cyclic change of the variables. An example is the function $f(x, y, z) = (x - y)(y - z)(z - x)$, which remains unchanged if x becomes y, y simultaneously becomes z, and z simultaneously becomes x. [107, 106, 126]

cyclotomic (adjective): the first component is from Greek *kuklos* "a circle," from the Indo-European root *kwel-* "to move around," as found in native English *wheel*. The second component is from Greek *temnein* "to cut." The Indo-European root is *tem-* "to cut." Related borrowings from Greek include *atom* (literally "not cut [into smaller pieces]") and *entomology* "the study of insects" (whose bodies are cut into distinctive sections). In mathematics the equation $x^n - 1 = 0$ is said to be cyclotomic; if all the solutions of the equation are graphed in the complex plane, they lie on a unit circle and "cut the circle" into equal arcs. [107, 225]

cylinder (noun), **cylindrical** (adjective): from Greek *kulindros* "a roller," from *kulindein* "to roll." The Indo-European root may be *skel-* "crooked," with subsidiary meaning "curved." While the cross-section of a mathematical cylinder may be any curve, unless otherwise qualified the word *cylinder* is usually taken to mean a circular cylinder. Similarly, although the parallel bases of a cylinder may form any angle with the axis of the cylinder, unless otherwise qualified the word *cylinder* is usually taken to mean a right cylinder. [195]

cylindroid (noun): the first component is *cylinder* (see previous entry). The ending is Greek-derived *-oid* (*q.v.*), "looking like." A cylindroid looks somewhat like a cylinder: it is a surface composed of all the straight lines that intersect each of two curves and that are parallel to a given plane. [195, 244]

cypher: see *cipher*.

D

damped (adjective): past participle of *to damp*, a native English word with cognates in other Germanic languages. In its earliest attested use, in the 14th century, the word was a noun meaning "vapor" or "noxious gas," as still seen in the line from the "Battle Hymn of the Republic" that talks about "the evening dews and damps." By the 16th century, the word had acquired the subsidiary meanings "fog, mist, humidity." Because foggy, rainy days can make people feel gloomy, the word also developed the meanings "depression, discouragement, stupor." Similarly, since a mist tends to reduce or extinguish a flame, *damp* developed into a verb meaning "to check, to discourage, to diminish." In mathematics, a damped wave is one whose amplitude diminishes over time, as typically represented by an equation like $y = ke^{-at}\cos(bt)$.

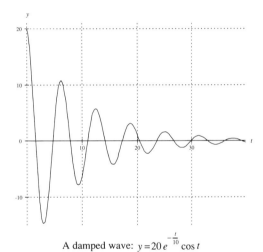

A damped wave: $y = 20\,e^{-\frac{t}{10}}\cos t$

dart (noun): from Old French, from a presumed Germanic *darodhaz* "spear." As the meaning of the word evolved, the spear lost its wooden shaft, so that a modern dart is invariably small. In mathematics a dart is a concave quadrilateral with two pairs of

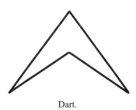

Dart.

adjacent, equal sides. In other words, a dart is a concave kite (*q.v.*). A dart is also known as a deltoid (*q.v.*).

data (plural), singular *datum* (noun): *datum* and *data* are past participles of Latin *dare* "to give." The *data* are literally the givens in a situation. The Indo-European root is *do-* "to give." Related borrowings from Latin are *tradition* (= what is given over from generation to generation) and *date*, since Roman letters often indicated the date by using the phrase "given (= *data*) on such and such a day." Borrowings from French include *endow* and *dowry*. Mathematicians often make conjectures after examining a lot of data and noticing patterns. [47]

day (noun): a native English word, from the Indo-European root *agh-* "day, span of time." How the Germanic family acquired the initial *d-* is unknown. Related cognates include *dawn* and *daisy*, originally *day's eye*, since the yellow center of the flower is revealed during the daytime and resembles an eye. The days in the Roman week were named for gods and goddesses, as are English days: Wednesday is Woden's day, Thursday is Thor's day, etc. The early Christian Church wanted to eliminate those vestiges of what it considered paganism, and advocated numbering rather than naming most days. Some European countries complied, though most later returned to pagan names. In Portuguese, however, the days from Monday through Friday are still *segunda-terça- quarta- quinta-* and *sexta-feira* ("second [day after the] fair," etc.) Hebrew and Arabic also number the days of the week.

de- (prefix): the Latin preposition *de* had many meanings, the most common being "from, down from, away from, out of." Those directional meanings are apparent in verbs like *depart*, *descend*, and *desert*. Less physical meanings of the Latin preposition *de* were "about, concerning, regarding, with respect to, as for," as seen in words like *define* and *denominator*. The Indo-European root may be *de-*, a demonstrative stem found in native English *to*. [32]

dea- (numerical prefix): according to the 1990 *Guinness Book of Records*, this prefix has the largest value in the metric system. It supposedly multiplies the unit to which it is attached by 10^{30}, which can be rewritten as $(10^3)^{10}$, or the tenth power of a thousand. *Dea-* is from Greek *deka* "ten," but the *-k-* has been deleted to avoid confusion with the prefix *deka-* (*q.v.*) that is already in use in the metric system. Structurally speaking, *dea-* is in accord with

the other recent "magnifying" prefixes added to the International System of Units: all have two syllables and end in an *-a*. The only problem is that the prefix *dea-* doesn't seem to exist. In a letter dated 1 December 1992, an editor at Guinness wrote that "[w]hile the prefixes dea and tredo did appear in earlier editions of the Guinness Book of Records, their origin appears to be a mystery. Subsequently, they were replaced by the officially recognised extremes, namely yotta and yocto which were officially adopted by the International Committee on Weights and Measures in 1991." [34]

deca-: see *deka-*

decade (noun): from Greek *deka* "ten," from the Indo-European root *dekm-* "ten." A decade is literally a group of ten things. Frequently the ten things are in sequence, as for instance the numbers from 10 through 19. Ten consecutive years whose tens digits are the same constitute a decade, as in the decade of the 1940's, which went from 1940 through 1949. [34]

decagon (noun): from Greek *deka* "ten" and the suffix *-gon* (*q.v.*) "angle." A decagon is a ten-angled (and of course also ten-sided) polygon. The ancient Greeks knew how to construct a regular decagon using only a compass and unmarked straightedge. [34, 65]

decahedron, plural *decahedra* (noun), **decahedral** (adjective): from Greek *deka* "ten" and *hedra* (prehistoric Greek *sedra*) "base"; the Indo-European root is *sed-*, as found in English *sit* and *seat*. A decahedron is a ten-based, i.e., ten-faced, polyhedron. [34, 184]

decakis- (numerical prefix): a Greek form meaning "ten times," based on *deka* "ten." The prefix is used as part of the name of a stellated regular solid. It indicates that each original face has become ten faces. A decakisdodecahedron, for example, is a polyhedron with ten times twelve, or 120, faces. [34]

December (noun): from Latin *decem*, from the Indo-European root *dekm-* "ten." In the Roman calendar, which began in March, December was the tenth month of the year. In spite of its name, it is now the twelfth month because January later became the first month of the year. [34]

deci-, abbreviated *d* (numerical prefix): an abbreviation of Latin *decimus* "tenth." The Indo-European root is *dekm-* "ten." In the metric system the prefix *deci-* indicates one-tenth of the following unit, as in *decimeter*. In 1791 the Paris Academy of Sciences chose the metric system prefixes for the first three negative powers of ten; all three (*centi-* and *milli-* being the last two) came from the appropriate Latin number words and all three ended in *-i*. In contrast, the prefixes for the first three positive powers of ten were chosen from Greek number words. In recent years the new submultiple prefixes have always ended with *-o* and have not necessarily been chosen from Latin. [34]

decidable (adjective): the first elements are from Latin *decidere* "to cut off," from *de* (*q.v.*) "off" and *cædere* "to cut down, to beat, to strike." The Indo-European root is *kaə-id-* "to strike." For an explanation of the suffix see *-able*. Even in Roman times *decidere* "to cut off," had developed a figurative sense "to cut short, to put an end to," and therefore "to settle, to determine," and finally "to decide." In mathematics the Austrian logician Kurt Gödel (1906–1978) proved that in any axiomatic system based on the natural numbers there will always be propositions that, while actually true, will be forever undecidable within that system. [32, 84, 69]

decile (noun): from Latin *decimus* "tenth." The Indo-European root is *dekm-* "ten." In statistics a decile is any of the ten equal parts into which a distribution may be broken up. Compare *fractile*, *quartile*, *quintile*, and *percentile*. [34]

decillion (numeral): patterned after *billion* by replacing the *b-* with Latin *dec-* "ten." Since in most countries a billion is the second power of a million, a decillion is defined as the tenth power of a million, or 10^{60}. In the United States, however, a billion is 10^9, and a decillion adds eight groups of three zeroes, making a decillion equal to 10^{33}. [34, 72, 151]

decimal (adjective, noun): from Latin *decimus* "tenth," the ordinal number corresponding to *decem* "ten," plus the adjectival suffix *-alis*, (as in Latin *anima* "breath, soul," and *animalis* "a creature that breathes"). From Latin *decimus* comes the verb *decimate*, which originally referred to the stern practice in the Roman army of randomly killing one-tenth of the soldiers in an unruly unit in order to re-establish discipline. With modern inflation, however, *decimate* is often used in the sense of "destroy a large part of," perhaps out of confusion with *devastate*. In mathematics a decimal system of numeration uses base ten. In common usage, *decimal* has also

come to be a noun meaning "a number between 0 and 1 written using a decimal point," as opposed to the equivalent number written as a fraction. Just as *decimate* has been abstracted away from the original sense of one-tenth, *decimal* is also sometimes used by analogy to refer to a number between 0 and 1 written in a base other than ten; in that way we avoid having to make up a term like "bimal" for base two, etc. Compare *octimal*, *hexadecimal*, and *vigesimal*. [34]

decouple (verb): via French, from Latin *de* (*q.v.*) "away from" and *couple* (*q.v.*). In mathematics, the coefficients in a system of equations are said to decouple when, say, given four equations in four unknowns, the values of the couple A and C can be found from one pair of equations, while the values of the couple B and D can be found from another pair of equations. [32, 103, 13, 238]

decrease (verb), **decrement** (noun): from Latin *de* (*q.v.*) "down from, away from," and *crescere* "to grow." The Indo-European root is *ker-* "to grow." Related borrowings from Latin include *procreate* and *concrete*. When a quantity decreases, it grows "down from" the original amount and becomes smaller. [32, 92]

deduce (verb), **deduction** (noun), **deductive** (adjective): from Latin *de* (*q.v.*) "down from," plus *ductus*, past participle of *ducere* "to lead." The Indo-European root is *deuk-* "to lead," as seen in native English *tug*. Related Italian terms are the *doge* of Venice and *il duce* "the leader." When you reason deductively, you start from certain first principles and rules that "lead down" to specific things that must follow as a consequence. Contrast *induction*. [32, 40]

defective (adjective): from Latin *de* (*q.v.*) "down from" and *factus*, past participle of *facere* "to make, to do." Something defective is "down from [being properly] done." In mathematics a defective equation is one that has fewer roots than another equation from which it has been derived; the number of roots is "down from" the original number. Contrast *extraneous* and *redundant*. [32, 41]

deficient (adjective): from Latin *de* (*q.v.*) "down from" and the present participle of *facere* "to make, to do." Something deficient is "down from [being properly] done"; in other words, something deficient falls short. A number like 8 is said to be deficient because when its proper divisors are added, the total falls short of the original number: $1 + 2 + 4 = 7$, which is 1 short of 8. Contrast *perfect* and *abundant*. [32, 41, 146]

define (verb), **definite** (adjective), **definition** (noun): from Latin *de* (*q.v.*), a preposition with many meanings, both physical and abstract, and *finis* "end, boundary, border," of unknown prior origin. When you define something you "put boundaries around" what it can mean. A good definition puts an end to confusion about what a term means. In calculus, a definite integral has numerical limits or "boundaries" on it, as opposed to an indefinite integral. [32, 60]

deform (verb), **deformation** (noun): from Latin *de* (*q.v.*) "down or away from" and Latin *forma* "contour, figure, shape." The Latin word seems to be a metathesis of Greek *morphe* "form, beauty, appearance," of unknown prior origin. In mathematics a deformation of a figure is any transformation that doesn't tear the figure. [32, 137]

degenerate (verb, adjective): from Latin *degeneratus*, past participle of *degenerare* "to become unlike one's race, to fall off, to degenerate." The prefix is *de-* (*q.v.*) "away from," and the root is *genus*, stem *gener-* "birth, descent, origin, race." The Indo-European root is *gen-* "to give birth," as seen in native English *kin* and *kind*. When something degenerates, it moves away from its genus or type, so that it is no longer general or typical. A degenerate conic section, for example, may be a point, a line, or two lines, none of which are what we generally think of as conic sections. [32, 64]

degree (noun): from Latin *de* (*q.v.*) "away from, down," and *gradus* "a step," from the verb *gradi* "to go, to walk." The Indo-European root is *ghredh-* "to walk, to go." As you descend a ladder, you take steps (= *gradi*) down (= *de*), so *degree* came to represent a rung on a ladder and, by analogy, a mark on a scale. The related word *grade* is also borrowed from Latin. The grade you make in a course—A, B, C, D, F—tells how high or low you stand in a course. If you think of polynomials being put on the steps of a ladder, with the higher powers at the top, then the degree of the polynomial corresponds to its highest power. (Compare the similar meaning of *dimension*, and notice how *degree* and *dimension* were similar for the ancient Greeks and Romans; for related concepts, also compare *squared* and *cubed*.) [32, 73]

deka- or **deca-** abbreviated *da* (numerical prefix): from Greek *deka* "ten," from the Indo-European root *dekm-* "ten." In the metric system, the prefix *deka-* multiplies the following unit by ten, as in *decaliter* "ten liters." The prefixes representing the first three positive powers of ten were chosen from Greek number words by the Paris Academy of Sciences in 1791. In contrast, the prefixes for the first three negative powers of ten were chosen from Latin number words. In the United States the preferred spelling is *deka-* rather than *deca-*. [34]

del (noun): short for *delta*. In calculus the operator *del*, equivalent to

$$\mathbf{i}\,\frac{\partial}{\partial x} + \mathbf{j}\,\frac{\partial}{\partial y} + \mathbf{k}\,\frac{\partial}{\partial z},$$

is symbolized by an upside-down Greek capital delta, ∇. The synonym *nabla* (*q.v.*) also refers to the shape of the symbol that represents the operator, rather than to any inherent property of the operator itself. See more under *atled*. [31]

deleted (adjective): from Latin *deletus*, past participle of *delere* "to blot out, to wipe out; to destroy, to annihilate." The word is of unknown prior origin. In mathematics a deleted neighborhood of a point c is a neighborhood of c that does not include c itself. Point c has been "wiped out."

delta (noun): the fourth letter of the Greek alphabet, written Δ as a capital and δ in lower case. It corresponds to the Hebrew letter *daleth*, ד, meaning "door," and to our Roman letter *d*. In calculus δ is used to set a tolerance for the amount that the independent variable of a function can change; δ is chosen because it represents the initial *d*-sound of the word *difference*. Similarly, the capital letter Δ is used in calculus to symbolize the word *difference*. From the shape of the capital Greek letter we have our meaning of *delta* as land that fans out at the mouth of a river. See further explanation under *aleph*. [31]

deltahedron, plural *deltahedra* (noun): a combination of *delta* and *polyhedron*. Because the Greek letter delta, Δ, is shaped like a triangle, a deltahedron is a polyhedron all of whose faces are triangles. An additional condition required of a deltahedron is that all the vertices be of the same type. [31, 184]

deltoid (noun): from *delta* and *-oid* (*qq.v.*) "looking like," and by extension "derived from." The term *deltoid* is used in two different ways in mathematics. (1) A deltoid is a three-cusped hypocycloid, a closed curve that has a shape like a capital Greek delta, Δ, except that the "sides" are concave rather than straight. The Swiss mathematician Leonhard Euler (1707–1783) first studied the deltoid as part of an investigation of caustic curves. (2) A deltoid is a concave quadrilateral with two pairs of adjacent, equal sides. The quadrilateral deltoid, which is also known as a dart, looks like the letter Δ, but with its base bent up in the middle. [31, 244]

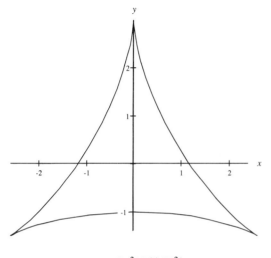

$$\text{Deltoid:}\quad \begin{aligned} x &= 2\cos t + \cos 2t \\ y &= 2\sin t - \sin 2t \end{aligned}$$

denary (adjective): from Latin *denarius* "containing ten," from the distributive adverb *deni* "every ten, ten each, by tens." The Indo-European root is *dekm-* "ten." The Roman *denarius* was a silver coin that originally contained 10 of the next lowest monetary unit, the *as* (hence our *ace* in a deck of cards). The *denarius* was in use throughout the Roman Empire. In Western Europe the word developed into Spanish *dinero* and Portuguese *dinheiro* "money." In Algeria, Bahrain, Iraq, Jordan, Kuwait, Libya, Tunisia, Yemen, and the countries that used to be known as Yugoslavia, Latin *denarius* has become the standard monetary unit, the *dinar*. In mathematics a denary system of numeration uses base 10. Nowadays *denary* has been largely replaced by the related word *decimal*. Compare *binary*, *octonary*, *quinary*, *senary*, *ternary*, and *vicenary*. [34]

denial (noun): from Old French *denier*, from Latin *denegare* "to say no, to refuse." The first component is from Latin *de* (*q.v.*), a demonstrative prefix. The second component is from Latin *negare* "to say no,

to deny." In logic the denial of a proposition is the same as the negation of that proposition. [32, 141]

denominate (adjective), **denominator** (noun): from Latin *de* (*q.v.*), a demonstrative prefix, and *nomen*, stem *nomin-*, "name," cognate to the English word with the same meaning. The Indo-European root is *no-men-* "name." A related borrowing from French is *misnomer* "a wrong name." In arithmetic, the denominator of a fraction names the kind of thing you're dealing with. In the fraction $\frac{3}{5}$, for example, you have three fifths, just the way you could have three sheep or three apples or three of any other thing. In fact, for a long time a fifth (of a gallon) was a common measure of alcoholic beverages in the United States. The denominator behaves like any noun: in fact, *noun* is just a modified version of the French word *nom* "name." Since the denominator names the sort of a thing you're dealing with, you must have a common denominator when you add fractions; otherwise it's like trying to add sheep and apples. Older names for the denominator included *inferior* and *base*. Denominate numbers are those with "names," i.e., units of measure, attached, as in 3 inches, 5 pounds, or 8 acres. As with fractions, denominate numbers can be added only when all the units are the same. [32, 145, 153]

dense (adjective), **density** (noun): from Latin *densus* "having its particles close together, crowded, thick." The word is used with its meaning virtually unchanged in English. In topology a set S is said to be dense in a space P if every point in P is a point or a limit point of S. In physical applications, density is defined as mass of matter per unit of volume.

denumerable (adjective): from Latin *de-* (*q.v.*), a prefix with demonstrative meaning, and *numerus* "number." Latin *numerus* may be from the Indo-European root *nem-* "to assign, to allot, to take." For an explanation of the suffix see *-able*. A denumerable set is one whose members can be put in one-to-one correspondence with the counting numbers. [32, 143, 69]

dependent (adjective): from Latin *de* (*q.v.*) "down from" and *pendent-*, present participial stem of *pendere* "to hang." The connection with Indo-European root *(s)pen-* "to draw, to stretch, to spin," is apparent in the fact that when a string is allowed to hang down with a weight attached, it is stretched tight. Native English cognates include *span* and *spindle*. In algebra the value of a dependent variable "hangs from" or "hinges on" the value of something else, namely the independent variable(s). In $z = 3x^2y$, for example, the value of z "hinges on" the values chosen for x and y. A system of two linear equations in two unknowns is said to be dependent if the coefficients in each equation are a fixed multiple of the corresponding coefficients in the other equation. Each of the equations depends on the other for meaning because the graph of one of the equations is identical to the graph of the other. [32, 206, 146]

depreciate (verb), **depreciation** (noun): from Latin *de* (*q.v.*) "down from" and *pretium* "price, worth, value." The Indo-European root is *per-* "to distribute, to traffic in, to sell." A related borrowing from Latin via French is *interpreter* (a go-between who "traffics in" meanings or negotiations). When the price or value of something goes down, it depreciates. Contrast *appreciate*. [32, 165]

depressed (adjective), **depression** (noun): from Latin *de* (*q.v.*) "down from" and *pressus*, past participle of *premere* "to press." The Indo-European root is *per-* "to strike." A depressed person feels emotionally pressed down. In mathematics a depressed equation is one that results from reducing ("pressing down") the number of roots of a given equation. In trigonometry, the angle of depression is measured by "pressing" a horizontal line downward until it joins the observer's eye to the object or point in question. [32, 166]

derive (verb), **derivative** (noun): from Latin *de* (*q.v.*) "off, from" and *rivus* "brook, stream," as seen in *rivulet*, probably borrowed from Italian. The Indo-European root is *rei-* "to flow, to run." The basic meaning of *derive* is "to flow from." In algebra a derived equation is one which "flows from" another after a series of manipulations. In calculus the derivative is a rate of flow or change. Looking at things picturesquely, the derivative "flows from" the difference quotient (see *differentiate*) when you take its limit as the denominator trickles down to zero. Linguistically speaking, the word *derivative* is derived from the word *derive*. [32, 180]

descending (adjective): from Latin *de* (*q.v.*) "down from" and *scandere* "to climb, leap." The Indo-European root is *skand-* "to climb." Surprisingly related is *scandal*, from Greek *skandalos* "a snare, trap, or stumbling block" (something you trip or leap over). In algebra, when terms are arranged in descending order the exponents "climb down" from term to term. [32, 194]

describe (verb): from Latin *de* (*q.v.*) "of, from, about" and *scribere* "to write." Underlying Latin *scribere* was the more physical Indo-European root *skribh-*, an extension of *sker-* "to cut, separate"; it is found in native English *shear*, *shard*, and *sharp*, as well as *scrap* and *scrape*, borrowed from Old Norse. Remember that in ancient times geometric figures were scratched on the ground or onto waxed tablets or other physical objects. In geometry, to describe an arc of a circle is to draw the arc from a given center. In nonmathematical English, to describe something is to write about it and tell what it is like. [32, 197]

detach (verb), **detachment** (noun): from French *dé* (*q.v.*) "away from, opposite to" and *attacher* "to fasten, to attach." The French word was built upon a presumed Germanic root *stag-*, as found in native English *stake*. The Indo-European root is *steg-* "a stick, a pole." Related borrowings include *stagger* (to be made to trip with a stick) from Old Norse, and *attack* (to fight against someone using sticks) from Italian. To *attach* originally meant "to drive a stick through." To *detach* is the opposite: "to remove a stake in order to set something free." In logic, the rule of detachment refers to implications: if an implication is true, and if its antecedent is true, then its consequent is true. That consequent can therefore be "detached" from the implication and stand alone as a true statement. In algebra, polynomials can sometimes be easily dealt with if the coefficients are detached from the variables and manipulated on their own, as in synthetic division, for instance. [32]

determine (verb), **determinant** (noun): from Latin *de* (*q.v.*) "of, from" and *terminus* "boundary." The Indo-European root is *ter-* "a peg, post, marker," and by extension "a boundary." Something which is determined has a definite boundary or value. In geometry, when we say that two points determine a straight line, we mean that those two points specify the line uniquely: the "boundaries" of the line have been clearly "marked." In algebra a determinant is a square array of numbers bounded by two parallel vertical lines. Furthermore, the value of a determinant is "determined" by following a set of rules. The French mathematician Augustin Louis Cauchy (1789–1857) apparently first used *determinant* in its modern sense in 1812; Carl Friedrich Gauss (1777–1855) had used the word in 1801, but more in the sense of what we would now call a discriminant. [32, 227, 146]

developable (adjective): from Latin *dis-*, a prefix that indicates undoing, and an assumed Celtic root *vol-* "to roll." The Indo-European root is *wel-* "to turn, to roll." For an explanation of the suffix see *-able*. In a play or a novel, when the plot develops, events "unroll." In mathematics a developable surface is one which can be unrolled, without distortion, onto a plane. The lateral surface of a cone is developable, but the surface of a sphere is not. [32, 248, 69]

deviate (verb), **deviation** (noun): from Latin *de* (*q.v.*) "away from" and *via* "a road, a way." The Indo-European root is *wegh-* "to go, to transport." Native English cognates include *weigh* and its homonym *way*. To deviate is literally to go out of one's way. In statistics, a standard deviation is a measure of how far the data "go out of their way" from the mean. [32, 243]

dextrorse or **dextrorsum** (adjective): the first component is from Latin *dexter* "right, on the right," from the Indo-European root *deks-* "right." The second component is from Latin *versus*, past participle of *vertere* "to turn," from the Indo-European root *wer-* "to turn." Latin *dexter* developed the extended meanings "handy, adroit, skillful," because most people are right-handed; the extended meaning is reflected in our borrowed word *dexterity*. In botany, the original directional sense is preserved: a dextrorse plant turns from left to right as it grows upward. In mathematics a dextrorse curve is a right-handed curve, i.e., one whose torsion at a given point is negative. Contrast *sinistrorse*. [251]

diabolic (adjective, noun): from Late Latin *diabolus* "the devil," in turn from the Late Greek compound *diabolos*. The first element is from Greek *dia-* "through, across." The second element is from Greek *ballein* "to throw," from the Indo-European root *gʷelə-* "to throw, to reach." The Greek compound was a translation of Hebrew *Satan*; it meant "accuser, slanderer," one who throws [untrue] words around, as the devil is supposed to do. In a magic square, the sum of every row and column is the same as that of each of the two diagonals. If each of the broken diagonals also has the same sum—a devilishly difficult feat—then the magic square is said to be diabolic. The word *diabolic* used in this way is a pun because it plays off the element *dia-* found in both *diabolic* and *diagonal*. A diabolic magic square is also called *panmagic* or *pandiagonal* (*qq.v.*). [45, 78]

diacaustic (adjective): a compound of Greek *dia-* "through, across," and *caustic* (*q.v.*). Whereas a caustic curve is the envelope of light rays emitted from a radiant source after reflection or refraction in a given curve, a catacaustic curve involves only refracted rays. When rays are refracted, they go through a prism or lens. [45, 97]

diagonal (adjective, noun), **diagonalize** (verb): from Greek *dia-* "through, across," and the root *gon-* "angle." In a polygon, a diagonal is a chord that "goes across" from one "angle" (vertex) to a non-adjacent "angle" (vertex). In a polyhedron, a diagonal is a chord that "goes across" from one "angle" (vertex) to another that's not in the same face. To diagonalize a matrix is to convert it to a matrix all of whose nonzero elements lie on the main diagonal of the matrix. [45, 65]

diagram (noun): from Greek *dia-* "across, apart," and the verb *graphein* "to write," as seen in the related borrowings *telegraph* and *phonograph*. The Indo-European root is *gerbh-* "to scratch," since writing was originally scratched or carved onto physical objects. In a diagram, you "write" or sketch things and spread them "apart" so you can see all the parts clearly. In geometry, a situation is frequently represented by a diagram, but a diagram isn't necessarily drawn to scale, and students shouldn't make unfounded assumptions based on the appearance of things in a diagram. [45, 68]

diakis- (numerical prefix): a Greek form meaning "two times," based on *di-* "two." The prefix is used as part of the name of a stellated regular solid; it indicates that for each original face there are now two faces. A diakisdodecahedron, for example, is a polyhedron with two times twelve, or 24, faces. [48]

dialytic (adjective): the first component is from Greek *dia-* "through, across," of unknown prior origin. The second component is from Greek *luein* "to set loose, to set free," from the Indo-European root *leu-* "to loosen, to cut up." Native English cognates include *lose*, *loose* and *forlorn*. The English mathematician James Joseph Sylvester (1814–1897) used the term *dialytic* to refer to a method of eliminating (= "loosening up") a variable from two algebraic equations. The related medical term *dialysis* refers to the separating of large molecules from small ones. [45, 116]

diameter (noun), **diametral** (adjective): from Greek *dia-* "across," of unknown prior origin, and *metron* "a measure." The Indo-European root is *me-* "to measure," as seen in native English *month* (a unit of time measured by the moon). A diameter (or conjugate diameter) of a conic section is a chord which bisects each member of a family of parallel chords; as the diameter cuts across each chord, it divides it into equal measures. Because a circle is the most familiar conic, the diameter of a circle is the most familiar type of diameter; in fact, most people know of no other type of diameter. [45, 126]

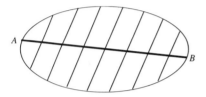

$\overline{AB}$ is a diameter of the ellipse
because it bisects each of the
ellipse's parallel chords.

diamond (noun): from the Late Latin stem *diamant-*, a metathesis of Vulgar Latin *adimant-*, from Classical Latin *adamant-*. The original Greek *adamant-* meant "hard metal, diamond," from *a-* "not" and the Indo-European root *dema-* "to break, to tame [a horse]." A hard piece of metal or jewel was originally conceived as being unbreakable. In geometry, a diamond is a lozenge, a non-square rhombus. It is named for its resemblance to the jewel. In logic, a diamond is one name for the symbol used to mean "it is possible that." In contrast to the diamond of geometry, the diamond symbol in logic is actually a square; its diagonals are horizontally and vertically oriented. [141, 36]

dichotomous (adjective), **dichotomy** (noun): the first component is from Greek *dikha* "in two (parts)," from the Indo-European root *dwo-* "two." The second component is from Greek *temnein* "to cut," from the Indo-European root *tem-* "to cut." Related borrowings from Greek include *tome*, a volume "cut" out of a larger set of books, and *entomology,* the study of insects, whose bodies are cut into distinct sections. In mathematics a dichotomous classification is one which allows only two mutually exclusive possibilities. An integer must be odd or even, for example; under "normal" circumstances a tossed coin must land heads or tails, a child must be born male or female, etc. Contrast *trichotomy*. [48, 225]

die, plural *dice* (noun): from Old French *de*, from Latin *datum* past participle of *dare* "to give." The Indo-European root is *do-* "to give." When the Romans played games that involved markers of various sorts, they used the verb "to give" whenever they played a stone or token or "man." The verb *dare* therefore took on the meaning "to play, to cast down." One of the most common objects cast down in games was a die, which came to be known by the word *datum* "cast down." Latin *datum* evolved into modern French *dé*, which English borrowed as *die*. Some people mistakenly say "a dice" rather than "a die." [47]

diffeomorphism (noun): a combination of *differentiable* and *homeomorphism* (*qq.v.*). A diffeomorphism is a one-to-one mapping for which both the mapping and its inverse are differentiable. [46, 23, 137]

difference (noun): the first element is from the Latin prefix *dis-* "apart, away," of unknown prior origin. The second element is from Latin *ferre* "to bring, to carry." The Indo-European root is *bher-* "to carry," as seen in Native English *to bear*. In subtraction, the difference is what remains when you take something away from something else. In algebra and calculus, the difference between two quantities is often symbolized by Latin *d* or Greek Δ, both representing the first letter of the word *difference*. That word, in the sense of "the result of subtraction," has seemed artificial to many speakers. Alternatives that have been used include *relic*, *excess*, and *balance*. Historically one of the most popular, perhaps because of its brevity, has been *rest*; it is still alive in a sentence like "I'll take three of these cookies and give you the rest." [46, 23]

differential (adjective, noun), **differentiate** (verb): the first element is from the Latin prefix *dis-* "apart, away," of unknown prior origin. The second element is from Latin *ferrent-*, present participial stem of *ferre* "to carry, to bring." The Indo-European root is *bher-* "to carry." Related borrowings from Latin include *defer* and *prefer*. In calculus one definition of the derivative is

$$\lim_{x \to a} \frac{f(x) - f(a)}{x - a};$$

the fraction contained in that limit is often referred to as the "difference quotient." That explains why, instead of a non-existent verb *to derivate*, we have *to differentiate*. After all, the derivative is a rate of change, and *difference* means the change in a quantity. [46, 23, 146]

digit (noun), **digital** (adjective): from Latin *digitus* "a finger," but more literally "a pointer," from the Indo-European root *deik-* "to show." A related borrowing from Latin is *indicate*. It is usually assumed that people first counted on their fingers. Since the kind of number that can be counted on fingers is a whole number, *digit* came to mean a whole number from one through ten (because people have only ten fingers). In Arabic notation, nine of those numbers from one through ten can be written with a single character, so *digit* eventually came to mean any single numerical character, including the late addition of zero, which is not normally countable on fingers. [33]

digon (noun), **digonal** (adjective): from Greek *di-* "two" and *gonia* "angle," from the Indo-European root *genu-* "angle, knee." In plane geometry a digon is a region enclosed by two curves meeting at two vertices. Probably the most commonly seen digon is the one in the shape of a stylized human eye. In spite of its name, this type of digon is not a polygon because its "sides" are not straight line segments. In solid geometry, if a sphere is cut by two distinct planes that have a diameter in common, the two planes divide the surface of the sphere into four spherical digons. The two circular arcs that bound each digon meet at opposite ends of the digon and form two equal "angles." The solid bounded by a digon and the two cutting planes is called a spherical *wedge* or *lune* (*qq.v.*). [48, 65]

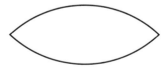

A two-dimensional digon

digraph (noun): a combination of *directed* and *graph* (*qq.v.*). A digraph is a graph consisting of nodes connected by directed line segments. The direction of a segment expresses a relation between the two nodes it joins. For example, if the nodes represent competing players in a round of games in which every player competes against every other player just once, then each directed segment might point from a winner to a loser. Because each segment points in one of two possible directions, some

people jump to the false conclusion that the *di-* in *digraph* means "two," as it does in *digon, dihedral,* and *dipyramid.* [46, 68]

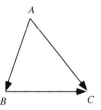

Digraph showing that
A defeated *B* and *C,*
and *B* defeated *C.*

dihedral (adjective); **dihedron**, plural *dihedra* (noun): from the Greek *di-* "two" and *hedra,* originally *sedra,* "base." The Indo-European root is *sed-* "to sit," as seen in native English *sit* and *seat* as well as *sedentary,* borrowed from Latin. In geometry, a dihedral angle is formed by two intersecting "bases," that is, planes, whereas a plane angle is formed by two intersecting lines. A dihedral angle is sometimes called a dihedron. The term *dihedron* is also used to refer to a solid with two edges and two faces meeting at each vertex; there may be any number (≥ 2) of vertices. Chinese wontons and Italian ravioli are approximate models of a dihedron with four vertices. In spite of its name, this type of dihedron is not a polyhedron because the two faces (= "bases") are curved surfaces rather than polygons. [48, 184]

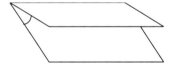

A dihedral angle.

dilation (noun): the first element is from the Latin prefix *dis-* "apart, away," of unknown prior origin. The second element is from Latin *latus* "wide," from the Indo-European root *stelə-* "to extend." When a person's pupils dilate they open wide. In mathematics a dilation is a type of two-dimensional transformation in which one dimension is magnified while the other stays the same. A dilation is also known as an *elongation* (*q.v.*). A dilation can be represented by the equations $x' = x$ and $y' = ky$, or $y' = y$ and $x' = kx$, where $k > 1$. Contrast *compression.* [46, 210]

dime (noun): from French *dîme* "tenth," from Old French *disme*, from Latin *decimus* "tenth." The Indo-European root is *dekm-* "ten." As first used in English, a dime was a tithe paid to the church; in fact the native English word *tithe* is a variant of the word *tenth*. In the American monetary system, the coin known as a dime is literally one-tenth of a dollar. [34]

dimension (noun): the first element is from the Latin prefix *dis-* "apart, away," of unknown prior origin. The second element is from Latin *mensura* "a measuring." The Indo-European root is *me-* "to measure." In geometry, the dimensions of a polygon are the measurements of its sides. In terms of physical reality, when you measure in one direction you are dealing with one dimension; when you start measuring in a perpendicular direction, you enter a second dimension, and so on. Compare the similar meaning of *degree*, and notice how *degree* and *dimension* were equivalent for the ancient Greeks and Romans; more specifically, compare *squared* and *cubed*. [46, 126]

dimetric (adjective), **dimetry** (noun): from Greek *di-* "two" and *metron* "a measure." The Indo-European root is *me-* "to measure." When a three-dimensional graph is projected onto a flat piece of paper, the positive *x*-, *y*- and *z*-axes are typically drawn radiating from a single point, and then a scale is placed on each axis. If the two-dimensional projections of two of the three axes have the same scale and the third one is different, the projection is said to be a dimetry. Dimetry is one type of axonometry (*q.v.*). Contrast *isometry* and *trimetry*. [48, 126]

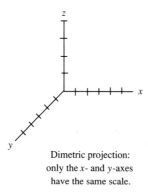

Dimetric projection:
only the *x*- and *y*-axes
have the same scale.

diminish (verb): the modern verb is a mixture of forms that resulted from two separate lines of development of the same Latin word. Classical Latin *deminuere* meant "to lessen by taking something

away." It was a compound of *de* "away from," and *minuere* "to make smaller, to lessen," from the Indo-European root *mei-* "small." The Latin compound led to French *diminuer*, from which Middle English borrowed the verb *diminue*. At the same time, English borrowed the verb *minish* "to lessen, reduce." It came from Old French *menuisier* "to cut into small pieces, to diminish." That verb had come from a presumed Vulgar Latin *minutiare*, from Classical Latin *minutus*, past participle of *minuere*. The verbs *diminue* and *minish* eventually merged into the modern *diminish*, which has long been used in the word problem sections of algebra books. [32, 131]

dipyramid (noun): from Greek *di-* "two," plus *pyramid* (*q.v.*). In solid geometry, a dipyramid is formed by joining the bases of two congruent pyramids. Although two congruent pyramids could be joined at corresponding faces or even along two edges or at their vertices, the resulting solids would not properly be dipyramids. [48, 176]

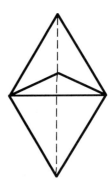

Dipyramid
with six faces.

direct (adjective), **directed** (adjective), **direction** (noun): the first element is from the Latin prefix *dis-* "apart, away from," of unknown prior origin. The second element is from Latin *rectus*, past participle of *regere* "to stretch, to keep in a straight line, to guide in the right direction." The Indo-European root is *reg-* "to move in a straight line." In mathematics, a directed line segment is considered positive or negative depending on which way it points. A vector is an example of a directed segment because it has not only length but direction, too. [46, 179]

director (noun): from *direct* (see above) and the Latin suffix *-or* "a man or masculine thing that does the indicated action." In analytic geometry, the director circle of an ellipse is the locus of the points of intersection of pairs of perpendicular tangent lines drawn to the ellipse. [46, 179, 153]

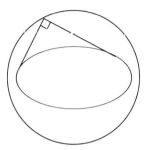

Director circle
of an ellipse.

directrix, plural *directrices*, (noun): a Latin feminine noun meaning "the female who directs"; the masculine version, *director*, is in common use today when speaking of movies (see *direct*). The feminine version of the word is used in mathematics because the Latin word for line, *linea*, was feminine. In a conic section, the directrix is the line that acts together with the focus to direct points to their proper location on the curve. In nonmathematical English, a female aviator is called an *aviatrix*, and a female executor of a will is an *executrix*. A female actor should be an *actrix*, but the word has been somewhat altered by French to *actress*. [46, 179, 236]

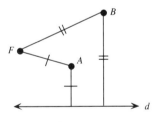

Line *d* is a directrix for
the parabola on which
points *A* and *B* lie.

disc or **disk** (noun): from Latin *discus* "a flat ring, a disk." The Latin word had been borrowed from the similar Greek *diskos*. The Indo-European root is *deik-* "to show," and by extension "to direct." That last meaning indicates [another cognate] that a discus was originally conceived of as an object meant to be directed or thrown. Because a discus was relatively flat, the word came to be applied to other flat objects or regions. One such extension led to *dish*, an object which is both round and flat. Another extension led to *desk*, which consists of a flat surface mounted on

legs. Other related words include *dais* and *token*. In mathematics, an open disc consists of all the points inside a circle, while a closed disc adds the points on the circle itself. Technically speaking a circle is a curve, not a region, though many people say *circle* when they mean *disc*. Compare *ball* in three dimensions. [33]

discrete (adjective): the first element is from the Latin prefix *dis-* "apart, away," of unknown prior origin. The second element is from Latin *cretus*, past participle of *cernere* "to sift, to distinguish, to discern." The Indo-European root is *krei-* "to sieve, to distinguish." A native English cognate is *to riddle* (as with holes, thereby creating a kind of sieve). A related borrowing from Latin is *crime* (which originally was an accusation; the truth of the charge still had to be "sifted out" or discerned.) In mathematics, a discrete variable is one whose values are discernible, distinct, "kept apart," as opposed to continuous values. In nonmathematical English someone who is discreet (different spelling but same etymology) is discerning enough to be prudent, to tone things down, or to keep them private. [46, 105]

discriminant (noun): the first element is from the Latin prefix *dis-* "apart, away," of unknown prior origin. The second element is from Latin *crimen*, stem *crimin-* "an accusation," the truth of the charge still having to be discerned (whereas the French-derived *crime* already assumes guilt). In fact *discern* is from the same Indo-European root *krei-* "to sieve, to distinguish." In mathematics the discriminant corresponding to a second-degree equation in one variable distinguishes among several possibilities: two distinct real roots, two identical real roots, or two complex conjugate roots. Different discriminants exist for more complicated polynomials as well. As used in nonmathematical English, the verb *discriminate* means "to make a distinction based on prejudice rather than on facts." [46, 105, 146]

disjoint (adjective): the first element is from Latin *dis-* "away, in two parts," often used as a negative; it is of unknown prior origin. The second element, *joint*, is the past participle of the French verb *joindre*, which is the source of and means the same as English *join*. The Indo-European root of the same meaning is *yeug-*. A native English cognate is *yoke*, which joins animals together. In mathematics, if each of two disjoint sets is represented pictorially by a circle, as in an Euler or Venn diagram, the two circles will stay apart from each other; they will not "join" anywhere because disjoint sets have no elements in common. [46, 259]

disjunction (noun): the first element is from the Latin prefix *dis-* "apart, away," of unknown prior origin. The second element is from Latin *iunct-*, past participial stem of *iungere* "to join." In mathematics a disjunction is a joining of two sets to produce a new set that contains every element that was in either of the original sets. Imagine that the two original sets are nonoverlapping intervals. Even though these two sets are "apart" from each other, combining them with a disjunction still produces a nonempty set; by comparison, a conjunction (*q.v.*) of the same two intervals would produce an empty set. In logic, a disjunction is another name for an or-statement: one thing may be true, or another thing (which may be "apart," i.e., mutually exclusive) may be true. A logical disjunction is often represented by the symbol ∨. Compare *union* and *vel*. [46, 259]

dispersion (noun): the first element is from the Latin prefix *dis-* "apart, away," of unknown prior origin. The second element is from Latin *sparsus*, past participle of *spargere* "to strew, to scatter, to sprinkle." Something sparse is "strewn far and wide." The Indo-European root is *(s)preg-* "to jerk, to scatter." Related Germanic borrowings include *sprinkle* and *freckles*. In statistics, dispersion is most commonly measured by the standard deviation. [46]

displacement (noun): the first element is from the Latin prefix *dis-* "apart, away," of unknown prior origin. The second element is from *place* (*q.v.*). A displacement is a translation, a type of rigid transformation in which a figure is moved without being rotated, stretched, or reflected. Sinusoidal functions are subject to a phase displacement. [46, 162]

dissect (verb), **dissection** (noun): the first element is from the Latin prefix *dis-* "apart, away," of unknown prior origin. The second element is from Latin *sectus*, the past participle of *secui* "to cut." The Indo-European root is *sek-* "to cut," as seen in the native English noun *saw*. When a geometrical figure is dissected, it is cut apart. As an interesting and certainly nontrivial example, imagine a square whose side is 112: it can be dissected into 21 other squares, each of a distinct size. [46, 185]

dissipate (verb), **dissipative** (adjective): from Latin *dissipare* "to spread, scatter, disperse." The first element is from the Latin prefix *dis-* "apart, away," of unknown prior origin. The second element is from the Indo-European root *swep-* "to throw, to sling."

In mathematics a dissipative system is one in which every motion leads eventually to a final resting position. As an example, consider M.J. Feigenbaum's transformation of the interval $[-1, 1]$ into itself, via the iteration $x_{n+1} = 1 - \mu x_n^2$, where x is in the stated interval and μ is between 0 and 0.75. For general x, successive iterates are "thrown away" from x and gradually dissipate to the equilibrium value $x_\infty = 0.73205\ldots$. [46]

distance (noun): the first element is from the Latin prefix *dis-* "apart, away," of unknown prior origin. The second element is from Latin *stare* "to stand." The Indo-European root is *sta-* "to stand," which is also a native English cognate. The distance between two points is literally "how far apart they stand from each other." [46, 208]

distribute (verb), **distribution** (noun), **distributive** (adjective), **distributivity** (noun): the first element is from the Latin prefix *dis-* "apart, away," of unknown prior origin. The second element is from Latin *tribu-*, borrowed from Etruscan, meaning "a division of the Roman people, a tribe." The related verb *tribuere* meant "to allot, to apportion," as when something was being divided up among the various tribes that made up ancient Italy. The related word *contribute*, borrowed from Latin, means literally "to work together with a group (= tribe)." In mathematics, one instance of the distributive property is given by the equation $a \cdot (b + c) = a \cdot b + a \cdot c$; the quantity a is allotted or distributed to b and also to c. Algebra textbooks almost invariably limit a discussion of distributive properties to the distribution of multiplication over addition or subtraction, but other distributive properties exist. The eight common distributive properties of algebra can be combined into what might be called THE distributive property, as illustrated in the order-of-operations chart below. A given operation can be distributed over the two operations that are found exactly one column to its left. For example, exponentiation can be distributed over multiplication or division, but not over addition or subtraction, nor over exponentiation or taking roots. [46]

Level 1	Level 2	Level 3
$+$	$\times$	x^n
$-$	$\div$	$\sqrt[n]{x}$

Distributive properties.

ditonic (adjective): from Greek *di-* "two" and *tonos* "sound, tone." The Indo-European root is *ten-* "to stretch." A native English cognate is *thin*, which describes an object that has been stretched. A musical tone (borrowed from Greek) is created when a stretched string is struck or plucked. The word *tone* is also used figuratively in the sense of "behavior, appearance." In mathematics the word *ditonic*, patterned after *monotonic*, stresses the fact that a phenomenon behaves in two ways. A ditonic function is initially strictly decreasing (or increasing), but beyond a certain point it becomes strictly increasing (or decreasing) and continues that way until it reaches a maximum (or minimum). [48, 226]

diverge (verb): **divergent** (adjective): from Latin *dis-* "apart, away from," and *vergere* "to turn, to verge." The Indo-European root is *wer-* "to turn, to bend." Native English cognates include *wrench* and *wring*. If an infinite series diverges, the running total of the terms eventually "turns away from" any specific value, so a divergent series has no limiting sum. Although the words *diverge* and *diverse* have now diverged somewhat in meaning, they are etymologically the same. [46, 251, 146]

divide (verb), **division** (noun), **divisible** (adjective), **divisibility** (noun): from Latin *dividere* "to separate, to divide." The prefix *di(s)-* meant "apart, in two." The Indo-European root underlying the second part of *dividere* is *weidh-* "to separate"; compare the English cognate *widow*, a woman separated from her dead husband. In arithmetic, when 5 is divided into 15 to give 3, the 5 and the 3 are the two parts or factors that the original 15 has been separated into. Our symbol "÷" for division first appeared in print in an algebra book by the Swiss mathematician Johann Heinrich Rahn (1622–1676). In old American textbooks, division was sometimes indicated by reversed parentheses,)(, which corresponded to the way the division algorithm used to be carried out. Where we now write

$$4 \overline{\smash{)}\, 8}\,^{\textstyle 2},$$

Americans in the early 19th century—and people in many countries even today—would write "4) 8 (2." For an explanation of the suffix in *divisible* see *-able*. [46, 245, 69]

dividend (noun): from Latin *dividere* "to divide," with the suffix *-nd-*, which creates a type of passive causative, so that Latin *numerus dividendus* meant "the number to be divided." In the statement 6 ÷ 3 =

2, the 6 is the dividend because it is the number that is to be divided by 3. In finance, the profits that a company makes are to be divided up and given to the shareholders, each of whom receives a *dividend*. Contrast *divisor*. [46, 245, 139]

divisor (noun): from Latin *divisus*, past participle of *dividere* "to separate, to divide," plus the suffix *-or*, which indicates a male person or thing that carries out a given action. In the same way that an actor is a man who acts, a divisor is a number that divides. In the statement 6 ÷ 3 = 2, the 3 is the divisor because it is the number that divides the 6. Contrast *dividend*. [46, 245, 153]

dodecagon (noun): the first part of the word is already a compound containing Greek *duo* "two" and *deka* "ten." Addition of the two components yields *duodeka*, or the shortened form *dodeka* "twelve." The other root is *gon-* "angle," so a dodecagon is a twelve-angled (and of course also twelve-sided) polygon. A dodecagon has also been called a duodecagon. [34, 48, 65]

dodecahedron, plural *dodecahedra* (noun): the first part of the word is already a compound containing Greek *duo* "two" and *deka* "ten." Addition of the two components yields *duodeka*, or the shortened form *dodeka* "twelve." The other root is *-hedra* "base." The prehistoric form was *sedra*, in which the relationship to the English cognates *sit* and *seat* is more obvious. A *dodecahedron* is a twelve-based, or twelve-faced polyhedron. A regular *dodecahedron* is one of the five regular polyhedra. Each of its twelve faces is a pentagon. A dodecahedron has occasionally been called a duodecahedron. [34, 48, 184]

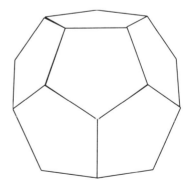

A regular dodecahedron.

dollar (noun): from Low German *daler*, from German *taler* or *thaler*, an abbreviation of *Joachims-*

thaler. The long word means "from Joachimstal," literally *Joachim's dale* (= valley). The *Joachims-thaler* was a silver coin first minted in Joachimsthal, Bohemia (now part of the Czech Republic), in 1519. Items of all sorts have been named for their places of origin: compare *china*, *damask* (from Damascus), *sherry* (from Jerez, Spain) and *suede* (from Sweden). The current spelling *dollar* was used in England sometime before the year 1600. The dollar was adopted as the official monetary unit of the United States on July 6, 1785. The new country had thrown off British rule and wanted to throw off the British monetary unit, the pound, as well. In 1782 Thomas Jefferson had advocated adopting the dollar because many Americans were already familiar with the Spanish dollar that circulated widely in the Spanish and British colonies of North America. However, since a bill establishing a mint wasn't passed until 1792, the first American dollars appeared only in 1794. As for the $ symbol used to represent the dollar, it is most commonly thought to be a combination of the letters *p* and *s* already in use in the abbreviation *ps.* that indicated pesos. Another theory holds that the symbol is a combination of the letters *U* and *S*, for *United States*. Still another theory traces the symbol to the Pillars of Hercules that appeared on the Spanish dollar. In computer mathematics a dollar sign is often used to indicate a number written in base sixteen. The German mathematician Ferdinand Joachimstal (1818–1861) could trace his ancestry to the same place that the dollar originally came from.

domain (noun): from Latin *domus* "house, home," as seen in our borrowed words *domicile* and *domestic*, as well as native English *timber* (the material used to build a house). The Indo-European root is *dema-* "home." In mathematics the domain of a function is the set of values for which the function "feels at home," in other words, the set of values that don't cause the function any trouble (such as zero in a denominator or a negative number inside a square root). Here's another way to look at the word: from Latin *domus* came *dominus* "lord, master," literally the person that ruled the home. A domain was the property owned by a lord. In that sense, the domain of a function $f(x)$ is the set of all the x's that the function "owns" or "has control of." Contrast *range*. [37]

dominant (adjective), **dominate** (verb): Latin *domus* meant "house, home." The head of the household was the *dominus*, or "lord" of the home. From that noun came the verb *dominari* "to be lord over, to dominate," and its present participial stem

dominant-. In mathematics one vector is said to be dominant over another if each of its components is greater than or equal to the corresponding component of the other vector. The term applies in a similar way to two infinite series. [37, 146]

dot (noun, adjective): from Old English *dott* "head of a boil." In the 16th century, the meaning was "small lump, clot." The word seems to be related to Old High German *tutto* "nipple" and therefore probably also to colloquial English *tit*. From the appearance of the anatomical things it designated, the word developed the abstract sense it has today "a small mark." In mathematics the dot product of two vectors is so called because it is indicated by a dot, as opposed to the cross product, which is indicated by the symbol ✕, a cross. The word *dot* is also commonly used to mean a decimal point.

double (noun, verb): a French word descended from Latin *duplus*, of the same meaning. The Latin word was a compound containing Indo-European *dwo-* "two" and *pel-*, meaning the same as its native English cognate "to fold." After a sheet of paper is folded over on itself, the number of layers has doubled. Compare *triple*, *quadruple*, *quintuple*, *sextuple*, *septuple*, and *octuple*. [48, 160]

down (adverb): a native English word, from the Indo-European root *dhuno-* "a fortified place." Since fortified places were often elevated, the word came to mean "hill." By the 12th century, the compound *of-dune* "off the hill," was in common use. With sound changes, that phrase was shortened to *adune*, and finally to *dune*, which gave modern *down*. The related *dune*, as in a sand dune, is from Dutch. Also cognate is the word *town*, since towns were often built on hills to give the inhabitants extra protection. In a standard rectangular coordinate system, the y-coordinate indicates how far up or down a point is located. [44]

dozen (noun): from French *douzaine* "a group of approximately twelve, a dozen," based on *douze* "twelve." The French word is from Latin *duodecim*, a compound of *duo* "two" and *decem* "ten." The suffix *-aine* is fairly generalized in French, and can be attached to many numbers; for instance, from *vingt* "twenty," comes *vingtaine* "a group of roughly 20." English has borrowed several such approximate number words from French: *quatrain* (a stanza of four lines), *dizain* (a poem of ten lines), and the more common *quarantine* "the period of roughly forty days during which a person suspected of being infected is kept under surveillance." [34, 48]

dram (noun): from Middle English *dragme*, ultimately from *drachma*, a silver coin used in ancient Greece; in fact the drachma is still the monetary unit of modern Greece. The word is from Greek *drakhme* "a handful," from the Greek root *drakh* "to grasp." In the American customary system of weights, a dram is one-sixteenth of an ounce.

dual (adjective, noun), **duality** (noun): via Latin *dualis* "containing two," from *duo* "two," from the Indo-European root of the same meaning, *dwo-*. In projective geometry, certain pairs (= twosomes) of elements are said to be duals. In a plane, for example, the elements *point* and *line* are such a pair. In the same way that two distinct points determine a line, two intersecting lines determine a point. [48]

duel (noun): from Latin *duellum* "war," perhaps from the Indo-European root *dwo-* "two," because a war is typically a fight between two opposing parties. Latin *duellum* was later altered to *bellum*, as seen in our borrowed word *bellicose* "warlike." In mathematics a duel is a two-person zero-sum game. See more under *noisy* and *silent*. [48]

dummy (noun, adjective): from *dumb*, a native English word with cognates in other Germanic languages. The original meaning seems to have been "stupid, without understanding," from which sense the word would have been mistakenly applied to people who couldn't hear or speak well. From the meaning "mute" the term *dummy* came to be applied to a fake person manipulated by a ventriloquist. A dummy has no intrinsic value, because everything it says or does comes from the ventriloquist. In mathematics a dummy variable is one that has no intrinsic interest: a slot must be filled, and any variable will do as well as any other to fill that slot.

duodecagon: see *dodecagon*.

duodecahedron: see *dodecahedron*.

duodecillion (numeral): patterned after *billion* by replacing the *bi-* with the root of Latin *duodecim* "twelve," a compound of *duo* "two" and *decem* "ten." Since in most countries a billion is the second power of a million, a duodecillion was defined as the twelfth power of a million, or 10^{72}. In the United States, however, a billion is 10^9, and a duodecillion adds eleven groups of three zeroes, making a duodecillion equal to 10^{39}. [48, 34, 72, 151]

duodecimal (adjective): from Latin *duodecimus* "twelfth," a compound made up of *duo* "two" and *decimus* "tenth." A duodecimal number system is one that uses base twelve. Some advocates— including Isaac Pitman (of Pitman shorthand fame), H.G. Wells, and George Bernard Shaw—have argued that we would be better off using base 12 for our calculations because 12 has more divisors (1, 2, 3, 4, 6, and 12) than 10 does (1, 2, 5, 10). Nevertheless, people persist in being born with ten fingers, and given the now almost universal adoption of the metric system of weights and measures, the duodecimal system will almost certainly remain just a curiosity. [48, 34]

duplation (noun): from Latin *duplus* "double," a compound of Latin *duo* "two" and the Indo-European root *plek-* "to plait," an extension of *pel-* "to fold." Although *duplation* means the same as the more common *duplication*, the term *duplation* is used almost exclusively in conjunction with a method of multiplying two numbers that involves the repeated doubling (duplation) of one number and the repeated halving (mediation) of the other. The method is sometimes called the Egyptian algorithm because the ancient Egyptians apparently used it; it has also been called Russian peasant multiplication for similar reasons. [48, 160]

duplicate (verb, noun), **duplication** (noun): from Latin *duplicare* "to double," from *duplex* "two-fold." *Duplex* is a compound of Latin *duo* "two" and the Indo-European root *plek-* "to plait," an extension of *pel-* "to fold." To duplicate something is to make it two-fold, i.e., twice as big as before. The duplication of the cube was one of the unsolved problems of classical geometry; with only a compass and ruler, the goal was to construct the base of a cube whose volume was to be twice the volume of a given cube. Only in modern times was the challenge proven to be impossible. In nonmathematical English a duplicate has come to mean a double in the sense of a copy. [48, 160]

dyad (noun), **dyadic** (adjective): from the Greek stem *duad-* "a pair," from the Indo-European root *dwo-* "two." In mathematics, a dyad consists of a pair of vectors written together without any indication of either scalar or vector multiplication. A dyadic numeration system is one that uses base two; in that context, Latin-derived *binary* is a more common synonym of Greek-derived *dyadic*. [48]

E

e (noun): most likely from the first letter of the word *exponential*, since the most "natural" exponential function is the one with base e. The first person to use e to symbolize the base of the natural logarithms was the Swiss mathematician Leonhard Euler (1707–1783). In an essay entitled "Meditations upon Experiments Made Recently on the Firing of Cannon," he wrote: "For the number whose logarithm is unity, let e be written, which is 2.718281 ..." By a happy coincidence, e is also the first letter in Euler's last name, so that e is now used to honor him. In 1873 Charles Hermite (1822–1901) proved that e is transcendental. The two most commonly encountered transcendental numbers are e and π.

eccentric (adjective), **eccentricity** (noun): *eccentric* is just a different spelling of *ex-centric*, from Latin *ex* "out of" and Greek-derived *center* (*q.v.*). Something eccentric is literally "off-center." A circle is the most "centric" plane curve because all points on the circle are an equal distance from the center. A circle happens to be an ellipse with both of its foci located at the center. As the two foci move in tandem away from the center, the resulting ellipse becomes more eccentric and appears more elongated. In nontechnical English a person is said to be eccentric when his behavior is off-center, or far from the norm. [49, 91]

echelon (noun): a French compound meaning "rung of a ladder," made up of *échelle* "ladder" and the augmentative suffix *-on*. The word is from Latin *scalæ* "ladder, stairs." The Indo-European root is *skand-* "to climb." Surprisingly related is *scandal*, from Greek *skandalos* "a snare, trap, or stumbling block" (something you trip or leap over). An echelon is an arrangement of parallel horizontal rows such that each row starts a bit farther to the right (or left) than the row above it, as if the beginning of each row were attached to a sloping ladder. When a matrix is in echelon form, the non-zero elements form an echelon, and the zero elements also form a complementary inverted echelon. [194, 151]

$$\begin{bmatrix} 1 & 3 & 7 & 2 \\ 0 & 5 & 4 & 9 \\ 0 & 0 & 6 & 1 \\ 0 & 0 & 0 & 8 \end{bmatrix}$$

A matrix in
echelon form.

ecological (adjective): the first element is from Greek *oikia* "dwelling," from *oikos* "house," from the Indo-European root *weik-* "clan." Related borrowings from Greek are *economy* and *diocese*. The second element is from Greek *logos* "reckoning, ratio, calculation," from the Indo-European root *leg-* "to gather." In nonmathematical English, ecology is the study (= reckoning) of organisms and their environment (= "house"). In statistics an ecological correlation is one based on rates or averages of groups (= "clans") rather than of individuals. [247, 111]

edge (noun): a native word, spelled *ecg* in Old English. The Indo-European root is *ak-* "sharp." Other native English cognates include *ear* (as in an ear of corn) and *hammer*. Borrowings from Latin include *acumen* and *eager* (= keen to do something). From Greek comes the prefix *oxy-* referring to *acid* (a Latin cognate), which is "sharp." In geometry, an edge of a polyhedron is a place where two faces make a "sharp ridge." [4]

eigenbasis (noun), **eigenfunction** (noun), **eigenvalue** (noun), **eigenvector** (noun): for the second components, see *basis*, *function*, *value* and *vector*. The first component, *eigen*, is a German adjective that means the same as the English cognate *own*. It is also translated into English as *characteristic*. The Indo-European root is *eik-* "to be master of, to possess," a meaning reflected in the fact that *own* in English can be not only an adjective, but a verb, *to own*. An eigenvalue of a matrix is a root of the *characteristic* equation of that matrix. Each square matrix has a *characteristic* equation all its own that it is guaranteed to satisfy. Compare *characteristic*.

eight (numeral): a native English word, from the Indo-European *okto*, of the same meaning. One hypothesis holds that the Indo-European word originally meant "two fours." From Latin *octo* English has borrowed words like *octet, octogenarian, octave,* and *octuple*. From the similar Greek *okto* English has

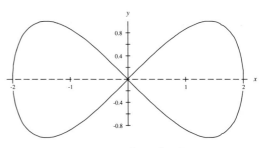

Eight curve: $x^4 = 4(x^2 - y^2)$

borrowed *octopus* and *octad* (a group or sequence of eight things). The word *eight* has also become the name of a curve whose shape is similar to a sideways figure 8, and whose equation is $x^4 = a^2(x^2 - y^2)$. [149]

either (pronoun): a native English word, from Old English *æghwæther*, whose three components have been so phonetically "worn down" that they are extremely difficult to recognize. The first element is from the Indo-European root *aiw-* "life, eternity." The second element is from the Indo-European root *kom-* "with." The third element is from the Indo-European root $k^w o$-, which appears in such relative pronouns as *who, what, where, which*, etc. The literal meaning of Old English *æghwæther* was "always together those [two]," i.e., "one or the other of the two." The English word *or* is ambiguous. In its inclusive sense it means "this thing or that thing or both"; in its exclusive sense it means "this thing or that thing but not both." In logic, to avoid confusion, the correlated pair *either . . . or* is used to indicate the exclusive or. The statement "either *p* or *q*" is defined as $(p \lor q) \land \sim (p \land q)$. [103, 109]

elastic (adjective), **elasticity** (noun): via New Latin, from Greek *elastikos* "impulsive," from *elaunein* "to drive." The Indo-European root is *el-* "to go." As first used in English, *elastic* described "the impulsive force of the atmosphere" that was believed to cause the expansion of gases. From that sense the word took on its modern meaning "returning to its original shape after being expanded." In mathematics the elasticity of a relationship is defined as the ratio of the percent of change in one variable to the percent of change in another variable.

element (noun), **elementary** (adjective): from the Latin plural *elementa* "first principles, rudiments, beginnings." The Latin singular *elementum*, of unknown prior origin, was a translation of Greek *stoikheon* "step, ground, base, element"; compare *stochastic*. Since elements are first principles, elementary school is supposed to expose students to basic knowledge. In mathematics the elements of a set are the rudimentary objects, so to speak, which may be further combined using union, intersection, etc. In solid geometry, an element is each of the straight lines making up a cone or cylinder.

elevation (noun): from Latin *ex*, reduced to *e* before a consonant, "away," and *levare* "to lighten." From the meaning "to lighten" came the sense "to raise," because when an object is sufficiently light you can

lift it up or it floats up by itself. The Indo-European root is *leg^w h-*, meaning the same as the native English cognate *light* (referring to weight, not brightness). Another native development is seen in *lung*, made up of lightweight tissue. Related borrowings from French include *relieve, lever*, and *levity*; from Gaelic comes the light, small creature known as a *leprechaun*. In trigonometry, the angle of elevation is measured from the horizontal upward to a line joining the observer's eye to the object or place in question. In geography the elevation of a point tells how high the point has been raised above sea level. [49]

eleven (numeral): a native English word with a surprising meaning that will delight the hearts of elementary school teachers who have to explain regrouping in subtraction. An early form of *eleven* was something like *endleofon*. The first component is from the Indo-European root *oi-no-* "one," while the second component is from the Indo-European root *leik^w-* "to leave." The compound, therefore, means "one-leave," or "one-left," because 1 is left over after you subtract away a group of 10. Compare the same sort of regrouping inherent in *twelve*. This "left-over" formation is common to the Germanic languages, which at one stage in their history presumably counted no higher than ten. The "left-over" formation was borrowed into Lithuanian, where it was extended to all the teens, but is not found in any other Indo-European languages. [148, 114]

eliminand (noun): from the stem of *eliminate* (*q.v.*) plus Latin *-nd*, which creates a type of passive causative: eliminands are things which are to be eliminated. In symbolic logic, the eliminands are the parts of the premises which are to be eliminated in the conclusion. Lewis Carroll provides this example: All cats understand French; some chickens are cats; [therefore] some chickens understand French. The two occurrences of *cats* are the eliminands. [49, 118, 139]

eliminant (noun): from Latin *eliminant-*, the present participle of *eliminare* (see next entry). Suppose you have two consistent linear equations in two unknowns and you wish to know if a third such equation is consistent with the other two. An eliminant is a determinant that can be used to find out. It gets its name from the fact that only the coefficients of the equations appear; the variables are eliminated. In older terminology *eliminant* was another name for a determinant in general. [49, 118, 146]

eliminate (verb), **elimination** (noun): from Latin *eliminare* "to turn out of doors," from *e(x)* "out of," and *limen*, stem *limin-*, "threshold, doorsill," of unknown prior origin. If you have chased something over the doorsill and out of your house, you have eliminated it, so the verb took on the more general meaning "to get rid of." In algebra a system of two linear equations is often solved for one of the variables by eliminating the other one. [49, 118]

ellipse (noun), **elliptical** (adjective): from Greek *en* "in" and *leipein* "to leave out," from the Indo-European root *leik*ʷ- "to leave." A related borrowing from Greek is *eclipse* (when the sun is "left out" of its usual role because it is blocked by the moon). Apollonius of Perga coined the term *ellipse* to describe the oval-shaped conic section because for an ellipse the square constructed on the ordinate has the same area as a rectangle whose height is equal to the abscissa and whose base lies along the latus rectum but is "left out," i.e., falls short of it. Although the name was originally chosen because of that rather complicated relationship, the name is also suitable in light of other properties of the ellipse. For example, since the eccentricity of an ellipse is less than 1, it is "left out from reaching" the cutoff value of 1 that distinguishes among an ellipse, a parabola, and a hyperbola. Similarly, given a vertical cone centered at the origin of a three-dimensional coordinate system, an ellipse results when the cutting angle falls short of (is less than) the angle between the *xy*-plane and an element of the cone. [52, 114]

ellipsis, plural *ellipses*, with last syllable pronounced the same as the verb *sees*, (noun): from Greek *leipein* "to leave out," from the Indo-European root *leik*ʷ- "to leave." The ellipsis, a punctuation mark made up of three periods arranged horizontally, is often used in mathematics. It indicates that part of a written expression (often a very large part) has been left out. For example, the set of even integers may be represented $\{\ldots, -4, -2, 0, 2, 4, \ldots\}$, where the ellipsis at the left indicates that the negative even integers less than -4 have been omitted, and the ellipsis at the right indicates that the positive even integers greater than 4 have been omitted. [52, 114]

ellipsoid (noun), **ellipsoidal** (adjective): from *ellipse* and *-oid* (*qq.v.*) "having the appearance of." An ellipsoid is a "three-dimensional ellipse." It looks like an ellipse in the sense that every cross-section perpendicular to an axis is an ellipse. The standard equation of an ellipse is

$$\frac{x^2}{a^2} + \frac{y^2}{b^2} = 1;$$

the standard equation of an ellipsoid looks a lot like it:

$$\frac{x^2}{a^2} + \frac{y^2}{b^2} + \frac{z^2}{c^2} = 1.$$

[52, 114, 244]

elongation (noun): from Latin *ex* "out of" and *longus* "long." The English adjective *long* is directly descended from Indo-European *del-* "long," and is borrowed neither from Latin *longus* nor from the resulting French *long*, though both of those are descended from the same Indo-European root. An elongation is a stretching out of something to make it longer. In mathematics, a simple elongation is a two-dimensional transformation represented by the equations $x' = x$ and $y' = ky$, or $y' = y$ and $x' = kx$, where $k > 1$. Compare *dilation*. Contrast *compression*. [49, 35]

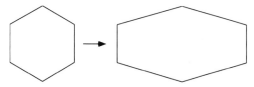

Elongation.

empirical (noun): from Latin *empiricus* "a physician whose art is founded solely on practice." The Latin word was borrowed from Greek *empeiria* "experience," from *en* "in" and *peira* "trial, experiment, attempt." The Indo-European root is *per-* "to try, to risk." A native English cognate is *fear*, which originally meant "sudden calamity, danger." Related borrowings from Latin include *experience, experiment,* and *expert*. In mathematics and science, an empirical solution is based on experience and experiments rather than on deduction. After an empirical solution is found, mathematicians often look for a deductive proof. [52, 165]

empty (adjective): the modern form of Old English *æmettig*, from the noun *æmetta* "leisure, rest." The Indo-European root is *med-* "to take appropriate measures." Related borrowings include the field of *medicine* (which takes appropriate measures to keep people healthy) and *modest* (taking appropriate measures to uphold decorum). In mathematics the empty set is the set with no members in it; also known as the null set, it is represented by the symbol $\emptyset$ or by an empty set of braces, {}. The intersection of two sets that have no elements in common is said to be empty. [128]

enantiomorph (noun), **enantiomorphic** (adjective), **enantiomorphous** (adjective): compounds made up of three Greek elements. The first component is from *en* "in," the second from *anti-* "opposite," and the third from *morph-* "shape." Two congruent solids are enantiomorphs if they are "oppositely shaped," i.e., if they are distinguishable mirror images of each other. A person's left and right hands being approximately enantiomorphic, we say that enantiomorphs have opposite "handedness." [52, 12, 137]

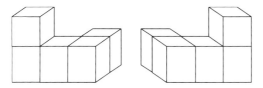

An enantiomorphic pair of block figures.

endecagon: see *hendecagon*.

endogenous (adjective): the first component is from Greek *endon* "within," from the Indo-European root *en-* "in." The second component is from Greek *-genes* "born," from the Indo-European root *gen-* "to give birth." Related borrowings from Greek include *genetics* and *Eugene* ("well born"). In mathematics an endogenous variable is one that a model seeks to determine or explain. An endogenous variable is intrinsic to the situation being modeled. Contrast *exogenous*. [52, 64]

endomorphism (noun): the first component is from Greek *endon* "within," from the Indo-European root *en-* "in." The second component is from Greek *morph-* "shape, form, beauty, outward appearance," of unknown prior origin. In mathematics an endomorphism is a homomorphism (*q.v.*) that "goes within" by mapping a set onto itself. [52, 137]

endpoint (noun): a compound of *end* and *point* (*q.v.*). The native English word *end* is from the Indo-European root *ant-* "front, forehead": your forehead is at one end of your body. The Indo-European root also developed the meaning "against" because what is in front of you pushes against your forehead if you move forward. That sense can be seen in the cognate prefix *anti-*, borrowed from Greek. The same sense is present in the native English prefix *un-* which acts "against" the meaning of the word it is attached to. Related borrowings from Latin include *ancient* and *antique*. In mathematics an interval which includes its endpoints is said to be closed. [12, 172]

enneacontahedron (noun): the first element is from the Greek compound *enneakonta*, from the Indo-European roots *newn-* "nine" and a not-easily-recognized form of *dekm-* "ten," so that *enneakonta* means "ninety." The second element is from Greek-derived *-hedron* "seat, base," used to indicate the face of a polyhedron. An enneacontahedron is a polyhedron with ninety faces; the most "common" enneacontahedron has 30 faces that are rhombs of one kind and 60 faces that are rhombs of another kind. [144, 34, 184]

enneagon (noun): the first component is Greek *ennea* "nine," from the Indo-European root *newn* "nine." The second component is Greek *gon-* "angle," from the Indo-European root *genu-*, with the same meaning as the English cognate *knee*. An enneagon is more commonly called a *nonagon*, a polygon with nine angles and therefore also nine sides. [144, 65]

enneahedron, plural *enneahedra* (noun), **enneahedral** (adjective): from Greek *ennea* "nine" and *hedra* (prehistoric Greek *sedra*) "base." An enneahedral solid has nine faces. Although the *Oxford English Dictionary* doesn't list the noun *enneahedron*, the fact that the adjective is attested there lends credence to the existence of *enneahedron*. Also listed in the large dictionary are *enneatic* (occurring once in nine times, nine days, nine years, etc.), *enneandrous* (having nine stamens), and *enneagynous* (having nine pistils). [144, 184]

entail (verb), **entailment** (noun): the verb *entail* is formed by adding the verbalizing prefix *en-* to the French-derived noun *tail* "the restriction of an inheritance to a specific person." The Old French source is *taillier* "to cut," as seen in our borrowed word *tailor*. The original Latin source is *talea* "a twig, a cutting," of unknown prior origin. When an estate is entailed, some heirs are "cut out of" the will, and others gain irrevocable title to the estate. By extension, *entail* came to mean "to have as an unavoidable consequence." In logic, entailment is a relationship in which a conclusion can be deduced from one or more premises; the conclusion is an unavoidable consequence of the premises. [52, 223]

entire (adjective): from French *entier* "whole," from Latin *integer*. Latin *integer* is a compound of *in-* "not" and the Indo-European root *tag-* "to touch," so that *integer* means literally "untouched, intact" (another cognate) and hence "whole." In mathematics an entire series is a power series which converges for all finite values of the independent variable; in other words, the domain is the entire real line or the entire complex plane. [141, 222]

entity (noun): from Latin *ent-* "being," the present participial stem of *esse* "to be." The Indo-European root is *es-* "to be," as seen in native English *is*. In logic, a statement that a certain type of thing exists is called an entity. In symbolic logic, entities are usually introduced using the existential quantifier symbol "∃." Contrast *nullity*. [56]

enumerate (verb), **enumeration** (noun), **enumerable** (adjective): from Latin *ex* "out," shortened to *e* before consonants, and *numerus* "number." The Indo-European root may be *nem-* "to apportion, to assign, to allot, to take," from which would have come the notion of a number, which measures the size of an allotment. In mathematics, enumeration is the process of establishing a one-to-one correspondence between the members of two sets. Historically speaking one of the sets has been the set of natural numbers, so that enumeration was originally just "counting out" or numbering. The adjective *enumerable* (also *denumerable*) means "capable of being enumerated." For an explanation of the suffix see *-able*. [49, 143, 69]

enunciation (noun): from Latin *enuntiare*, a compound of *ex* "out," shortened to *e* before consonants, and *nuntiare* "to announce." The verb is based on Latin *nuntius* "a messenger," from the Indo-European root *neu-* "to shout." Related borrowings from French include *announce*, *denounce*, and *renounce*. In classical geometry, the enunciation of a proposition is the part that "shouts out" or announces (at the beginning, of course) what is to be proved. [49]

envelope (noun): a French word, from Latin *in*, with the same meaning as the identical English cognate, and a hypothetical *volup-* or *velup-* "to wrap up," of unknown prior origin. The linguist Calvert Watkins suggests that the unknown element may be related to the Indo-European root *wel-* "to turn, to roll," as found in *evolute*, *helix*, *involute*, and *volume*. In the same way that a paper envelope is used to wrap up or "surround" a letter, a mathematical envelope is a curve that "surrounds" (in the sense of having a common tangent with) every member of a family of curves or lines. [52, 248]

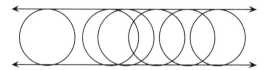

The two parallel lines form an envelope
for the family of congruent circles
whose centers lie on a horizontal line.

épi (noun, adjective): a French word meaning "ear of grain, spike of a flower"; it comes from a neuter variant of Latin *spica* "an ear or spike of grain." The Indo-European root is *spei-* "sharp point." English cognates include *spire*, *spoke*, and *spit* (a stick for roasting meat). In mathematics, an épi spiral is defined by the polar equation

$$r = \frac{a}{\sin n\theta},$$

for integral $n \geq 2$. The curve has n branches if n is odd, and $2n$ branches if n is even. Each branch has two asymptotes radiating out from the center, giving the curve a "spiky" appearance that someone must have thought resembles an ear of grain. An épi spiral doesn't wind around its center like other spirals, although the pieces are circularly arranged about the center. An épi spiral is the inverse with respect to the pole of the corresponding rose curve. [203]

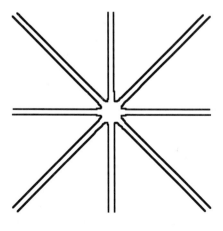

Épi spiral: $r = \dfrac{1}{\sin 4\theta}$

epicycloid (noun): from Greek *epi-* "over," *kuklos* "circle, wheel," and *-oid* (q.v.) "having the appearance of." The Indo-European word underlying Greek *epi* is also *epi* "near, at, against." As is typical with prepositions, there can be great changes in meaning temporally and/or spatially, so that Greek *epi* has diverged in meaning from the Indo-European *epi*. The Indo-European root underlying Greek *kuklos* is $k^w el$- "to revolve, go around," as found in native English *wheel*. In mathematics an epicycloid is a curve traced by a point on the circumference of one circle that rolls "over" (= around the outside of) another circle. See also *epitrochoid*, *trochoid*, and *hypotrochoid*. [53, 107, 244]

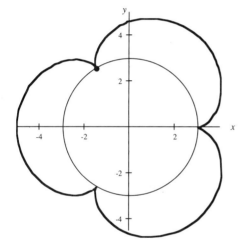

An epicycloid of three cusps:

$$\begin{cases} x = 4\cos\theta - \cos 4\theta \\ y = 4\sin\theta - \sin 4\theta \end{cases}$$

epimorphism (noun): a combination of Greek *epi-* "on, over," and *homomorphism* (q.v.). In abstract algebra, an epimorphism is a surjective homomorphism. Loosely speaking, one set is mapped "over onto" another. [53, 137]

epitrochoid (noun): from Greek *epi-* "over," *trokhos* "a wheel," and the suffix *-oid* (q.v.) "having the appearance of." The Indo-European root of the principal component is *dhregh-* "to run." From the Greek word, via Latin, comes English *truck*. In mathematics, a trochoid is the locus of a point on the (possibly extended) radius of a circle (= wheel) as that circle rolls "over" (= around the outside of) another circle. If the tracing point is on the circumference of the rolling circle, the resulting locus is called an *epicycloid* (q.v.). [53, 43, 244]

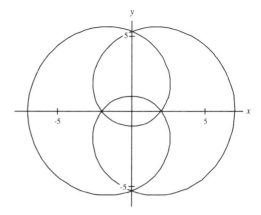

Epitrochoid:
$$x = 3\cos t - 4\cos t 3$$
$$y = 3\sin t - 4\sin 3t$$

epols (noun): a reversal of the order of the letters in *slope* (q.v.). In a linear equation of the type $y = mx + b$, the slope m is defined as $\Delta y/\Delta x$. On the other hand, in a linear equation of the type $x = ny + a$, the epols n is defined as $\Delta x/\Delta y$. Compare *atled* for another instance of backward spelling corresponding to a mathematical reversal.

epsilon (noun): the fifth letter of the Greek alphabet, written E as a capital and ϵ in lower case. The Greek name is actually *e-psilon*, meaning "simple *e*," as opposed to the other kind of *e* represented by the Greek letter *eta* (as when we say *double u*, to distinguish *w* from single *u*). In calculus ϵ is used to set a tolerance for the amount by which the dependent variable of a function can change.

equal (adjective), **equation** (noun): from Latin *æquus* "even, level," of unknown prior origin. Two quantities are equal when they are "even" with each other. If you think of two amounts being weighed in a balance scale, they are equal in weight when the balance scale is level. If you think of liquids poured into two identical glasses, the amounts are equal when the liquids reach the same level in both glasses. Our modern symbol "=" for equality first appeared in an algebra book by the English mathematician Robert Recorde (c. 1510–1558), although the parallel lines were longer than they are in the current version. The parallel lines symbolize equality because, in Recorde's words "no two things can be more equal." Although the word *equality* represents the abstract idea of things being equal, a mathematical relation like $2 + 3 = 5$ is called an equation rather than an equality. Contrast *inequality*. [2]

equant (noun): from Latin *æquant*-, present participial stem of *æquare* "to be equal," from *æquus* "even, level," therefore "equal." A noncentral point E inside a circle is called an equant if a point moving on the circumference of the circle appears from E to subtend equal angles in equal times. [2, 146]

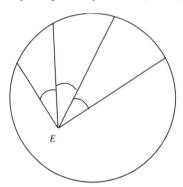

Point E is an equant.

equator (noun): from Latin *æquatus* "made equal," from *æquus* "even, level," of unknown prior origin. The ending *-or* is an agental suffix meaning "the male person or thing that performs the indicated action." An equator is a circle that "equates" a surface, i.e., divides it into two equal parts. In particular, the equator of an ellipsoid of revolution is the largest of all the ellipsoid's circular cross-sections. [2, 153]

Equator of a
prolate spheroid.

equiangular (adjective): from Latin *æquus* "even, level," of unknown prior origin, and *angle* (*q.v.*). In mathematics anything having equal angles can be called equiangular. An equiangular spiral, for instance, is one which makes a constant angle with every ray emerging from the center of the spiral. [2, 11, 238]

equiareal (adjective): from Latin *æquus* "even, level," of unknown prior origin, and *area* (*q.v.*). A mapping that preserves areas is said to be equiareal. [2, 16]

equiconjugate (adjective): from Latin *æquus* "even, level," of unknown prior origin, and *conjugate* (*q.v.*). A diameter, or conjugate diameter, of a conic section is a chord which bisects each member of a family of parallel chords. Two such diameters that are equal are said to be equiconjugate. [2, 103, 259]

equidistant (adjective): from Latin *æquus* "even, level," of unknown prior origin, and *distant* "standing apart." The Indo-European root is *sta-* "to stand." Two points located the same (= equal) distance from a third point are said to be equidistant from that third point. [2, 46, 208, 146]

equilateral (adjective): from Latin *æquus* "even, level," and *latus*, stem *later-*, "side," both of uncertain origin. Related borrowings from Latin are *bilateral* and *multilateral*. In geometry an equilateral triangle is one in which all sides are equal in length. [2, 110]

equilibrium, plural *equilibria* (noun): the first element is from Latin *æquus* "even, level," of unknown prior origin. The second element is from Latin *libra* "a balance scale"; the word may be of Mediterranean origin. When something is in equilibrium, it is "evenly balanced" and therefore won't change without external influence. In a mathematical model of a dissipative system, equilibria are possible rest positions of the system. As an example, consider M.J. Feigenbaum's transformation of the interval $[-1, 1]$ into itself, via the iteration $x_{n+1} = 1 - \mu x_n^2$, where x is in the interval and μ is between 0 and 0.75. For any such x, the system approaches the equilibrium value $x_\infty = 0.73205\ldots$. [2, 120]

equimultiple (noun): from Latin *æquus* "even, level," therefore "equal," of unknown prior origin, and *multiple* (*q.v.*). Equimultiples are equal multiples. For example, given a common fraction $\frac{a}{b}$ reduced to lowest terms, and a different-looking common fraction $\frac{c}{d}$ equal to it, then c and d are said to be equimultiples of a and b. Compare *homologous*. [2, 132, 160]

equipollent (adjective): the first component is from Latin *æquus* "even, level," of unknown prior origin. The second component is from Latin *pollent-*, present participial stem of *pollere* "to have weight, to be able, powerful," also of unknown prior origin. Classical Latin *pollere* was a somewhat archaic verb that was often doubled up with its synonym *posse* (later *potere*) for rhetorical effect, much as English uses redundant, old-fashioned expressions like *kith and kin* or *might and main*. In mathematics two sets are said to be equipollent (or equipotent) when their members can be placed in one-to-one correspondence. The two sets are, metaphorically speaking, equally powerful. [2, 146]

equipotent (adjective): the first component is from Latin *æquus* "even, level," of unknown prior origin. The second component is from Latin *potens*, stem *potent-* "able, having power." The Indo-European root is *poti-* "powerful; lord." In mathematics, two sets are said to be equipotent (or equipollent) when their members can be placed in one-to-one correspondence. The two sets are, metaphorically speaking, equally powerful. [2, 173, 146]

equitangential (adjective): from Latin *æquus* "even, level," of unknown prior origin, and *tangent* (*q.v.*). An equitangential curve is one for which the lengths of all tangential segments, when appropriately defined, are equal. The tractrix is such a curve; if the tractrix is in standard position with its cusp on the y-axis and the x-axis acting as an asymptote, then segments measured along tangent lines from the point of tangency to the x-axis are equal. [2, 222, 146]

equivalence (noun), **equivalent** (adjective): from Latin *æquus* "even, level," of unknown prior origin, and *valere* "to be strong, to have value." The Indo-European root is *wal-* "to be strong." A native English cognate is *wield* (originally "to use power"). Two things that are equivalent have equal value; one of them can be substituted for the other. The difference in usage between *equal* and *equivalent* is sometimes subtle. Equations like $2x + 3 = 5$ and $4x + 6 = 10$ cannot be said to be equal because $5 \neq 10$; those equations are equivalent, however, because $x = 1$ is their common and only solution. [2, 241, 146]

ergodic (adjective): the first element is from Greek *ergon* "work, action," from the Indo-European root *werg-* "to do," as seen in native English *work*. The second element is from Greek *hodos* (prehistoric root *sod-*) "way, course," from the Indo-European root *sed-* "to go." A related borrowing is *odometer*, which keeps track of how many miles your car goes. In mathematics, ergodic theory has to do with the study of measure-preserving transformations. In an ergodic Markov chain, for example, you can "work your way" from one state to any other. [183]

error (noun): a Latin word meaning "a wandering, a straying." The Indo-European root *ers-* "to be in motion" also appears in English *race*, borrowed from Old Norse. From chivalry comes the term *knight errant*, a person who wanders in quest of worthy enterprises. Because an error is literally a straying from something true or correct, in nonmathematical English *error* is a synonym of *mistake*. In statistics, however, errors are an inevitable consequence of measuring and are not mistakes in the nonmathematical sense.

escribe (verb): from Latin *ex* "out of," shortened to *e* before a consonant, and *scribere* "to write." Although in Classical Latin *scribere* meant "to write," the earlier meaning of the Indo-European root *skribh-*, an extension of *sker-*, was more physical: "to cut, to separate." In ancient times geometric figures were drawn on (= cut into) the ground or scratched into waxed tablets. In geometry an escribed circle is a circle which is tangent to one side of a triangle and also to the extensions of the other two sides; the circle is "written outside of" the triangle. [49, 197]

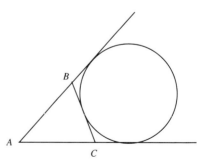

Triangle *ABC* and an escribed circle.

essential (adjective): from Latin *essentia* "the being or essence of a thing," from *essent-*, present participial stem of *esse* "to be." The Indo-European root *es-*

"to be," appears in native English *is*. Because the essence of a thing is its most basic property, *essential* has come to mean "fundamental, necessary." In the equation $y = mx + b$, the m and the b are called essential constants because no fewer than those two constants are required to distinguish one straight line from another. [56, 146]

estimate (verb, noun): from Latin *æstimare* "to appraise, to value." Although the origin of the Latin verb isn't certain, one plausible hypothesis is that the word is based on *æs*, stem *ær-*, "copper ore, bronze," since Roman coins were made of copper and bronze. The Indo-European root *ayos-* "copper, bronze," is found in native English *ore* and perhaps also in native English *iron*. From the use of *æstimare* with the meaning "to appraise" we have the mathematical sense "find an approximate value." From the more metaphorical sense, via French, we also have *esteem*.

ethnomathematics (noun): from Greek *ethnos* "people" and *mathematics* (*q.v.*). The Greek word—whose original initial *s-* has been lost—is from the Indo-European root *s(w)e-*, which refers back to the subject of the sentence, and might be translated "oneself" (in fact English *self* is a cognate, as is the *sib-* in *sibling*). When Greeks used the word *ethnos*, they were saying, in effect "we ourselves, we the people." Ethnomathematics is the study of mathematical concepts and their manifestations among ethnic groups around the world. Since most of our mathematics has evolved from Greek foundations, ethnomathematics deals primarily with non-Western cultures. [219, 135]

eureka (verb, exclamation): a first-person past tense of the Greek verb *heuriskein* "to find, to discover," from the Indo-European root *wer*- "to find." When Archimedes was in his bath one day, he discovered that a submerged object displaces its own weight in water; he is then supposed to have taken to the streets naked, shouting triumphantly "Eureka!", in other words "I have found [it]!" The exclamation is now quoted to indicate any breakthrough, not just one in mathematics or science. [252]

evaluate (verb): via French *evaluation*, from Latin *ex* "out of," shortened to *e* before a consonant, and *valere* "to be worth, to have value." The Indo-European root is *wal-* "to be strong." Related borrowings from French are *avail* and *prevail*. In mathematics, to evaluate an expression is to "get value out of" the expression by substituting specific

values for all the variables and then combining and simplifying as far as possible. [49, 241]

even (adjective): a native English word with cognates in other Germanic languages. *Even* means "level, having no variation." A whole number is even if it can be divided into two "level" or "uniform" amounts. For example, 8 pennies can be divided into two even (level) stacks of 4, whereas with an odd number of pennies like 7 the two stacks are necessarily uneven and don't reach the same level. The ancient Greeks considered even integers to be female. An even function $f(x)$ is one for which $f(-x) = f(x)$; an even function gets its name from the fact that when it is expressed as a power series, all the exponents are even. The derivative of an even function is odd, and vice versa.

event (noun): the prefix is Latin *ex*, shortened to *e* before a consonant "out of"; the main part of the word is from *ventus*, the past participle of Latin *venire* "to come." The Indo-European is $g^w a$- or $g^w em$-, with the same meaning as the English cognate *come*. Related borrowings from Latin include *invent* (= to come upon) and *convene* (= come together). In the study of probability, an event is literally the way something "comes out" or happens. If an event is random, it comes "out of the blue," so to speak. Notice that the English word *outcome* is an exact translation of *event*. [49, 77]

evolute (noun), **evolution** (noun): the prefix is Latin *ex* "out of," shortened to *e* before a consonant; the main part of the word is from *volutus*, past participle of *volvere* "to roll." The Indo-European root is *wel-* "to turn, to roll." In mathematics the evolute of a curve is the locus of the centers of curvature of that curve. The circle of curvature can be imagined rolling "out" along the given curve, and as it rolls its center traces the evolute. The evolute bears a complementary relationship to the involute, as indicated in the fact that Latin *e* is the antonym of *in*. In nineteenth century mathematics books the term *evolution* meant "extracting a root of a number," which is the inverse of what was then called *involution* "raising a number to a power." The "rooting" sense of evolution is rarely used now, but in biology, evolution is still the process by which species are "rolled out" over time. [49, 248]

ex- (prefix): this prefix appears in many English words borrowed from Greek and Latin. The Indo-European root is *eghs* "out." In Latin, the preposition *ex*, generally shortened to *e* before consonants, meant "out of, away from." A comparative form, *exterus*, underlies English borrowings like *external*, *extra*, and *exterior*. In Greek, the basic form was *ex* or *ek*, as seen in English borrowings like *exotic*, *exocentric*, *ectoplasm*, and *ectomorph*. [49]

exa-, abbreviated *E* (numerical prefix): the prefix *exa-* multiplies the unit to which it is attached by 10^{18}, which can be rewritten as $(10^3)^6$, the sixth power of a thousand. *Exa-* is from Greek *hex* "six," with the final *-a* added for uniformity: all of the numerical prefixes in the International System of Units have two syllables and end in a vowel, which in the case of magnifying prefixes created in recent times is always an *-a*. The initial *h-* of the Greek root was dropped to avoid confusion with *hexameter* "a kind of verse with six feet," as well as with other words like *hexagon* and *hexafoil* in which the prefix refers to six of something rather than to raising something to the sixth power. The prefix *exa-* became a part of the International System of Units in 1975. [220]

exact (adjective): from Latin *exactus* "precise, accurate, exact." The Latin adjective was the past participle of *exigere*, a compound of *ex-* "out of" and *agere* "to drive, to do," so that *exigere* meant "to drive out." When Robin Hood exacted sums of money from people, he drove the money out of their pockets and into his own. Latin *exigere* also developed the subsidiary meaning "to bring to an end, to perfect." In mathematics an exact value is "perfect" in that it is correct, as opposed to an approximate value. [49, 3]

example (noun): from Latin *ex* "out of" and *emere* "to take." The Indo-European root is *em-* "to take, to distribute." If you are exempt from something, you have been taken out of the list of potential candidates or victims. If you have a group of things that are of the same type or follow a common pattern, an example is one of those things that you have taken out to show people so they can understand the general pattern. The word *example* was also shortened to *sample*, which has approximately the same meaning. [49, 51]

exceed (verb), **excess** (noun): from Latin *excedi* and its past participle *excessus* "to go out, to go away," from *ex-* "out" and the Indo-European root *ked-* "to go, to yield." Related borrowings from French include *ancestor* (a relative who has gone before) and *deceased* (someone who has permanently gone away.) When one quantity exceeds another, its size has "gone out" past the size of the other. In spherical trigonometry, the spherical excess of a triangle

is the amount by which the sum of the angles of the triangle "goes out beyond" 180°. [49, 88]

excenter (noun): from Latin *ex* "out of" and *center* (*q.v.*). In geometry, the excenter is the center of a triangle's escribed (*q.v.*) circle. The excenter is always outside of the triangle. Contrast *circumcenter*, *incenter*, and *orthocenter*. [49, 91]

except (preposition), **exception** (noun): from Latin *ex* "out" and *capere* "to take, to seize." The Indo-European root is *kap-* "to grasp," as seen in the English cognates *have* and *haft*. If the domain of a function is all real numbers except 0, the value 0 has been taken out of the domain. An exception is a special case that has been "taken out of" a certain rule. For example, in algebra it is generally true that $x^0 = 1$; the one exception occurs for $x = 0$. [49, 86]

exchange (verb, noun): from a presumed Vulgar Latin *excambiare*, a compound of *ex* "out, away," hence "different," and *cambiare* "to trade, barter, exchange." The Indo-European root is *(s)kamb-* "to curve, to bend"; when you bend something, you change its shape or location or, mathematically speaking, its value. The exchange value of a symbol in the Hindu-Arabic system of numeration is what it can be "traded in for." For example, the exchange value of the 3 in 365 is three hundred, whereas the exchange value of the 3 in 36 is thirty. The exchange value is not the same as the place value, which in the same two examples would be 100 and 10. [49, 193]

excircle (noun): from Latin *ex-* "outside of," and Greek-derived *circle* (*q.v.*). With regard to a triangle, an excircle is the same as an escribed (*q.v.*) circle, i.e., a circle tangent to one side of the triangle and to the extensions of the other two sides. See more under *excenter*. Contrast *incircle*. [49, 198, 238]

exclude (verb), **exclusive** (adjective): from Latin *ex* "out of" and *claudere* "to shut, to close." The Indo-European root is *kleu-* "a hook, a peg," so that the Latin verb would have meant originally "to lock with a hook or a bolt." Related borrowings from French include *cloister*, where religious people are "closed up," and *cloisonné* enamel, where each color is "closed" in a little frame of its own. In mathematics, if certain values are excluded from the domain of a function, those values are "shut out." If two sets are mutually exclusive, all the members of each are "shut out" of the other. If an interval extends from 1 to 3, exclusive, then the endpoints 1 and 3 are "shut out of" the interval. [49, 101]

exhaustion (noun): from Latin *ex* "out of" and *haustus*, past participle of *haurire* "to draw up water." The Indo-European root is *aus-* "to draw water." The literal meaning of *exhaust* is "to draw out or use up." Metaphorically, if your strength is used up, you feel exhausted. In mathematics the method of exhaustion is a type of limiting process in which the difference between a theoretical value and a measured or calculated value gradually "gets used up," i.e., approaches zero. Archimedes approximated the value of π by using the method of exhaustion with a 96-sided polygon inscribed in a circle. [49]

exist (verb), **existence** (noun): from Latin *exsistere*, also spelled *existere*, "to step out, emerge, appear, arise, become, exist." The components of *exsistere* are *ex* "out of" and *sistere*, a reduplicated form of the more basic *stare* "to stand." The Indo-European root is *sta-* "to stand," as is apparent in the native English cognate of the same meaning. When something exists, it "stands out" against the nothingness that would be there without it. In mathematics the symbol ∃, which is the first letter of *Exist* turned backwards, is used to mean "there exists." An existence proof establishes that a certain thing exists without necessarily indicating what the thing is like, what value it has, or how to locate it. [49, 208]

exogenous (adjective): the first component is from Greek *exo-* "outside of," from the Indo-European root *eghs-* "out." The second component is from Greek *-genes* "born," from the Indo-European root *gen-* "to give birth." Related borrowings from Greek include *genealogy* and *genotype*. In mathematics an exogenous variable is one that is determined outside of a situation. An exogenous variable is extrinsic to the situation being modeled. Contrast *endogenous*. [49, 64]

exotic (adjective): from Greek *exotikos* "outside," from the Indo-European roots *eghs* "out." Something exotic is "outside" our normal experience. In abstract mathematics a four-dimensional manifold is said to be exotic if it is topologically but not differentially equivalent to R^4. The existence of an exotic manifold was first shown by Simon Kirwan Donaldson (b. 1957). [49]

expand (verb), **expansion** (noun): from Latin *ex* "out of" and *pandere* "to spread out." The Indo-European root is *petə-* "to spread." Related borrowings from Latin include *pace* (the span made by stretching your legs) and *patent* (something wide open, obvious). In algebra, when you expand

an expression like $(a + b)^3$, you "spread it out" to the equivalent form $a^3 + 3a^2b + 3ab^2 + b^3$. [49, 169]

expected (adjective): from Latin *ex* "out" and *spectare* "to look at, behold." When you expect company, for example, you may keep looking out the window to see if your guests have arrived. The Indo-European root is *spek-* "to observe." Related borrowings from Latin include *despise* (to look down on), *inspect*, and *specimen*. A related and metathesized borrowing from Greek is *skeptical* (= looking askance at things). In the mathematics of probability, the expected value is the average number that you can look forward to seeing turn up in a certain situation. [49, 205]

explicit (adjective): from Latin *ex-* "out(side)" and *plicare* "to fold." The Indo-European root is *plek-* "to plait," an extension of *pel-* "to fold." Something explicit is literally "folded out" so you can see it clearly. When an equation is in explicit form, one variable has been "folded out" so that it's all alone on one side of the equation. The opposite is *implicit* (*q.v.*). In nonmathematical English, something explicit is openly stated or shown. [49, 160]

exponent (noun), **exponential** (adjective), **exponentiation** (noun): from Latin *ex* "away, out," and *ponent-*, present participial stem of *ponere* "to put." (See more under *component*.) When you expose something, you put it out so it can be seen. Similarly, in mathematical notation the exponent is the small number or letter that is "put out" to the right and above the base when that base is being raised to a power. An exponent is therefore named after its physical appearance in writing rather than its mathematical significance. In algebra the technical name for the operation of raising a number to a power is exponentiation. An exponential function is one like 2^x in which the variable is in the exponent while the base is a constant. René Descartes' (1596–1650) book *La Géométrie*, published in 1637, was one of the first to use our modern system of writing exponents. [49, 14, 146]

expression (noun): from Latin *ex* "out" and *pressus*, past participle of *premere* "to press." To express something is to "press" it out in words and therefore to convey meaning. A mathematical expression is a group of numbers, variables and operations; a mathematical expression means something, but since an expression doesn't contain an equal sign or an inequality symbol, no claim is being made about the expression. For that reason you can simplify, but not solve, an expression. [49, 166]

exsecant (noun): a contraction of *external secant*. The first word is from Latin *externus* "outward, external, from the more basic *ex* "outside of." For the second word see *secant*. The exsecant function of an angle θ is defined as exsec θ = sec θ − 1. The name can be explained in this way: for a first-quadrant angle θ in a unit circle, sec θ is the length of a secant line segment beginning at the center of the circle and ending where the segment intersects the vertical tangent line. The external secant is the part of the secant segment outside the unit circle, and is therefore 1 unit shorter than the secant. In proving trigonometric identities, the quantity sec θ − 1 is sometimes multiplied by its conjugate; the resulting $\sec^2 \theta$ − 1 moves "outside of" the realm of secants and can be rewritten as $\tan^2 \theta$. Compare *versed* (sine). [49, 185, 146]

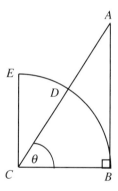

$$CD = CB = CE = 1,$$
$$CA = \sec\theta,$$
$$\text{exsec } \theta = DA.$$

extend (verb), **extension** (noun), **extent** (noun): from Latin *ex* "out of" and *tendere* "to stretch, to extend." The Indo-European root is *ten-* "to stretch," as seen in the English cognate *thin*. If a side of a triangle is extended, it is "stretched out" and therefore made longer. When a set is defined by extension, the list of its members is "stretched out" until all the members have been mentioned or implicitly indicated. For example, the set of positive even integers may be defined by extension as $\{2, 4, 6, 8, \ldots\}$. Compare *roster*; contrast *intension*. In plane analytic geometry, the extent of a curve is how far the curve "stretches" horizontally and vertically; the extent embraces both the domain and the range. [49, 226]

exterior (adjective, noun): from Latin *exterus* "on the outside, external," the comparative degree of the more basic *ex* "outside of." As if that comparative weren't enough, a second comparative ending, *-ior*, was later added, giving *exterior* the literal meaning "more more outside." (Compare English *lesser*, which is also a double comparative.) The exterior of a closed figure is whatever is outside the figure. An exterior angle of a polygon is outside the polygon. [49]

extract (noun): from Latin *ex* "out of" and *tractus*, past participle of *trahere* "to draw, drag, haul." The Indo-European root is *tragh-* "to draw, drag." An Indo-European variant *dhragh-* is the source of native English *draw* and the Old Norse borrowing *drag*. The English use of *draw* in the sense "sketch a picture" comes from the fact that you "drag" your pencil around in the process of making the picture. In mathematics, if you extract a root of a number you "drag it out" of the number in the same way you might metaphorically pull a plant out by its roots. [49, 234]

extraneous (adjective): from Latin *extraneus* "outside, external, strange," from *extra* "on the outside," from the more basic *ex* "out of." The Indo-European root is *eghs* "out." A related borrowing from Greek is *exotic* "outside of our familiar experiences." In algebra an extraneous solution is "outside" the set of true solutions. An extraneous solution typically appears if both sides of an equation have been squared one or more times, especially when radicals are involved at the beginning. Compare *redundant*. [49]

extrapolate (verb), **extrapolation** (noun): from Latin *extra* "outside of" and *polire* "to smooth, to furbish, to polish." The Indo-European root is *pel-* "to thrust, to strike, to drive." In the long-ago days B.C. (before calculators), values of trigonometric, exponential, logarithmic, and other important functions were listed in tables. Of course only a finite number of values of each function could be included, and it sometimes happened that the argument you wanted to look up fell outside the interval between adjacent arguments in the table. You then had to "go outside the gap" by pretending you were dealing with a linear function and estimating, by the use of a simple proportion, how far beyond the corresponding function values the needed value fell. In statistics a regression line can be generated from a set of points, but extrapolating along the line to the left of the leftmost point or to the right of the rightmost point may lead to unwarranted conclusions. [49, 161]

extreme (adjective, noun), **extremum**, plural *extrema* (noun): Latin *extremus* is the superlative degree of the adjective *exterus* "outside," which is already a comparative of *ex* "outside of." The literal meaning of *extreme* is therefore the extremely redundant "the most more outside." In mathematics the extreme terms in a proportion are the first and fourth ones, the ones farthest out from the center when the proportion is written on one line; in the proportion $a : b : : c : d$, the extremes are a and d. In calculus an extremum is the "farthest out" value of a function, whether a maximum or a minimum. [49]

F

face (noun), **facet** (noun): via French, from Latin *facies* "shape, form." The Indo-European root is *dhe-* "to set, to put," so a face means literally the shape put on something, and especially the shape taken on by the front of the head. In geometry, each of the polygonal shapes that bounds a polyhedron is called a face, by analogy with the face of a person. The word *facet* is a diminutive of French *face*. Each facet of a jewel, for example, is like a little face on the surface of the jewel. In mathematics a facet of a convex polytope is a face that is not contained in any larger face. [41, 57]

factor (noun, verb), **factorize** (verb), **factorable** (adjective): *factor* is a Latin word meaning "a maker, doer, performer, perpetrator," from Latin *factus*, past participle of *facere* "to do, make," with the addition of the agental suffix *-or*. A factor contributes to an act or process. Mathematically speaking, let the process in question be multiplication: each of the factors then contributes something toward the product. In early 19th century American textbooks, factors were called "component parts." A nonmathematical borrowing related to *factor* is *factory*, a place where products are made. Some English-speaking countries use the verb *factorize* rather than *factor*. For the suffix in *factorable*, see *-able*. [41, 153, 69]

factorial (noun): from *factor* "a number that is part of a product" (see previous entry). A factorial is a special product containing as its factors all the integers from 1 up to a given positive whole number. (By exception to the general definition, zero factorial

is defined to be 1.) The exclamation mark used to represent a factorial is a reflection of the fact that the factorial of even a relatively small number can be very large. The exclamation mark was first used to represent factorials in 1808 by Christian Kramp (1760–1826), although as recently as the beginning of the 20th century American mathematics books used the symbol ⌐ instead. For example, ⌐3 meant 3 factorial. [41, 153]

fall (verb): a native English word, from the Indo-European root *p(h)ol-* "to fall." In mathematics a portion of a curve is said to be falling if it takes on lesser values when moving from left to right. A falling portion of a curve is also said to be *decreasing*.

fallacy (noun), **false** (adjective): from Latin *falsus*, past participle of *fallere* "to deceive, trick, dupe, cheat, lie under oath," of unknown prior origin. A derivative sense "to disappoint expectation, to be defective," led to the related borrowing *fail*, borrowed from French. In logic, a false statement "cheats" us out of the truth. [59]

family (noun): via French, from Latin *familia* "family." The original meaning was "all the slaves belonging to one master," from *famulus* "a slave, a servant." By extension, *family* came to mean an entire household, including both slaves and masters. Eventually the word *family* was used to describe only the people living in a house who were related by blood or marriage. Latin *famulus* seems to have been a common Italic word; it may ultimately have been borrowed from Etruscan. In mathematics a family is a group of curves that are related because they share similar features or definitions. For example, the ellipse, parabola, and hyperbola are the three members of a family called the conic sections.

fathom (noun): a native English word, from the Indo-European root *peta-* "to spread out." The original Germanic meaning appears to have been "the length of two arms spread out." A fathom is now defined as a length of six feet. Because a fathom is used principally to measure the depth of water, a verb *to fathom* has developed, with the meaning "to get to the bottom of." [169]

feasible (adjective): via French *faisible* "capable of being done," from *faire* "to do, to make," from Latin *facere* "to do, to make." The Indo-European root is *dhe-* "to set," as seen in the native English cognate *do*. For an explanation of the suffix see *-able*. The adjective *feasible* means literally "capable of being done," or more succinctly "doable." In mathematics a feasible solution is one which not only satisfies a particular equation or system of equations but also satisfies all the constraints that apply to the problem under consideration. For example, in a question involving the heights of people, an "answer" like −28 ft. "won't do." It isn't feasible to have negative numbers in that problem; even disregarding the minus sign, it isn't feasible to consider people 28 feet tall. [41, 69]

femto-, abbreviated *f* (numerical prefix): from Danish and Norwegian *femten*, cognate to and meaning the same as native English *fifteen*. In 1964 *femto-* became part of the set of submultiple prefixes used in the International System of Units. *Femto-* multiplies the unit to which it is attached by 10^{-15}. [164, 34]

festoon (noun): via French, from Italian *festone* "a big festival," from *festa* "feast" and the augmentative suffix *-one*. Vulgar Latin *festa* was the plural of Classical Latin *festum* "holiday, festival." The Indo-European root is *dhes-* "having to do with religion." A festoon is a type of decoration used in festivals: it is a garland of flowers or leaves hanging on a string that is suspended from two equally high points. In mathematics a festoon is a parabola whose equation is of the type $y = cx^2$, where $c > 0$. Such a parabola approximates the shape of a flowers-and-leaves festoon. [151]

field (noun): a native English word, from the Indo-European *pelə-* "flat" or "to spread," as seen in native English *floor* as well as Greek-derived *planet* (which "spreads out" in its orbit). Because a field may stretch out for quite a distance, the word has come to be used metaphorically to mean "realm, area of study." In the physical world, for example, a magnetic field "spreads out" from a magnet. By even further extension, a field in mathematics has come to mean a set of elements that can be combined with two operations that obey certain properties. [162]

figurate (adjective): from Latin *figura* "shape, form, figure." Figurate numbers are integers that can be represented by dots arranged into polygonal figures. The lowest rank of figurate numbers is the triangular numbers, so-called because they can be arranged into triangles. The following orders of figurate numbers are square, pentagonal, hexagonal, heptagonal, octagonal, etc. The first number in each rank is always 1, which represents a degenerate triangle, square, pentagon, etc. [42]

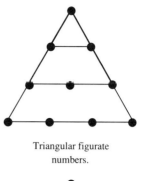

Triangular figurate numbers.

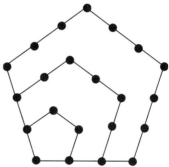

Pentagonal figurate numbers.

figure (verb, noun): from Latin *figura* "shape, form, figure," from the Indo-European root *dheigh-* "to form, to build." An English cognate is *dough*, which is built up into mounds called loaves. Even more surprisingly related is *lady*, from Old English *hlæfdige*, literally a "loaf kneader." In mathematics or even more generally, a figure is a simplified shape or diagram that represents an object, a concept, or a situation. The digits that are used to represent numbers are examples of simple, common figures, so that "to figure" became an English synonym for "to calculate." [42]

filter (noun): from Medieval Latin *filtrum* "a piece of felt," borrowed from the Germanic languages, as seen in native English *felt*. The original meaning of *felt* was "compressed wool," from the Indo-European root *pel-* "to strike, to drive." Because of felt's tightly compressed fibers, a piece of felt was often used to strain impurities out of a liquid. That use accounts for the modern meaning of *filter*. In mathematics, a filter is a family of nonempty subsets of another set that fulfill certain conditions. [161]

finite (adjective): from Latin *finis* "end, boundary, border," of unknown prior origin. Something finite has an end; it doesn't keep on going forever. To use a related word that comes from French, it *finishes* somewhere. Another borrowing from French is *fine*: if something is of fine quality, all the necessary work on it has ended. The old noun *finity*, with its French-derived suffix, has now given way to *finiteness*, which bears a native English suffix. [60]

first (ordinal number): a native English compound. The Indo-European root underlying the first part of the word is *per-* "forward, through, in front of"; there are many other extended meanings. The *-st* suffix is the same one that normally appears in English superlatives such as *best*, *most*, *scariest*, etc. The first object in a series is literally the one "most in front." All but two of the English ordinal numbers are formed from their corresponding cardinals. The first exception is *first*. The second exception is *second* (q.v.). To the ancient Greeks, 1 wasn't considered a number, but rather the generator of all numbers. The first ordinal, therefore, couldn't be formed from the corresponding cardinal because there was no corresponding cardinal. [165]

fit (verb), **fitting** (noun): the descendant of the Middle English verb *fitten* "to marshal forces," perhaps from earlier meanings "arrange, adjust, match." In mathematics, curve fitting involves the arranging or matching up of points and functions.

five (cardinal number): a native English word. When compounded with two different forms of English *ten*, *five* yields *fifteen* and *fifty*. Related English words are *finger*, of which there are five on each hand, and *fist*, which contains five folded fingers. The Indo-European root is *penk^{w}e* "five." In Latin the word developed into *quinque*, with the initial *qu-* influenced by the beginning of the previous Latin numeral *quattuor* "four," as well as by the *-que* at the end of *quinque*. Related borrowings from Latin are *quintuplet* and *quintet*. In Greek the original Indo-European root developed into *pente*, as in *pentathlon*, a set of five athletic events in the Olympics. A related borrowing from Sanskrit is *punch*, a drink originally made from five ingredients. From Yiddish comes *fin*, a slang term for a five-dollar bill. The ancient Pythagoreans believed that the number 5 represents marriage because it is the sum of the first even (= female) number, 2, and the first odd (= male) number, 3. [164]

fixed (adjective): from Latin *fixus*, past participle of *figere* "to fix, to fasten, to attach." The Indo-European root is *dhig^w-* "to stick, to fix." Native

English cognates include *dike* and *ditch*. In mathematics a fixed point is one which stays in the same place after a transformation acts on it. For example, if a circle and all the points inside it rotate about the circle's center, that center is a fixed point.

flat (noun): from Old Norse *flatr* "flat," from the Indo-European root *plat-* "to spread." As a liquid spreads, it flattens out. The native English *flat* "a floor of a building," therefore also "an apartment," is from the same root. In geometry, a flat angle is the same as a straight angle, i.e., an angle of 180°. Many mathematical references to flat things use a Latin- or Greek-derived cognate rather than the native English term; see, for example, *platykurtic* and *plane*. [162]

flecnode (noun): from Latin *flectere* "to bend, curve, turn," of unknown prior origin; plus *node* (q.v.). A flecnode is a node that also happens to be a point of inflection. [61, 142]

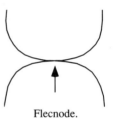

Flecnode.

flexagon (noun): the first component is from Latin *flexus* "bent," of unknown prior origin. The second component is the same suffix found in *pentagon* and *hexagon*; borrowed from Greek *gonia*, it means "angle." The first flexagon was created by Arthur E. Stone in the fall of 1939. It was a trihexaflexagon made by folding a strip of paper into the shape of a hexagon: when pinched and opened in various ways, it revealed three different faces. [61, 65]

flexion (noun): from Latin *flexus* "bent," of unknown prior origin. The term *flexion* is a synonym of second derivative. The second derivative of a function corresponds to the "bending" or concavity of a curve. [61]

fluxion (noun): from Latin *fluxus*, past participle of *fluere* "to flow." The Indo-European root is *bhleu-* "to swell, overflow," an extension of *bhel-* "to thrive, to bloom." A native English cognate is *bloat*, but not *flow*, even though *flow* coincidentally has the same meaning as Latin *fluere*. Related borrowings from Latin include *fluctuate*, *fluid*, and *superfluous*. Isaac Newton's word for *derivative* was *fluxion*, because a rate of change measures the "flow" of a function. Newton indicated a fluxion by placing a dot over the variable in question; he wrote $\dot{x}$ where we now more commonly write x'. [21]

focal (adjective); **focus**, plural *foci* (noun): *focus* is a Latin word meaning "fireplace, hearth," of unknown prior origin. The hearth was an important gathering place for the family, the center of home life. As a result, *focus* came to mean "center of attention." In a conic section, a focus is an important point from which distances are measured. A focal chord is one which passes through a focus. A derivative of the word *focus* is *fuel*, which feeds the home fires. [62]

folium, plural *folia* (noun): a Latin word meaning "leaf." The Indo-European root is *bhel-* "to thrive, to bloom." Related English cognates are *blossom* and *blade*, as in a blade of grass. From Old Norse comes *bloom*. Related borrowings from French include *foil* and *flower*. In mathematics the folia are a family of curves defined by the polar equation $r = \cos\theta(4a\sin^2\theta - b)$. Depending on the relative sizes of *a* and *b*, a given member of the family may have one, two, or three "leaves" or loops. See pictures under *bifolium* and *trifolium*. The folium of Descartes is a curve whose rectangular equation is $x^3 + y^3 = 3axy$. Named for René Descartes (1596–1650), who studied it, it always has one "leaf." [21]

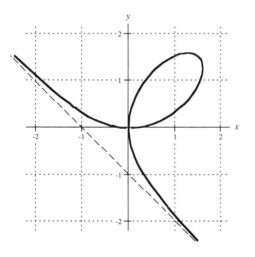

Folium of Descartes: $x = \dfrac{3t}{1+t^3}, \quad y = \dfrac{3t^2}{1+t^3}$.

foot (noun): a native English unit of length so named because somebody's foot was that long. The Indo-European root is *ped-*"foot." Native English cognates include *fetter* (which binds your feet) and *fetch* (to go on foot for something). Related borrowings from Latin include *pedestrian* and *pawn* (a footsoldier). From Greek come *podium* (which you place your feet upon) and *octopus*. In mathematics the place at which a perpendicular dropped from a point to a line meets that line is called the foot of the perpendicular; the word is used by analogy to a human foot, which is the lowest part of a person's body. In the U.S. Customary System a foot is a unit of length equal to one-third of a yard. [158]

force (noun): a French word, from a presumed Vulgar Latin *fortia*, from Classical Latin *fortis* "strong," as seen in our borrowed words *fort* and *fortitude*. The literal meaning of *force* is "strength (to do work)." The Indo-European root may be *bhergh-* "high." The connection between the notions "high" and "strong" is that in ancient (and not-so-ancient) times, people built fortified towns on hills because they could more easily defend those towns against enemies attacking from below. The Indo-European root may also be *dher-* "to hold firmly, support," as seen in our borrowed words *firm* and *confirm*. In mathematics, forces are frequently represented by vectors.

forest (noun): an Old French word, from Latin *forestis (silva)* "outside (woods)," meaning "royal woods reserved for hunting." The more basic Latin word is *foris* "at the door, from outside," from the Indo-European root *dhwer-* "door," which is an English cognate. A related borrowing from French is *foreign*. By analogy with the botanical world, in graph theory a forest is a graph whose components are trees (*q.v.*).

form (noun), **formal** (adjective), **formalism** (noun): from Latin *forma* "contour, figure, shape." The Latin word appears to be a metathesis of Greek *morphe* "form, beauty, appearance," of unknown prior origin. The island known in Chinese as Taiwan was named *Formosa* by the Portuguese because it had a fine form, in other words, because it was beautiful, which is what *formosa* means in Portuguese. Much of secondary mathematics involves matching algebraic expressions to certain standard forms. A formal mathematical statement is one which adheres to the rules for writing and manipulating mathematical statements and symbols, i.e., it uses the appropriate forms, but it need not have any connection to things in the physical world. In a similar way, formalism is the belief that mathematics consists of "forms" and symbols that can be manipulated by certain rules but that have no intrinsic meaning nor any necessary connection to the "real" world. [137]

formula (noun): from Latin *forma* "contour, figure, shape, form (*q.v.*)," plus the diminutive ending *-ula*. A formula is literally "a little form." In Classical Latin, *formula* meant "a fine form, an established mode of expression (as in judicial proceedings), a rule, a principle." We now use the word in the sense of a rule; for instance, the formula $C = 2\pi r$ is a rule that can be used to calculate the circumference of a circle when the radius is known. In traditional English usage *formula* retained its Latin plural *formulæ*, but in modern usage the plural is almost always *formulas*. [137, 238]

four (numeral): a native English word, from the Indo-European root $k^w etwer$-, of the same meaning. When compounded with two forms of the native English word *ten*, the same root yielded *fourteen* and *forty*. Evidence from various languages suggests that at one time four was the highest Indo-European number. In Latin, the basic word developed into *quattuor* and many related Latin words beginning with the variant *quadr-*. Borrowings from Latin include *quadratic* and *quadrilateral*. The corresponding ordinal number was *quartus* "fourth," as seen in *quart* and *quartic*. In Greek, the Indo-European root developed into *tetra*, with variant form *tessares*. Borrowings from Greek include *tetrahedron* and *tesseract*. [108]

fractal (noun): coined by the mathematician Benoit Mandelbrot (who was born in 1924 and whose last name means "almond bread" in German). The word is a variant of *fractional*, since a fractal is a type of fractional dimension found in many naturally occurring physical phenomena. The Indo-European root is *bhreg-*, which is more recognizable in the native English *break* than in the Latin cognate *fractus* "broken," which is the source of *fractal*. [24]

fraction (noun), **fractional** (adjective): from Latin *fractus*, past participle of *frangere* "to break," which is the native English cognate. The Indo-European root is *bhreg-*, of the same meaning. Related borrowings from Latin include *fragile* (= breakable), *diffraction* (breaking up into colors) and *fragment*. A fraction is literally a piece broken off something. In fact, in 16th century English mathematics books, fractions were sometimes referred to as "broken numbers." Compare the meaning of *algebra*. [24]

fractile (noun): from Latin *fractus*, past participle of *frangere* "to break," which is the native English

cognate. The Indo-European root is *bhreg-*, of the same meaning. In statistics a fractile is a generic term for any of the equal parts into which a distribution may be consecutively divided. Commonly used fractiles are *quartiles*, *quintiles*, *deciles*, and *percentiles* (*qq.v.*). [24]

free (adjective), **freedom** (noun): *free* is a native English word whose original meaning was "dear, beloved," from the Indo-European root *pri-* "to love." The change to the modern meaning resulted from the fact that in a household family members loved other family members but didn't usually love slaves: the beloved people were the free ones. The original meaning of the word can still be seen in the native English cognate *friend*, a person you love. Also related is *Friday*, the day devoted to the beloved Norse goddess Frigg. In mathematics the term *degrees of freedom* refers to the number of independent variables involved in a statistic. [41]

frequency (noun): from Latin *frequens* "repeated, frequent, crowded." The Indo-European root may be *bhrek^w-* "to cram together." In statistics, the frequency of a value is the number of times it is repeated in a distribution. With regard to sinusoidal functions, the frequency is the reciprocal of the period; it tells how many cycles are "crowded together" in a certain interval.

frontier (noun): via French, from Latin *frons*, stem *front-*, "forehead, foremost part," of unknown prior origin. Since the foremost parts of a region are on its fringes, *frontier* came to mean "border, boundary." In mathematics a frontier point is the same as a boundary point.

frustum, plural *frusta* (noun): a Latin noun meaning "a piece." The same Indo-European root *bhreus-* "to

The unbroken curves
are the boundaries of a
frustum of a cone.

cut, to break" is found in native English *bruise*. Popular etymology mistakenly connects *frustum* to *frustrate*, just as people confuse *prostrate* with *prostate*: it's easy to conjecture that when you are cut off from what you desire you feel frustrated. In geometry a frustum is a piece of a cone or pyramid "cut off" between two parallel planes. Contrast *truncated*.

function (noun), **functional** (adjective), **functor** (noun): from Latin *functus*, past participle of the verb *fungi* "to perform" (not the same as the *fungi* meaning yeasts and molds). The Indo-European root is *bheug-* "to enjoy." Related borrowings from Latin include *defunct* (no longer performing because dead) and *perfunctory* (going through the motions of performing but without any feeling). A mathematical function can be conceived of as a set of operations that are performed on each value that is put into it. The German mathematician Gottfried Wilhelm Leibniz (1646–1716) invented the term *function*. The Swiss mathematician Leonhard Euler (1707–1783) used the notation $f(x)$ to represent a function because the word *function* begins with the letter *f*. Well into this century the term *function* was still used for relationships that return more than one value for an element of the domain. For example, when the square-root function operated on 4, the result was $+2$ or -2. Such a relation used to be called a multiple-valued function. To indicate the modern sense of a function, mathematicians used to use the term *single-valued function*. A functional equation is an equation in which the unknown is a function rather than a variable. For example, the functions $f(x) = \sec^2 x$ and $g(x) = \csc^2 x$ are solutions of the functional equation $f(x) + g(x) = f(x) \cdot g(x)$. A functor is a mapping from one category into another that is compatible with it; the Latin word means literally "performer." [153]

furlong (noun): from Old English *furh*, modern *furrow*, and *long* (q.v.). A furlong was originally a furrow long, the furrow being the length of a square field whose area was ten acres. In modern usage, a furlong is one-eighth of a mile. Despite having similar pronunciations, the *fur-* of *furlong* and the number *four* aren't related. [35]

fuzzy (adjective): In the 17th century the word meant "spongy," and in the 18th century "fluffy." The word appears to be related to the Dutch *voos* "spongy." In that case, the Indo-European root would be *pu-* "to rot, decay," and related borrowings would include *foul*, *putrid*, and *pus*. Something spongy or fluffy is "not precisely delineated," and that meaning is

now used in set theory and logic. In a fuzzy set, there is a relaxation of the criteria that determine membership in the set. It is no longer the case that a given element clearly does or doesn't belong to the set. As an example consider the set of intelligent human beings. While people like Newton or Einstein would almost universally be accorded membership in that set, there are many other people who only marginally might be considered members. Contrast *crisp*.

G

gallon (noun): from Old North French *galon*, an augmentative form based on the root found in Medieval Latin *galleta* "a liquid measure, a jug" perhaps of Celtic origin. In the U.S. Customary System a gallon is a liquid measure equal to four quarts. British and Canadian gallons contain different amounts. [151]

game (noun): from Old English *gamen*, with the final consonant now lost except in the related word *backgammon*. In Gothic, a related Germanic language, *gaman* meant "fellowship." The word may have contained *man*, meaning "person"; the original meaning would then have been "to enjoy [the company of other people]." In the 20th century a branch of mathematics has been developed dealing with game-playing strategies.

gamma (noun): third letter of the Greek alphabet, written Γ as a capital and γ in lower case. It corresponds to the Hebrew letter *gimel*, ג, meaning "camel" (notice how similar our borrowed word *camel* still sounds). See more information under *aleph*. Along with α and β, γ is often used in geometry to designate an angle in a triangle; in triangle *ABC*, γ is the angle at vertex *A*. The gamma function, $\Gamma(x)$, is defined to be

$$\int_0^\infty t^{x-1} e^{-t} \, dt.$$

gelosia: see *jalousie*.

gematria or **gematry** (noun): from Rabbinical Hebrew *gematriya*, from Greek *geometria* "geometry," taken in the general sense of "arithmetic" or "mathematics." Neither ancient Hebrew nor Greek had numerals, but instead used letters of the alphabet to represent numbers. Letters used in that way wouldn't generally form real words, of course. Gematria is an extension of the system in which a real word acquires a numerical value equal to the sum of the numerical values of its letters. Words or phrases of equal value are considered equivalent. Much time has been spent over the centuries using gematria to "demonstrate" that a certain thing or person is exceptionally good or bad. Like the word *get*, *gematria* is pronounced with a "hard g." [63, 126]

general (adjective), **generalize** (verb), **generalization** (noun): from the Latin noun *genus*, stem *gener-*, "birth, origin, race." The Indo-European root is *gen-* "to give birth," as seen in native English *kin* and *kind*. Related borrowings from Greek are *gene* and *genetic*. Things that are all born in one family or have the same origin are therefore of the same type or race; compare *genus* as used in biology. Something *general* is native to all the members of a given genus, race, type or class. In mathematics the general term of a sequence or series gives the pattern that all the individual terms follow; all the terms are, so to speak, members of the same family. A general term or formula makes it easy to see the main or most important characteristics of the group. From this sense of "main" and "most important" we get the use of *general* as the head of an army. *To generalize* is to make a statement that is true of all members of a group. In mathematics and other disciplines, a generalization is often first discovered by noticing patterns in data; the conjectured generalization must then be proved. [64]

generator (noun): from the Latin noun *genus*, stem *gener-*, "birth, origin, kind." The Indo-European root is *gen-* "to give birth." The Latin agental suffix *-or* indicates a masculine person or thing that performs a certain action. For example, Charles Lindbergh was an aviator because he was involved in aviation. In mathematics a group may have one or several generators that give rise to all the elements in the group. [64, 153]

generatrix (noun): from the Latin noun *genus*, root *gener-*, "birth, origin, kind." The Latin agental suffix *-atrix* indicates a female person or thing that performs a certain action. For example, Amelia Earhart was an aviatrix because she was involved in aviation. In mathematics a generatrix (feminine because the word for *line* was feminine in Latin) of a ruled surface is the line which, by following a certain pattern, generates the surface. [64, 236]

genus, plural *genera* (noun): a Latin noun meaning originally "birth, descent, origin," and by extension "race, class, sort, species, kind." The Indo-European

root is *gen-* "to give birth," as seen in the native English cognates *kin* and *king* (= born into the royal line). In biology a genus is a classification of plants or animals. In mathematics the genus of a surface is a number that distinguishes among the many possible kinds of surfaces. [64]

geodesic (noun, adjective): from Greek *geo-* "the earth" and *daiesthai* "to divide." While the first component is of unknown prior origin, the second is from the Indo-European root *da-* "to divide," which is also found in Greek-derived words such as *democracy* (having to do with divisions of society). Geodesy is a branch of science that deals with the size and shape of the earth. As used in mathematics, *geodesic* refers to a curve on a surface such that the principal normal at any point on the curve is the same as the principal normal of the surface. The word *geodesic*, normally quite technical, was popularized in recent years by the inventor and designer Buckminster Fuller (1895–1983), whose geodesic dome received a lot of attention in the 1960's and 70's. [63, 30]

geometry (noun), **geometric** (adjective), **geometer** (noun): from Greek *geo-* "earth," of unknown prior origin, and *metron* "a measure." The Indo-European root is *me-* "to measure." As indicated by the etymology, geometry must originally have dealt with measuring land. Although geometry gradually grew more abstract, people assumed until the beginning of the 19th century that the axioms and postulates of geometry naturally corresponded to the physical world as they knew it on earth. In modern terms, however, geometry need have no physical referent at all. A geometer is a mathematician who specializes in geometry. [63, 126]

giga-, abbreviated *G* (numerical prefix): from Greek *gigas* "giant," of unknown prior origin. Related borrowings from Greek include *giant* and *gigantic*. In the metric system the prefix *giga-* multiplies the following unit by one billion. The capacity of commonly available computer hard drives began to be measured in gigabytes in the 1980's.

gill (noun): from Old French *gille*, from Late Latin *gillo* "a water pot," of unknown prior origin. In the U.S. Customary System a gill is a unit of measurement equal to four fluid ounces.

girth (noun): from Old Norse *gyorth* "a girdle," akin to the native English verb *to gird*. Compare the way other verbs such as *die* and *bear* are transformed, respectively, into the nouns *death* and *berth* by the use of the *-th* suffix. The Indo-European root is *gher-*

"to grasp, to enclose." It is found in native English *yard* (not the unrelated unit of length, but an enclosed space like a back yard); it is also found in *garden*, which English borrowed from French and which French had earlier borrowed from Germanic. In solid geometry the girth is the distance around the cross-section of a surface, provided that all such cross-sections have the same perimeter. The United Parcel Service requires that the length plus girth of a box not exceed 130 inches.

glide (verb): a native English word, from Germanic *glid-* "to glide." A related borrowing, probably from Yiddish, is *glitch* "a slip, mistake." In mathematics a glide reflection is one type of transformation. [74]

glissette (noun): a French word, from *glisser* "to slip, to slide," plus the diminutive ending *-ette*. *Glisser* was altered from Old French *glier,* apparently under the influence of the word *glacier.* Old French *glier* had been borrowed from Germanic *glidan,* a cognate of English *glide.* In mathematics a glissette is the locus of a point (or, alternatively, the envelope of a curve) carried by a curve or segment which slides on other given curves or points. The first systematic study of glissettes was published by W. H. Besant in 1869. Compare *roulette.* [74, 57]

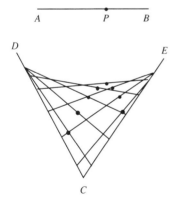

$\overline{AB}$ is a rod of fixed length. As *A* slides on $\overline{CD}$ and *B* slides on $\overline{CE}$, the locus of point *P* is a glissette.

global (adjective): from Latin *globus,* perhaps from the Indo-European root *gel-* "to form into a ball." If so, related English cognates include *clump, club,* and even *cloud,* which originally meant "hill." The huge ball that we live on, the earth, is also called the globe. The adjective *global* came to mean "referring to or found everywhere on the globe." In calculus,

global does not refer to the shape of a globe but to the universal scope of something like a maximum or minimum. In that sense *global* is contrasted with *local.*

gnomon (noun). a Greek word meaning "carpenter's square," but literally "someone who knows," from the Indo-European root *gno-* "to know," which is the English cognate. The word *gnomon* was used to refer a carpenter's square because it "knows" how to measure right angles. The word was also used to name the vertically positioned part of a sundial that resembles a carpenter's square and whose shadow lets us know what time it is. The term *gnomon* was later transferred to any object having a shape similar to that of a carpenter's square or a sundial's gnomon. In particular, in geometry a gnomon is a region which, when added to a polygon, produces a new polygon that is geometrically similar to the original one. [75]

The shaded regions are gnomons.

golden (adjective): a native English word, from the Indo-European root *ghel-* "to shine." Other English cognates include *gleam, glow, yellow* and *gall,* the latter two because of their "shining" color. In mathematics the golden mean or golden ratio, now often symbolized by the Greek letters τ or ϕ, refers to the number

$$\frac{1 + \sqrt{5}}{2}.$$

It is called "golden" because the ancient Greeks believed that a rectangle whose sides are in the ratio of $\phi : 1$ is the most precious, most esthetically pleasing,

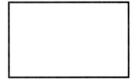

Golden rectangle.

of all possible rectangles. The Greeks didn't, however, use the term *golden*. In 1835 a German writer, Georg Simon Ohm, was apparently the first person to use the term *golden section* in print (though in German, of course).

-gon (suffix): from Greek *gonia* "angle, corner," from the Indo-European root *genu-* "knee, angle," as seen in native English *knee* and *kneel*. The connection between "knee" and "angle" is one of shape: at the knee, the upper and lower parts of a leg come together to form an angle. Related borrowings from Latin include *genuflect* and *geniculate*. From Greek come *gonion* "the point at the rear of the joint in a jaw," and *goniometer* "an instrument for measuring angles in crystals." In geometry a polygon is a closed figure made up of straight line segments; each pair of adjacent segments meets at an angle. [65]

googol (numeral), **googolplex** (numeral): a word invented by a nine-year-old boy when his uncle, the American mathematician Edward Kasner (1878–1959), asked him to think of a name for a very large number. To be exact, a googol is the made-up name for 10^{100}, and a googolplex is 10^{googol}. (Is there any significance to the fact that the two *o*'s in *googol* resemble the two 0's in 100?) The suffix *-plex* is from Latin words like *duplex* "double" and *triplex* "triple." The Indo-European root is *plek-* "to plait," an extension of *pel-* "to fold." [160]

grad (noun): from Latin *gradus* "a step," from *gradi* "to go, to walk." From that same root, via French, comes our word *degree*. If you do something by degrees, you do it in steps or stages. Similarly, the grade you make on a test indicates what "step" you are on, usually assuming that the metaphorical ladder has 100 steps. In America, following Babylonian tradition, a right angle is divided into 90 equal "steps" or degrees. In Europe, which is heavily committed to the metric system and powers of 10, a right angle has been divided into 100 grads. The term *grad* is a short form of *centigrade*, in which the division into 100 parts is more obvious. [73]

gradient (noun): from Latin *gradient-*, present participial stem of *gradi* "to go, to walk." The grade of a road is a measure of the rate at which the road "walks" uphill. If a rising road makes an angle of θ with the horizontal, the gradient is defined as $\sin \theta$, or sometimes $100 \sin \theta$. In this sense a gradient is similar to a slope (the slope being equivalent to $\tan \theta$). In calculus the gradient extends the concept to functions of two or more variables: for example, the gra-

dient vector $\nabla f(x, y)$, defined to be

$$\frac{\partial f}{\partial x}\mathbf{i} + \frac{\partial f}{\partial y}\mathbf{j},$$

gives an indication of the "tilt" of a plane tangent to the surface given by $z = f(x, y)$. [73, 146]

grain (noun): via French, from Latin *granum* "grain." The Indo-European root is *gre-no-* "grain," recognizable in the English cognate of the same meaning, *corn*. (In England, *corn* is still a generic term referring to grain, rather than to maize, as in America.) In the U.S. Customary System, a grain is a weight equal to about $\frac{1}{7000}$ pound, presumably because 7000 grains of wheat or some similar crop weigh approximately one pound. [76]

gram (noun): the Greek verb *graphein* meant "to scratch, to carve," and later "to write," since writing was originally scratched on wood or the ground, or else carved on tablets or stone slabs. The derived word *gramma* referred to anything used for or in writing, including letters of the alphabet. A small weight of the type used on a scale often bore a *gramma* on it, one or more little letters designating how much it weighed. From there *gramma* came to refer to the weight itself, rather than the letter designating that weight, and that is why in the metric system a gram is a basic unit of mass. Because of the etymological connection between *gram* and *graph*, *gram* is sometimes used in place of *graph* as the final component in compounds. For example, a histogram might more properly be called a histograph. [68]

graph (noun, verb): from the Greek verb *graphein* "to write," from the Indo-European root *gerbh-* "to scratch." Writing was originally scratched on wood, stone, tablets or the ground itself. Early geometers scratched pictures on the ground, too, and that is how we acquired the word *graph* in the sense of a diagram or picture of a relation among variables. A related native English word is *crab*, since crabs scratch the ground as they walk along. [68]

great, comparative **greater**, superlative **greatest** (adjective): a native English word, from the Indo-European root *ghreu-* "to rub, grind," which is an extension of *gher-* "to scrape, scratch." English cognates include *grit* and *groats*. In Old English *great* meant "thick, coarse." As time passed, *great* took on the extended meanings "bulky" and "of considerable size." In solid geometry, a great circle of a sphere is the intersection of the sphere with a plane passing through its center. In retailing, a great gross is a dozen gross (= 12 × 144, or 1728), just as the father of a grandfather is called a great grandfather. In algebra, the "greater than" symbol ">" was first used by the British mathematician Thomas Harriot (1560–1621). The greatest common divisor of two or more integers, abbreviated *gcd*, is the largest integer that exactly divides all of them. [71]

gross (noun): from French *grosse douzaine* "a big or fat dozen (q.v.)." (Compare the literal meanings of *thousand* and *million*.) French *grosse* is from Late Latin *grossus*, which may be from the Indo-European root *g^w res-* "thick, fat." In the 15th century, *gross* meant "large, bulky"; although those meanings have disappeared from English, they are still current in French. From the sense "large(r)" comes our use of expressions such as *gross profits*. In the 15th century the word also meant "dense, thick, coarse," and "concerned with large masses." From those meanings comes the modern use of *gross* as "vulgar, obscene, disgusting." A gross is a dozen dozen, and a dozen gross (= 12 × 144, or 1728) is known as a great gross, just as the father of a grandfather is called a great grandfather.

group (noun), **grouping** (adjective): via French *groupe*, from Italian *gruppo*, borrowed from Germanic. The original source is the Indo-European root *ger-* "curving, crooked." When things are heaped up, they make a somewhat rounded, and therefore curving, heap. That heap of things came to be called a group. English has many words indirectly related to *group*, including *crock*, *creep*, *cripple*, *grapple*, *cramp*, *crimp*, *crochet*, and *crumb*. In mathematics, a group is a "heap" of things called elements, which may be combined under a given operation subject to certain other stipulations. In algebra, grouping symbols such as brackets, braces, and parentheses are used to put terms together in a "heap." Many American students mistakenly believe that the required order for grouping symbols is parentheses inside brackets inside braces, but in fact there is no required order. There is also no limit to the number of times the same grouping symbol may be repeated. [66]

groupoid or **gruppoid** (noun): from English *group* or Italian *gruppo* (see previous entry) plus the Greek-derived suffix *-oid* (q.v.) "looking like," therefore "resembling." In mathematics a groupoid resembles a group but lacks some of the properties of a group such as associativity. To be a groupoid, all that is required is a set and a binary operation which is closed in that set. [66, 244]

H

hailstone (noun): a native English compound. The first element is from the Indo-European root *kaghlo-* "pebble, hail"; the second element is from the Indo-European root *stei-* "stone," so the compound is somewhat redundant. In number theory hailstone numbers are sequences of integers generated in the following way. (a) Choose a positive integer. (b) If it is odd, triple it and add 1, but if it is even, divide it by 2. (c) Reapply the rules of part (b) to each new result. For example, a starting number of 7 produces the hailstone sequence 7, 22, 11, 34, 17, 52, 26, 13, 40, 20, 10, 5, 16, 8, 4, 2, 1, 4, 2, 1, ...; the final 4, 2, 1 is bound to repeat forever. These sequences get their name from the fact that the values within a hailstone sequence typically rise and fall like a hailstone inside a cloud. Just as a real hailstone eventually becomes so heavy that it falls to earth, every starting integer ever tested has produced a hailstone sequence that eventually drops down to the number 1, at which time it "bounces" into the small loop 4, 2, 1, ... The first person known to have discussed the hailstone numbers—though not necessarily under that name—is one Lothar Collatz, while he was a student in the 1930's.

half, plural *halves* (noun), **halve** (verb): native English words, from the Indo-European root *(s)kel-* "to cut." English cognates include *shale* (which is "cut" into layers) and *shield* (originally cut from boards). In Old English, *half* referred to one of the two opposite sides of an object when the object was metaphorically cut in two. That early sense is maintained in the expression "on behalf of," as when a lawyer acts on behalf of his client. In modern usage, a half is one of the parts that result when an object is cut into two equal pieces. The verb *to halve* means "to cut in half, to make into halves," just as *to calve* means "to produce one or more calves." The verb *halve* is seldom used now, in part because for some dialects of American English *halve* is pronounced the same as *have*. Perhaps the way to have your cake and eat it too is to halve the cake, then eat one of the halves and have the other left over. [196]

handle (noun): a compound of native English *hand* plus a diminutive suffix, so that a handle is "a little hand" or "something to be grasped by a hand." The word *hand* has cognates in many Germanic languages but does not seem to be of common Indo-European origin. In topology the cutting of holes in and adding of handles to a surface changes the genus of the surface.

happy (adjective): from the obsolescent English noun *hap* "fortune, chance," which is still alive in *mishap*, *hapless*, and *happen*. English *hap* was borrowed from Old Norse *happ* "chance, good luck." The Indo-European root is *kob-* "to suit, fit, succeed." The English adjective happy originally meant "fortunate, lucky," a sense which is preserved when we speak of a happy coincidence. The most common meaning of *happy* now, however, is "glad, joyous." In mathematics, let the sum of the squares of the digits of a positive integer s_0 be represented by s_1; in a similar way, let the sum of the squares of the digits of s_1 be represented by s_2, and so on. If some $s_i = 1$, for $i \geq 1$, then the original integer s_0 is said to be happy. Although common superstition holds 13 to be an unlucky number, mathematically speaking 13 is a happy number because $1^2 + 3^2 = 10$, and $1^2 + 0^2 = 1$. Contrast *lucky*.

harmonic (adjective): from Greek *harmos* "joint," came *harmonia* "joint, agreement, concord," since a state of harmony results when things are joined together well. The Indo-European root is *ar-* "to fit together"; a joint is a place where bones fit together. A related borrowing from Greek is *arthritis*, a disease of the joints. Also related is native English *read* "to fit words together to make sense out of them." When the ancient Greeks studied music, they became aware that strings whose lengths were in certain ratios had pleasant sounds that seemed "to fit together" well when played at the same time; such ratios were $\frac{1}{2}$, $\frac{1}{4}$, $\frac{2}{3}$ (which equals $\frac{1}{2} + \frac{1}{6}$), etc. The study of music was not distinct from that of arithmetic, and so the term *harmonic* came to refer to a sequence of numbers whose reciprocals are in arithmetic progression. That notion of reciprocals was also extended to the harmonic mean of two numbers r_1 and r_2, given by

$$\frac{2}{\dfrac{1}{r_1} + \dfrac{1}{r_2}}.$$

If r_1 and r_2 are the average rates of speed traveling from point A to point B and back from point B to point A along the same route, then the harmonic mean gives the average rate of speed for the round trip. [15]

hat (noun): a native English word, from the Indo-European root *kadh-* "to shelter, to cover," as seen also in *hood* and *heed*. In mathematics the term *hat* is commonly used as the name of the accent mark

formally known as a circumflex (*q.v.*). In statistics $\hat{p}$ is a quantity that approximates the actual proportion p of some subgroup in a larger population. It is usually called "p hat" rather than the more unwieldy "p circumflex" because the circumflex is like a little hat placed over the letter p. The fact that p is usually unknown led one professor to remark that $\hat{p}$ is called "p hat" because "it's cold out there in the unknown world."

haversine (noun): a contraction of "half the versed sine," so that the haversine of an angle θ, abbreviated hav θ, is defined as

$$\text{hav } \theta = \frac{1 - \cos \theta}{2}.$$

Haversines are used in solving oblique spherical triangles. Compare *coversine*. See explanation under *versed*. [196, 251, 192]

hecto-, abbreviated h (numerical prefix): from Greek *hekaton* "one hundred." In prehistoric Greek, the first part of the word was *se-*, rather than *he-*, from the Indo-European root *sem-* "one." The second part of the word is from the latter portion of the Indo-European root *dekm* "ten." In the International System of Units the prefix *hecto-* multiplies the following unit by 100. In that same system, the hectare, a unit of area equal to 100 ares, has become more common and more "basic" than the are. The metric prefixes representing the first three positive powers of ten were chosen from Greek number words by the Paris Academy of Sciences in 1791. In contrast, the prefixes for the first three negative powers of ten were chosen from Latin number words. [188, 34]

-hedron (nominal suffix): from Greek *hedra* "base, seat." The prehistoric Greek form was *sedra*, in which the Indo-European root *sed-* "to sit" can be recognized. The base of a polyhedron is the part that the solid "sits" on; it is more commonly known as a face of the polyhedron. In mathematics when *-hedron* is used as a suffix after a Greek number word, the resulting compound represents a geometric solid with the given number of faces or "bases." Examples are *tetrahedron*, *pentahedron*, etc. [184]

height: see *high*.

helicoid (noun), **helicoidal** (adjective): from *helix* and Greek-derived *-oid* (*qq.v.*) "looking like." In mathematics a helix is a three-dimensional spiral. If the helix is rotated about an axis and is simultaneously translated in the direction of that axis so that the ratio of the two rates is constant, the resulting surface is called a helicoid. A physical example is a propeller screw. [248, 244]

helix, plural *helices* (noun), **helical** (adjective): *helix* is a Latin word taken from Greek, where it referred to a kind of slender, flexible willow (which is an English cognate), and, more abstractly, to any spiral-shaped object. The Indo-European root is *wel-* "to turn, to roll," as seen in the English cognate *wallow*. In mathematics, a helix may be thought of as a three-dimensional spiral. Two common types of helix are conical and cylindrical. The root of *helix* is in common use in the word *helicopter*, literally "spiral wing." [248]

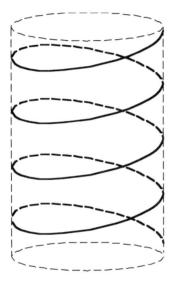

Cylindrical helix.

hemicycle (noun): the first component is Greek *hemi* (prehistoric *semi*) "half," from the Indo-European root *semi-* "half." The second component is from Greek *kuklos* "circle," from the Indo-European root *kwel-* "to move around," as found in native English *wheel*. A hemicycle is half of a circle. The word *hemicycle* is an all-Greek compound that was later translated into the all-Latin *semicircle* that we now more commonly use. [189, 107]

hemicylinder (noun), **hemicylindrical** (adjective): the first component is Greek *hemi* (prehistoric *semi*) "half," from the Indo-European root *semi-* "half." The second component is from Greek-derived *cylinder* (*q.v.*). A hemicylinder is the shape that results when a right circular cylinder is cut in half by a plane containing the axis of the cylinder. [189, 195]

hemisphere (noun), **hemispherical** (adjective): the first component is Greek *hemi* (prehistoric *semi*) "half," from the Indo-European root *semi-* "half." The second component is from Greek *sphaira* "ball, sphere," of unknown prior origin. A hemisphere is half of a sphere. [189, 207]

hendecagon, also erroneously but persistently *endecagon* (noun): the first component is from a presumed Greek *hens*, from the Indo-European root *sem-* "one, as one." The second component is from Greek *deka* "ten," from the Indo-European root *dekm-* "ten." Together the first two components produce *hendeca-* meaning "one [plus] ten," or "eleven." When the suffix *-gon* (*q.v.*) "angle" is added, the compound means "a polygon of eleven sides." [188, 34, 65]

hendecahedron, (also erroneously but persistently *endecahedron*), plural *hendecahedra* (noun): the first component is from a presumed Greek *hens*, from the Indo-European root *sem-* "one, as one." The second component is from Greek *deka* "ten," from the Indo-European root *dekm-* "ten." Together the first two components produce *hendeca-* meaning "one [plus] ten," or "eleven." When the suffix *-hedron* (*q.v.*) "base" is added, the compound means "a polyhedron of eleven faces." [188, 34, 184]

heptadecagon (noun): the first component is from Greek *hept-*, from prehistoric Greek *sept-* "seven," where the relation to Latin *septem* and English *seven* is more obvious. The second component is from Greek *deka-*, from the Indo-European root *dekm-* "ten." The ending is from Greek *gon-*, from the Indo-European root *genu-* "angle, knee." A heptadecagon is a seventeen-angled (and therefore also seventeen-sided) polygon. Carl Friedrich Gauss (1777–1855) was the first person to discover how to construct a regular heptadecagon using only a compass and unmarked straightedge. [191, 34, 65]

heptagon (noun), **heptagonal** (adjective): the first component is from Greek *hept-*, from prehistoric Greek *sept-* "seven," where the relation to Latin *septem* and English *seven* is more obvious. The second component is from Greek *gon-*, from the Indo-European root *genu-* "angle, knee." A heptagon is a seven-angled (and therefore also seven-sided) polygon. [191, 65]

heptahedron, plural *heptahedra* (noun), **heptahedral** (adjective): the first component is from Greek *hept-*, from prehistoric Greek *sept-* "seven," where the relation to Latin *septem* and English *seven* is more obvious. The second component is *hedra*, from prehistoric Greek *sedra* "base." *Sedra* is from the Indo-European root *sed-* "to sit," as seen in the English cognates *sit* and *seat*. A heptahedron is a seven-based, i.e., seven-faced, polyhedron. [191, 184]

heptiamond (noun): see *polyiamond*.

heptomino (noun): from Greek *hept* "seven" and all but the first letter of Latin *domino*. For the second component and further explanation see *polyomino*. [191, 37]

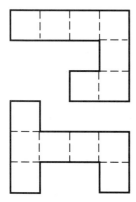

Two heptominoes.

hermit (noun, adjective): via Late Latin and French, from Greek *eremites* "a person of the desert," from *eremos* "deserted, empty." The Indo-European root is *erə-* "to separate." A related borrowing from Latin is *rare*. A hermit is a person who lives alone, though not necessarily in a desert. In mathematics a hermit point of a curve is one that is separated from all the other points on the curve. A hermit point is now more commonly known as an isolated point. [55]

heterological (adjective): the first component is from Greek *heteros*, originally "one of two," a compound based on the Indo-European root *sem-* "one." *Heteros* later acquired the extended meanings "other" and "different." The second component is from Greek *logos*, in the sense of "speech, discourse." The Indo-European root is *leg-* "to gather." The term *heterological* was coined by the mathematician Hermann Weyl (1885–1955) as part of his studies of paradoxes. He defined an adjective as being heterological if it doesn't describe itself. For example, in English the adjective *French* is heterological. A paradox arises in trying to decide whether the adjective *heterological* belongs in the set of heterological adjectives. [188, 111]

heteroscedastic (adjective), **heteroscedasticity** (noun): the first component is from Greek *heteros*, originally "one of two," a compound based on the Indo-European root *sem-* "one"; *heteros* later acquired the extended meanings "other" and "different." The second component is from Greek *skedasis* "a scattering," of uncertain origin. One hypothesis traces the word to the Indo-European root *skei-* "to cut, to split," an extension of *sek-* "to cut," as seen in the split personality of a *schizophrenic*. Another possible link is with a different extension of *sek-*, *sked-* "to split, to scatter," as seen in native English *shatter*. A third hypothesis traces the word to the Indo-European root *skel-* "crooked," as seen in *scoliosis* and *isosceles*, borrowed from Greek. In statistics a distribution is said to be heteroscedastic if its variations (= "scatterings") are not all the same. The word *heteroscedastic* may be pronounced as if the first *c* were a *k* or as if the first *c* were omitted. [188, 185, 195]

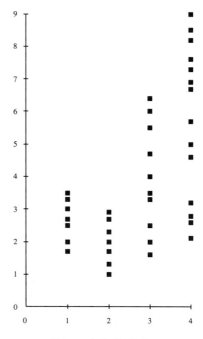

Heteroscedastic distribution.

heterosquare (noun): the first element is from Greek *heteros*, originally "one of two," a compound based on the Indo-European root *sem-* "one." *Heteros* later acquired the extended meanings "other" and "different." The second element is *square* (*q.v.*). A heterosquare of order *n* is a square array of the numbers from 1 to n^2 such that the sum of each row, column, or diagonal is different from all the others. A heterosquare is the "opposite" of a magic square, in which all sums are the same. The smallest possible heterosquare is of order three. [188, 108]

1	2	3
8	9	4
7	6	5

Heterosquare
of order 3.

heuristic (noun): from Greek *heuriskein* "to find, discover." The Indo-European root is *werə-* "to find." In a heuristic approach to learning mathematics, students are encouraged to explore and discover things on their own. Strict advocates of that approach might do well to discover that it takes more than a little bit of prompting to get a student to become a new Archimedes and "discover" two thousand years of mathematical advances in a few hours or even a few months. [252]

hex (adjective): from Greek *hex-* (prehistoric Greek *sex-*) "six," from the Indo-European root of the same meaning, *s(w)eks-*. The numbers 1, 7, 19, 37, ..., are called hex numbers because they can be arranged in patterns of concentric hexagons. In 1942 Piet Hein invented a game that he named *hex* because it is played on a diamond-shaped board made up of eleven rows of hexagonal cells. The word *hex* "an evil spell" is from a different source. [220]

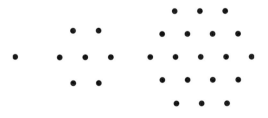

The first three hex numbers.

hexacontahedron, plural *hexacontahedra* (noun): the first element is from the Greek compound *hexakonta*, from the Indo-European roots *(s)weks-* "six," and a not-easily-recognized form of *dekm-* "ten," so that *hexakonta* means "sixty." The second element is from Greek-derived *-hedron* (*q.v.*) "seat, base," used to indicate the face of a polyhedron. A hexaconta-

hedron is a polyhedron with sixty faces. [220, 34, 184]

hexadecagon (noun): the first component is from Greek *hex-*, via prehistoric Greek *sex-*, from the Indo-European root *s(w)eks-* "six." The second component is from Greek *deka-*, from the Indo-European root *dekm-* "ten." The ending is from Greek *gon-*, from the Indo-European root *genu-* "angle, knee." A hexadecagon is a sixteen-angled (and therefore also sixteen-sided) polygon. [220, 34, 65]

hexadecimal (adjective): the first component is from Greek *hex-*, via prehistoric Greek *sex-*, from the Indo-European root *s(w)eks-* "six." The second component is from Latin *decimus* "tenth. "The hexadecimal system of numeration, which uses base sixteen, is common in computer science. The digits from ten through fifteen are usually represented by the capital letters from A through F. Since *hexadecimal* is a rather long word, it is sometimes abbreviated *hex*. The word *hexadecimal* is unusual because Greek and Latin elements are combined; the expected purely Latin form would be *sexadecimal*, but then computer hackers would be tempted to shorten the word to *sex*. Nevertheless, a similar Latin-based form is *sextodecimo*, used in the printing industry to refer to the size of a page obtained from a large sheet by folding it in half four times. Compare *octimal*, *decimal*, and *vigesimal*. [220, 34]

hexafoil (noun): the first component is from Greek *hex-*, via prehistoric Greek *sex-*, from the Indo-European root *s(w)eks-* "six." For the second component and further explanation see *multifoil*. [220, 21]

hexagon (noun), **hexagonal** (adjective): the first component is from Greek *hex-*, via prehistoric Greek *sex-*, from the Indo-European root *s(w)eks-* "six." The second component is from Greek *gonia* "angle." A hexagon is a six-angled (and therefore also six-sided) polygon. Hexagons are found in nature in the honeycombs of bees. [220, 65]

hexagram (noun): the first component is from Greek *hex-*, via prehistoric Greek *sex-*, from the Indo-European root *s(w)eks-* "six." The second component is from Greek *gramma* "letter, figure." A hexagram is a plane figure made up of two overlapping equilateral triangles, as in one version of the Mogen David or so-called Jewish star. In his famous "mystic hexagram" theorem, Blaise Pascal (1623–1662) used *hexagram* as a synonym of (a possibly reflex) *hexagon*. [220, 68]

hexahedron, plural *hexahedra* (noun), **hexahedral** (adjective): the first component is from Greek *hex-*, via prehistoric Greek *sex-*, from the Indo-European root *s(w)eks-* "six." The second component is from Greek *hedra*, from prehistoric Greek *sedra* "base." The Indo-European root is *sed-* "to sit," as seen in the English cognates *sit* and *seat*. A hexahedron is a six-based, i.e., six-faced, polyhedron. A regular hexahedron, commonly known as a cube (*q.v.*), is one of the five regular polyhedra. [220, 184]

hexakis- (numerical prefix): a Greek form meaning "six times," based on *hex-*, from prehistoric Greek *sex* "six." The prefix is used as part of the name of a stellated regular solid. It indicates that each original face became six faces. A hexakisicosahedron, for example is a polyhedron with six times twenty, or 120, faces. [220]

hexiamond (noun): see *polyiamond*.

hexomino (noun): from Greek *hex* "six," plus all but the first letter of Latin *domino*. For the second component and further explanation see *polyomino*. [220, 37]

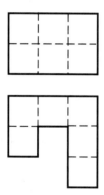

Two hexominoes.

high (adjective), **height** (noun): the native English *high* originally referred to something arched, a meaning that reflected the underlying Indo-European root *keu-* "to bend." Since something arched is inherently high, the meaning of English *high* shifted from the bending of an object to its height. The noun *height* should be *heighth*, just like *width* and *length*, but the form with a final *-t* rather than *-th* began in the north of England and eventually became the standard form. Some southern Americans still say *heighth* even though they write *height*. In pairs of opposite adjectives like *high* and *low*, the "positive" member of the pair also serves

as a generic term for the entire range of values: the question "how high?" can be answered with a measurement that is high or low. By contrast, the question "how low?" already presumes that the answer will involve something low. In a standard rectangular coordinate system, the *y*-value indicates the height of a point. In calculus a high point on a curve is the same as a relative maximum. [96]

hippopede (noun): the first component is from Greek *hippos* "horse," from the Indo-European root *ekwo-* "horse." A related borrowing from Greek is *hippopotamus*, literally "water horse." A related borrowing from Latin is *equestrian*. The second component is from the Indo-European root *ped-* "foot," which is the English cognate. A hippopede is literally a horse's foot. In mathematics it is a curve first studied by Proclus around 75 B.C. Also known as the horse fetter (an English cognate of *ped-*), it is represented by the polar equation $r^2 = 4b(a - b\sin\theta)$. For appropriate values of a and b, the curve looks like the infinity symbol, ∞. That shape apparently resembles the fetters that used to be put on horses to hinder their movement. [158]

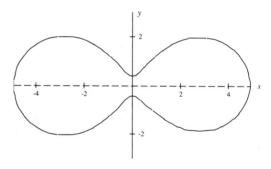

Hippopede: $(x^2 + y^2)^2 + 12(x^2 + y^2) = 36x^2$

histogram (noun): the second element, *-gram*, is indisputably from Greek *gramma* "piece of writing, picture," from the Indo-European root *gerbh-* "to scratch," because diagrams were originally scratched on earth, clay, etc. Reference books explain the first element, *histo-*, in two ways. (1) It may be from Greek *histos* "anything upright," but particularly "the upright beam of a loom," and then, by extension "anything woven, a web, a tissue." If Greek *histos* is the source, then the Indo-European root is *sta-* "to stand," as seen in the native English cognate *stand*. According to this explanation, the upright bars of a histogram account for its name. (2) *Histo-* may be a contraction of *history*, from Latin *historia*, in turn from Greek *istoria* "inquiry, observation." Greek *his-*

tor "a learned man," represents a presumed *wid-tor*, from the Indo-European root *weid-* "to see." Compare the native English cognate *wise*. Etymologically speaking, history is "what has been seen (and presumably also understood)." According to this explanation, a histogram is a "picture history" of a statistical distribution. (3) Whoever coined the term *histogram* may have had both of the above associations in mind, since each is plausible. [208, 244, 68]

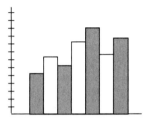

Histogram.

hodograph (noun): the first component is from Greek *hodos* "way," from the Indo-European root *sed-* "to go." Related borrowings include *exodus* (a going out) and *odometer*. The second component, *graph*, is from Greek *graphein* "to write, draw." In mathematics, a hodograph is a curve formed by the tips of the velocity vectors of a moving particle, provided the initial points of all the vectors coincide. [183, 68]

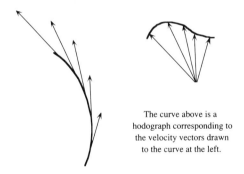

The curve above is a hodograph corresponding to the velocity vectors drawn to the curve at the left.

hole (noun), **hollow** (adjective): closely related native English words, from the Indo-European root *kel-* "to cover, to conceal." It may seem paradoxical that *hole* should come from a root meaning "to cover," but the human response to a hole is commonly to cover it. As for the meaning "to conceal," if a hole is deep enough it conceals whatever is at its bottom. Related borrowings from Germanic are the *holster* which conceals a gun and the *helmet* which covers a

head. In mathematics the graph of

$$y = g(x) = \frac{f(x)}{f(x)}$$

typically has a hole in it wherever $f(x) = 0$. In the graphing of an interval on a number line, or of a curve in two-dimensional coordinate system, a "hollowed-out" dot is used to indicate such a hole. In topology a hole can be of two types: a cavity or a channel (*qq.v.*). [89]

holomorphic (adjective): the first element is from Greek *holos* "whole," from the Indo-European root *sol-* "whole." The second element is from Greek *morph-* "shape, form, beauty, outward appearance," of unknown prior origin. In mathematics a single-valued function $f(z)$ is holomorphic at $z = a$ if and only if $f(z)$ is differentiable (i.e., the derivative "maintains its appearance," so to speak) throughout a (= for the whole) neighborhood of $z = a$. [201, 137]

homeomorphic (adjective), **homeomorphism** (noun): the first element is from Greek *homoios* "similar," in turn from *homo-* (archaic *somo-*), with the same meaning as the English cognate "same." The Indo-European root is *sem-* "one, as one." The second element is from Greek *morph-* "shape, form, beauty, outward appearance," of unknown prior origin. In mathematics a homeomorphism is a one-to-one correspondence, continuous in both directions, between the points in two geometric figures. Topologically speaking, the two figures have the same shape. [188, 137]

homogeneous (adjective), **homogeneity** (noun): the first element of the Greek compound is *homo* (prehistoric *somo-*) "same," from the Indo-European root *sem-* "one." An English cognate is *same*; when things are the *same*, they are all of *one* type. The second element is from *genos* "kind," from the Indo-European root *gen-* "born," "created." In physics, if an object is *homogeneous*, all the matter in it has "the same kind of" (i.e., equal) density. In linear algebra, if a system of linear equations is homogeneous, the constant on one side of each equation is always the same, namely 0. Although in nonmathematical English the opposite of *homogeneous* is *heterogeneous*, a system of linear equations that isn't homogeneous is called *nonhomogeneous* rather than *heterogeneous*. [188, 64]

homologous (adjective), **homology** (noun): the first component is from Greek *homos*, archaic *somos*, with the same meaning as the English cognate "same." The Indo-European root is *sem-* "one, as one." The second component is from *logos* "word, proportion." The Indo-European root is *leg-* "to gather," because words are "gathered together" when a person speaks. In mathematics homologous elements play the same or similar roles in two figures or relationships. For example, the vertices of a triangle and the corresponding vertices of a projection of that triangle are homologous points. The numerators or denominators of two equal fractions are also said to be homologous. Compare *equimultiple*. [188, 111]

homomorphic (adjective), **homomorphism** (noun): the first component is from Greek *homo-* (archaic *somo-*), with the same meaning as the English cognate "same." The Indo-European root is *sem-* "one, as one." The second component is from Greek *morph-* "shape, form, beauty, outward appearance," of unknown prior origin. A homomorphism is a correspondence between two sets, a domain and a range, in which each element of the domain has a unique image in the range, and each element in the range is the image of at least one element in the domain. Metaphorically speaking, the two sets have the same form. [188, 137]

homoscedastic (adjective), **homoscedasticity** (noun): the first component is from Greek *homo-* (archaic *somo-*), with the same meaning as the English cognate "same." The Indo-European root is *sem-* "one, as one." The second component is from Greek *skedasis* "a scattering," of uncertain origin. One hypothesis traces the word to the Indo-European root *skei-* "to cut, to split," as seen in the split personality of a *schizophrenic*. Another hypothesis traces the word to the Indo-European root *skel-* "crooked," as seen in the related borrowings *scoliosis* and *isosceles*. In statistics, a distribution is said to be homoscedastic if its variations (= "scatterings") are the same. The word *homoscedastic* may be pronounced as if the first *c* were a *k* or as if it were omitted. [188, 185, 195]

homothetic (adjective), **homothesis** (noun): the first component is from Greek *homo-* (archaic *somo-*), with the same meaning as the English cognate "same." The Indo-European root is *sem-* "one, as one." The second component is from Greek *tithenai* "to put," from the Indo-European root *dhe-* "to put." The related *thesis*, used by itself in English, is a document that you "put forth" in order to get a degree. In geometry, homothetic figures have been put

in the same relative position, so that lines through corresponding points all meet in a single point and are divided in a single, constant ratio by that point. Contrast *copolar*. [188, 41]

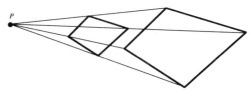

The two quadrilaterals are homothetic with respect to point *P*.

homotopic (adjective), **homotopy** (noun): the first component is from Greek *homo-* (archaic *somo-*), with the same meaning as the English cognate "same." The Indo-European root is *sem-* "one, as one." The second component is from Greek *topos* "place, region," of uncertain origin. In topology, two mappings are homotopic if each can be continuously deformed into the other; metaphorically speaking, the two mappings describe the same place. [188, 232]

horizontal (adjective): from Greek *horizein* "to divide, to separate," from *horos* "to bound, to limit," of unknown prior origin. The noun *horizon* means "the line that divides the sky from the land"; it is the place that marks the limit or boundary of the land that a person can see. Since the horizon goes across a person's field of vision, *horizontal* came to be applied to anything that goes from side to side, particularly a straight line. A surprisingly large number of English speakers don't connect the words *horizon* and *horizontal*, perhaps in part because the two words are stressed on different syllables. [79]

horopter (noun): the first element is from Greek *horos* "boundary, limit." The second element is Greek *opter* "one who looks," from the Indo-European root *okw-* "to see," as found in native English *eye* and the second part of *window*. Related borrowings from Greek include *optic* and *autopsy*. A horopter is a curve or surface in space containing all points such that images fall on corresponding points of the two retinæ. More specifically, a horopter is also defined as a space curve resulting from the intersection of a cylinder and a hyperbolic paraboloid. [79, 150]

hosohedron, plural *hosohedra* (noun): the first component is from Greek *hosos* "how great, as great, how many, as many." The second component is from Greek *hedra* "base, seat," from the Indo-European root *sed-* "to sit." Normally the *-hedron* suffix indicates some sort of polyhedron, a solid with faces that are planar and edges that are straight line segments. In spite of its name, a hosohedron isn't a polyhedron. Each face of a hosohedron is a digon, a figure whose one face is a non-planar surface and whose two edges are curves rather than line segments. There may be as many (= *hosos*) faces (= *hedra*) as desired. [184]

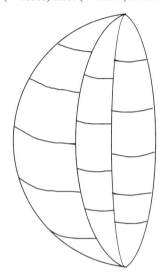

Three consecutive digonal
faces of a hosohedron.

hour (noun): via French, from Latin *hora* "an hour," though the Roman hour varied in length during the year because an hour was considered to be one-twelfth of the time from sunrise to sunset. Latin *hora* had been borrowed from Greek *hora*, which meant not "hour" but "season." Surprisingly, the underlying Indo-European root *yer-* appears in the native English cognate *year*, which seems to have been the original meaning. As the Indo-European root passed through Greek and Latin, the amount of time it designated became shorter and shorter, providing a linguistic example of inflation. [258]

hull (noun): a native English word, from the Old English verb *helan* "to cover." The Indo-European root is *kel-* "to cover, conceal." Related cognates include *hall* (a covered place), Scandinavian *Valhalla* (the hall reserved for those who died in battle), and even *hell* (a "covered" place beneath the earth). In mathematics the convex hull of a set is the smallest convex set containing (= "covering") all the points in the set. [89]

hundred (numeral): a native English compound. The first element, *hund*, actually means "ten." It

comes from *dekm-tom*, an extension of the more basic Indo-European root *dekm* "ten." The second element is from Old English *rad* "number," so that *hundred* means literally the "tens-number" in the sense that it is ten times ten. Old English *rad* is from the Indo-European root *ar-* "to fit together." That root is found in *arm* (both the anatomical and military kinds), *art*, Greek *arthros* "joint," and *arithmos* "number." Also related is native English *read*, to fit words together to make sense out of them. Although *hundred* means "tens number," the cognate Old Norse word *hundrath* could mean either ten tens or ten twelves, an indication that the ancient Scandinavians used both base ten and base twelve. A vestige of that ambiguity is found in our word *hundredweight*; while currently defined in the United States as 100 lbs, in England it has at various times meant 100 lbs., 112 lbs., and 120 lbs. See more under *ten*. [34, 15]

hyper- (prefix): from Greek *huper* "over." The Indo-European root *uper*, of the same meaning, is also found in the native English cognate *over* and in the Latin cognate *super*. The related words are sometimes interchangeable: we can speak of a child as being hyperactive or overactive, though not "superactive." [239]

hyperbola (noun), **hyperbolic** (adjective): the prefix *hyper-* (*q.v.*) is from Greek *huper-* "*over*," which is in fact the native English cognate. The main element *-bola* is from the Greek verb *ballein* "to cast, to throw." The Indo-European root is $g^w el \partial$- "to throw, reach." A related borrowing from Greek is *ballistic*. In mathematics, a hyperbola is a conic section with eccentricity greater than 1; in other words, the eccentricity is "thrown over" the cutoff value of 1 that distinguishes among an ellipse, a parabola, and a hyperbola. Similarly, given a vertical cone centered at the origin of a three-dimensional coordinate system, a hyperbola results when the cutting angle goes over (is greater than) the angle between the *xy*-plane and an element of the cone. In trigonometry, the hyperbolic functions are based on a unit hyperbola rather than on a unit circle, as are the more familiar circular functions. In nonmathematical English, *hyperbolic* is the adjective corresponding to *hyperbole*, meaning "speech that is 'thrown over' the normal amount, exaggeration." [239, 78]

hyperboloid (noun): from *hyperbola* and Greek-derived *-oid* (*qq.v.*) "looking like." A hyperboloid is a three-dimensional figure that looks like the two-dimensional hyperbola in the sense that every cross-section parallel to the axis of the hyperboloid is a hyperbola. A hyperboloid of revolution is generated when a hyperbola is revolved about one of its axes. When revolution is about the transverse axis, the resulting hyperboloid has two sheets. When revolution is about the conjugate axis, the resulting hyperboloid has one sheet. [239, 78, 244]

hypercomplex (adjective): from Greek-derived *hyper-* "over, beyond," and *complex (qq.v.)*. Whereas a complex number is of the binary form $a(1) + b(i)$, a hypercomplex number can have any number of terms, each consisting of a coefficient that multiplies a distinct unit. The German Sanskrit scholar Hermann Günther Grassmann (1809–1877) developed the general algebra of hypercomplex numbers. [239, 103, 160]

hypercube (noun): from Greek-derived *hyper-* "over, beyond," and *cube (qq.v.)*. A hypercube, which exists in four or more dimensions, goes "beyond" the three dimensions in which a conventional cube resides. A hypercube in four dimensions is called a tesseract (*q.v.*). [239, 96].

hyperellipse (noun), **hyperellipsoid** (noun): from Greek-derived *hyper-* "over, beyond," and *ellipse* or *ellipsoid (qq.v.)*. A hyperellipsoid, which exists in four or more dimensions, goes "beyond" the three dimensions in which a conventional ellipsoid resides. The standard equation of an ellipsoid is

$$\left(\frac{x}{a}\right)^2 + \left(\frac{y}{b}\right)^2 + \left(\frac{z}{c}\right)^2 = 1.$$

By contrast, the standard equation of a hyperellipsoid is

$$\left(\frac{x_1}{a_1}\right)^2 + \left(\frac{x_2}{a_2}\right)^2 + \cdots + \left(\frac{x_n}{a_n}\right)^2 = 1, \quad \text{for } n \geq 4.$$

A hyperellipse is the same as a hyperellipsoid, provided that a hyperellipse isn't construed as a three-dimensional ellipsoid (which is one dimension beyond an ellipse). [239, 52, 114, 244]

hypergeometric (adjective): from Greek-derived *hyper-* "over, beyond," and *geometric (qq.v.)* A hypergeometric series is so named because it "goes beyond" the complexity of a simple geometric series. [239, 63, 126]

hyperplane (noun): from Greek-derived *hyper-* "over, beyond," and *plane (qq.v.)*. A hyperplane, which exists in four or more dimensions, goes "beyond" the three dimensions in which a conventional plane resides. The standard equation for a plane

in three dimensions is $ax + by + cz - d = 0$. In contrast, the standard equation of a hyperplane is $a_1x_1 + a_2x_2 + \cdots + x_nx_n - x_{n+1} = 0$, for $n \geq 4$. [239, 162]

hyperspace (noun): from Greek-derived *hyper-* "over, beyond," and *space (qq.v.)*. Hyperspace is analogous to conventional three-dimensional space, but hyperspace contains four or more dimensions (at least one of which might be called "hyperthetical"). [239, 202]

hypersphere (noun): from Greek-derived *hyper-* "over, beyond," and *sphere (qq.v.)*. A hypersphere, which exists in four or more dimensions, goes "beyond" the three dimensions in which a conventional sphere resides. The standard equation of a sphere is $x^2 + y^2 + z^2 = 1$. In contrast, the standard equation of a hypersphere is $x_1^2 + x_2^2 + \cdots + x_n^2 = 1$, for $n \geq 4$. [239, 207]

hypo- (prefix): from Greek *hupo* "under." The Indo-European word *upo* meant "under," but also "up from under," and as a result the word developed opposite meanings in different Indo-European languages. English shows the upward meaning in *up* and the related verb *to open (up)*. Like Greek *hupo*, the Latin cognate *sub* retains the meaning "under." Therefore the cognate prefixes *hypo-* and *sub-* both mean "under," as in *hypodermic* and *subtcutaneous*, which are synonymous borrowings from Greek and Latin. [240]

hypocycloid (noun): from Greek-derived *hypo-* (*q.v.*) "under(neath)" and *kuklos* "circle." A hypocycloid is the curve traced by a point on the circumference of one circle that rolls around "underneath," that is, inside, another circle. [240, 107, 244]

hypotenuse (noun): from Greek-derived *hypo-* (*q.v.*) "under" and *teinein* "to stretch." The Latin cognates, which may be more familiar, are *sub* for "under" and *tendere*, to stretch," as combined in *subtend*. The Indo-European root underlying Greek *teinein* is *ten-* "to stretch." A Native English cognate is the adjective *thin*, which describes an object that has been stretched. When a right angle is inscribed in a circle, the diameter of the circle subtends that right angle. The diameter automatically becomes the hypotenuse of the right triangle thus formed. Even when a right triangle is not inscribed in a circle, the hypotenuse is the side that "stretches" from one leg to the other. [240, 226]

hypothesis (noun): from Greek-derived *hypo-* (*q.v.*) "under" and *thesis* "a putting." A thesis is a docu-

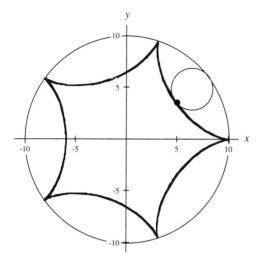

Hypocycloid: $\begin{cases} x = 8\cos t + 2\cos 4t \\ y = 8\sin t - 2\sin 4t \end{cases}$

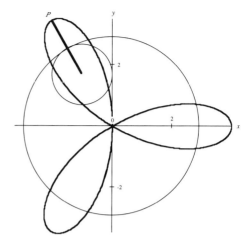

Hypotrochoid: $\begin{cases} x = 2\cos t + 2\cos 2t \\ y = 2\sin t - 2\sin 2t \end{cases}$

Point *P* on the extended radius of the small circle traces a hypotrochoid as the small circle rolls inside the large circle. This hypotrochoid happens to be a rose of three petals.

ment that a college student "puts forward" in order to get a degree. When you make a *hypothesis*, your idea is put under consideration to see if it holds up under scrutiny. The same Indo-European root *dhe-* that appears in *thesis* appears in native English *do* and *deed*: your deeds are what you put into practice. Compare the meanings of the components in *supposition*. [240, 41]

hypotrochoid (noun): from Greek-derived *hypo-* (*q.v.*) "under," *trokhos* "a wheel," and the suffix *-oid* (*q.v.*) "having the appearance of." The Indo-European root underlying Greek *trokhos* is *dhregh-* "to run." (English *truck* is ultimately borrowed via Latin from the same Greek word.) In mathematics, a hypotrochoid is the locus of a point on the (possibly extended) radius of a circle (= wheel) as that circle rolls "under" (= inside of) another circle. If the tracing point is on the circumference of the rolling circle, the resulting locus is called a *hypocycloid*. Compare *trochoid* and *hypotrochoid*. [240, 43, 244]

I

icosagon (noun): from Greek *eikosi* "twenty" and the root *gon-* "angle." Greek *eikosi* is from Indo-European *wi* "in half," hence "two," and a not easily recognizable *dkmt-* (with the initial *d-* lost) "ten." An icosagon is a polygon with two times ten, or twenty, angles, and therefore also twenty sides. [253, 34, 65]

icosahedron, plural *icosahedra* (noun): from Greek *eikosi* "twenty" and *hedra*, from prehistoric Greek *sedra* "base," as seen in the English cognates *sit* and *seat*. Greek *eikosi* is from Indo-European *wi*, meaning "in half," hence "two," and a not easily recognizable *dkmt-* (with the initial *d-* lost) "ten." An icosahedron is a solid with two times ten, or twenty "bases," that is, faces. A regular icosahedron, whose twenty faces are equilateral triangles, is one of the five possible regular polyhedra. [253, 34, 184]

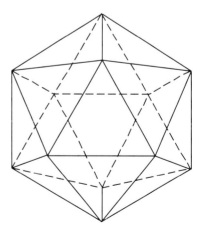

Icosahedron.

icosidodecahedron, plural *icosidodecahedra* (noun): a compound made up of *icosahedron* and *dodecahedron* (*qq.v.*). An icosidodecahedron is a semiregular polyhedron with thirty-two faces. Like the icosahedron and the dodecahedron, all of its faces are either equilateral triangles (20) or pentagons (12). [253, 34, 48, 184]

icositetrahedron, plural *icositetrahedra* (noun): from Greek *eikosi* "twenty," *tetra* "four," and *hedra* "base." An icositetrahedron is a polyhedron with twenty-four faces. In the type of icositetrahedron that actually occurs in crystals, there are three mutually orthogonal planes of symmetry and every face is a quadrilateral with no parallel sides. [253, 34, 108, 184]

ideal (noun): via French, from Late Latin *idealis*, an adjective corresponding to the noun *idea*. *Idea* had been borrowed from the similar Greek *idea*, from an Indo-European root *weid* "to see." (Prehistoric Greek lost its initial *w-*.) From the literal meaning of the root, Classical Greek developed many subsidiary meanings: "look, semblance, form, kind, nature, model." An idea, in our modern sense, was conceived of as a vision abstracted away from the physical sense of sight. The native English cognate *wise* shows a similar development: a wise person can "see" the truth. Related borrowings from Latin include *vision* and *visualize*. The nonmathematical adjective *ideal* refers to ideas of a lofty or positive nature. In projective geometry, an ideal point and an ideal line are located at infinity; they can't actually be seen, but must be visualized or imagined. [244]

idemfactor (noun): from Latin *idem* "same," plus *factor*. Latin *idem* is formed on the Indo-European pronominal stem *i-* which is also found in the archaic English words *yon* and *yea*. In mathematics the dyadic **ii** + **jj** + **kk** is called the idemfactor because in each term both factors are the same. Furthermore, its scalar product with any vector leaves the vector the same as before. [80, 41, 153]

idempotent (adjective): from Latin *idem* "same," plus Vulgar Latin *potent-*, present participial stem of *potere* "to be able, to have power." Latin *idem* is formed on the Indo-European pronominal stem *i-* which is also found in native English *yet* and *yes*. Vulgar Latin *potere* is from the Indo-European root *poti-* "powerful; lord." In mathematics an idempotent element is one which is unchanged (= has the same "power") after being multiplied by itself. Examples are the numbers 0 and 1 and the identity matrix. [80, 173, 146]

identity (noun): from Latin *identidem* "several times, over and over, repeatedly," from *idem* "same," formed on the Indo-European pronominal stem *i-* which is also found in native English *yonder*. In mathematics an identity is an equation for which both sides repeatedly have the same value for each value of the variables involved. An identity element is a number which, when combined with any quantity using a given operation, leaves that quantity the same. The identity element for multiplication is 1, and for addition it is 0. [80]

image (noun): a French word with the same meaning as in English. The original Latin noun *imago*, stem *imagin-*, meant not only "an image" but also "an imitation, a copy, a likeness." The word is related to *imitate*, likewise borrowed via French from Latin, but the ultimate source is still unknown. In a mathematical mapping, each point in one set is "copied" to a corresponding point, known as its image, in another set. [81]

imaginary (adjective): from Latin *imaginarius*, an adjective corresponding to the noun *imago* "image" (*q.v.*). From the sense "a copy," the word *image* took on the sense "portrait, picture." The meaning was then further extended from a painting or drawing as a physical object to an image as a strictly mental thing. That led to the meaning "which exists only in the mind." Imaginary numbers were so named because initially people couldn't find a "real," physical counterpart for them. While they may be imaginary, such numbers can give very real results in science and engineering. The Swiss mathematician Leonhard Euler (1707–1783) introduced the use of the letter *i* to stand for the imaginary unit. As recently as the end of the 19th century imaginary numbers were also called *impossible* numbers. [81]

im grossen (phrase): the first German word is a contraction of *in dem* "in the." Just like its identical English cognate, German *in* comes from the Indo-European root *en-* "in." German *dem* is cognate to native English *the*, from the Indo-European demonstrative pronoun *to-*, as also found in *that* and *their*. The German adjective *gross*—not related to English *gross* (*q.v.*)—means "great," which is the English cognate, though the earlier meaning was "thick, coarse." The Indo-European root is *ghreu-* "to rub, grind," which is an extension of *gher-* "to scrape, scratch." Other English cognates include *grit* and *groats*. The German phrase *im grossen* refers to the behavior of a function over an interval or in its entirety, as contrasted with the behavior of the function in the neighborhood of a point; contrast *im kleinen*. [52, 231, 71]

im kleinen (phrase): the first German word is a contraction of *in dem* "in the." Just like its identical English cognate, German *in* comes from the Indo-European root *en-* "in." German *dem* is cognate to native English *the*, from the Indo-European demonstrative pronoun *to-*, as also found in *then* and *there*. The German adjective *klein* currently means "small," though in older German the word meant "clean, pure, delicate, fine, neat." The Indo-European root is *gel-* "bright, pure," as seen in native English *clean*. The German phrase *im kleinen* refers to the behavior of a function in the neighborhood of a point, as contrasted with the behavior of the function over an interval or in its entirety. Contrast *im grossen*. The German mathematician Felix Klein (1849–1925) lived up to the Indo-European rather than the modern German meaning of his last name. [52, 231]

implication (noun), **implicit** (adjective), **imply** (verb): from Latin *in-* "in(side)" and *plicare* "to fold." The Indo-European root is *plek-* "to plait," an extension of *pel-* "to fold." Something implicit is literally "folded in" so that you can't see it clearly. In other words, when you fold something in, it becomes an intrinsic, often inextricable part of the whole, as when flour is folded into dough. When an equation is in implicit form, the variables have been "folded up" or mixed together in such a way that no one variable is all by itself on one side of the equation. The opposite is *explicit*. In logic, an implication is the name given to an if-then statement: if it rains, then I'll get wet. The getting wet is implied by (= is an unavoidable consequence of) the raining. In nonmathematical English, something implicit is understood but not stated openly. [52, 160]

imprimitive (adjective): from Latin *in-* "not," assimilated to *im-*, and *primitive* (q.v.), in its literal meaning of "first, earliest." In number theory, an integer *d* is said to be a primitive divisor of a function $f(n)$ if and only if *d* divides $f(n)$ but doesn't divide $f(x)$ for integers *x* such that $0 < x < n$; in other words, *n* is the first value of *x* that allows $f(x)$ to be divided by *d*. If *d* divides $f(n)$ but also divides $f(x)$ for another integer *m* between 0 and *n*, then *d* is said to be an imprimitive divisor of $f(x)$ when $x = n$. [141, 165]

improper (adjective): from Latin *in-* "not," assimilated to *im-*, and *proper* (q.v.). In calculus what makes an integral improper is the fact that it involves infinity, either as an explicit limit or because

the integrand becomes infinite. Although improper integrals are part of the standard curriculum, somehow we don't normally speak of proper integrals. In a discussion of fractions, however, it is proper to speak of both proper and improper fractions. [141, 165]

in- (negative prefix): a negative Latin prefix corresponding to native English *un-* and Greek *an-*. The Indo-European root is *ne-* "not," which used to appear in Old English *ne*, a word now replaced by the compound *not*. Latin *in-* usually assimilated wholly or partially to certain following consonants, as seen in borrowed words such as *imbalance*, *illegal*, *immutable*, *impractical*, and *irregular* rather than *inbalance*, *inlegal*, *inmutable*, *inpractical*, and *inregular*. If a word with the negative prefix *in-* doesn't appear in this dictionary, try looking up the positive version instead (for example *perfect* or *regular*.) [141]

in- (prepositional prefix): a native English word, from the Indo-European root *en-* "in." The original preposition with unchanged meaning appears in the Latin prefixes *in-* and *en-* as well as Greek *en-*. The *-n-* of Latin *in-* usually assimilated wholly or partially to the following consonant, as seen in *imbibe*, *illuminate*, *immerse*, *impact*, and *irradiate*, rather than *inbibe*, *inluminate*, *inmerse*, *inpact*, and *inradiate*. Latin also developed extended versions of *in-* such as *inter*, *intra*, and *intro*, which may also appear as prefixes of words in this dictionary. [52]

incenter (noun): from *in* and *center* (qq.v.). With regard to a triangle, the incenter is the center of the circle (known as the incircle) that can be inscribed in that triangle. Because the incenter is the point at which the bisectors of the three angles of the triangle meet, the incenter is necessarily inside the triangle. Contrast *circumcenter*, *excenter*, and *orthocenter*. [52, 91]

inch (noun): from Latin *uncia*, a unit of weight. The Latin word was derived from *unus*, from the Indo-European root *oi-no-* "one." The *uncia* took its name from the fact that it was one subunit within a larger unit; specifically, each *uncia* was one of the twelve parts of the *libra*, or Roman pound. Although English changed the *inch* from a unit of weight to one of length, it kept the fractional value of one-twelfth of the next highest unit, which is why there are twelve inches in a foot. Compare *ounce*. [148]

incidence (noun): from *incident-*, present participial stem of Latin *incidere* "to fall upon." The components are *in* "in, on," and *cadere* "to fall." The Indo-European root is *kad-* "to fall." Related borrowings from Latin include *accident* and *decadent*. An incident is something that befalls a person. In physics, when a ray of light falls on a reflective surface, the angle of incidence equals the angle of reflection. In mathematics, an incidence matrix is one that tells which points lie (= "fall") on which arcs of a network. [52, 83, 146]

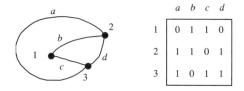

The array of 0's and 1's inside the square
forms an incidence matrix corresponding to
the three nodes and four arcs at the left.

incircle (noun): from *in* and *circle* (*qq.v.*). With regard to a triangle, the incircle is the circle that can be inscribed in that triangle. See more under *incenter*. Contrast *excircle*. [52, 198, 238]

inclination (noun): from Latin *inclinare*, consisting of *in* "in" and *clinare* "to lean." The Indo-European root is *klei-* "to lean." Related native English words are *lean* and *lid* (older *hlid*, the covering that "leans" over a pot.) Related borrowings include *clinic* (from the beds on which patients recline); *climax* (from the Greek word for ladder, since a climax was originally a series of events building up to a conclusion); and *climate* (from Greek *klima* "the sloping surface of the earth"). In mathematics the angle of inclination of a line (not to exceed 180°) is measured counterclockwise from the positive *x*-axis to the given line. [52, 100]

include (verb), **inclusive** (adjective): from Latin *in* "in" and *claudere* "to close." The Indo-European root is *kleu-* "a hook, a peg," so that the Latin verb would have meant originally "to lock with a hook or a bolt." Related borrowings include *seclude*, *closet*, and *close* itself. Something included is literally "closed in" or enclosed. In the geometry of triangles, SAS indicates that an angle is included or "enclosed" between two adjacent sides of a triangle. If an interval stretches from 3 to 4, inclusive, the endpoints 3 and 4 are included in the interval. [52, 101]

increase (verb, noun), **increment** (noun): from Latin *in* "in, on" and *crescere* "to grow." The Indo-

European root is *ker-* "to grow." Related borrowings from Latin include *create*, *cereal* (from Ceres, goddess of grain), and *crescendo*. When a quantity increases, it grows a little bit ("onward"). In calculus a curve is said to increase when it grows higher from left to right. An increment is a small increase in the value of a variable. The opposite of *increment* is not *excrement* (which comes from a different source), but *decrement*. [52, 92]

indegree (noun): the first element is English *in*, from the Indo-European root *en-*, of the same meaning. The second element is degree (*q.v.*). In graph theory, the indegree of a point *P* in a graph is the number (= degree) of arcs in the graph whose terminal point is *P*. The prefix *in-* indicates that each arc goes into point *P*. Contrast *outdegree*. [52, 32, 73]

independent (adjective): from Latin *in-* "not," *de* "from," and *pendent-*, present participial stem of *pendere* "to hang." The Indo-European root is *(s)pen-* "to draw, to stretch, to spin." Native English cognates include *spin* and *spider*. In algebra an independent variable doesn't depend or "hang" on the value of anything else. In $z = 3x^2y$, for example, both *x* and *y* are independent variables; they can take on any value. [141, 32, 206, 146]

indeterminate (adjective): from Latin *in* "not" and *determinatus*, past participle of *determinare* "to bound, to limit, to determine." The main Latin root is *terminus* "boundary line, limit," from the Indo-European root *ter-* "peg, post, boundary." Something determined has a boundary, a distinct end, and hence a definite value. An indeterminate form like $\frac{0}{0}$, on the other hand, does not have a definite value. Depending on the variable expressions that approach zero in the numerator and denominator, however, $\frac{0}{0}$ may approach a determinable value. Although we speak of indeterminate forms, we somehow rarely speak of "determinate" forms. [141, 32, 227]

index plural *indices* (noun): a Latin word meaning literally "pointer," in particular the second finger on a person's hand. The prefix *in-* means "in, to, at," and the Indo-European root is *deik-* "to show." A summation sign, which is actually the capital form of the Greek letter *sigma*, has two indices: the lower index points out or indicates the lowest value of the variable to use in the summation, while the upper index points out or indicates the highest value to use in the summation. For example, in

$$\sum_{n=3}^{5} n,$$

the lower index is 3 and the upper index is 5; the indicated sum is $3 + 4 + 5$. With regard to radicals, an index points out which root is being taken. For example, the index 3 in $\sqrt[3]{n}$ indicates the cube root of n. [52, 33]

indicator (noun): from *indicate*, a verb based on *index* (*q.v.*), plus the Latin agental suffix *-or* "the masculine person or thing that does the indicated action." In number theory, the indicator of a positive integer is the number of smaller positive integers that are relatively prime to it. The indicator "points out" how many smaller, relatively prime integers there are. Of course there are many functions that point out something; why this one should bear the name *indicator*, and not any of the others, isn't clear. In fact the indicator is now more commonly known as Euler's ϕ function (see *phi*). [52, 33, 153]

indicatrix, plural *indicatrices* (noun): the first element is from Latin *indicare* "to point out," based on *index* (*q.v.*) "a pointer." The ending is the Latin feminine suffix *-atrix* "the female person or thing that performs the action in question." In mathematics an indicatrix is a locus of the extremities of a line segment, in particular a radius, relative to a space curve. The feminine suffix is used because the Latin word *linea* "line" was feminine. [52, 33, 236]

induction (noun), **inductive** (adjective): from Latin *in*, meaning the same as its identical English cognate, plus *ductus*, past participle of the verb *ducere* "to lead." The Indo-European root is *deuk-* "to lead." A native English cognate is *to tug*. If you are inducted into the army, you are "led into" military service. In mathematics, when you reason by induction, a few specific examples "lead you in" to the general pattern. Compare the etymology of *inference*. Contrast *deduction*. [52, 40]

inequality (noun): from Latin *in-* "not" and *æquus* "even, level," of unknown prior origin. The corresponding adjective is *unequal*, in which the negative prefix is native English *un-* rather than Latin *in-*, though both are from the Indo-European root *ne-* "not." (Most people think the Declaration of Independence contains the words "inalienable rights" but the actual text is "unalienable Rights.") Although a mathematical relation like "$2 + 3 = 5$" is called an equation, a relationship like "$2 + 3 > 4$" is almost always called an inequality. The word *inequation* is listed in *The Oxford English Dictionary*, but it has fallen out of use. [141, 2]

infer (verb), **inference** (noun): via French, from Latin *in*, cognate to English *in*, and *ferre* "to bring."

The Indo-European root underlying Latin *ferre* is *bher-* "to carry," as seen in native English *to bear*. Metaphorically speaking, when you infer something you are "carried in" to a conclusion based on the evidence. Of course you still have to prove that your inference is correct. Compare the etymology of *induction*. [52, 23]

inferior (adjective): from Latin *inferus* "below, underneath, lower." A related borrowing from Latin is *infernal* "having to do with hell," which was conceived of as being a place below the ground. The Indo-European root is *ndher-*, which is more recognizable in the English cognate *under* than in Latin *inferus* or the related *infra*. In mathematics, the term *limit inferior* refers to the smallest of the accumulation points of a sequence. Contrast *superior*. [140]

infimum, plural *infima* (noun): a Latin superlative of *inferus* "below, underneath, lower," so that *infimum*, meaning literally "the lowest," corresponds to the comparative *inferior* (*q.v.*). The Indo-European root is *ndher-*, which is more recognizable in the English cognate *under* than in the related Latin *inferus* or *infra*. Mathematically speaking, and contrary to the etymology of the word, an infimum isn't the smallest of anything; the infimum is the *greatest* value that is still "under" (or equal to) the value of every member of the set. In other words, an infimum is the same as the greatest lower bound of a set. Contrast *supremum*. [140]

infinite (adjective), **infinity** (noun), **infinitary** (adjective): from Latin *in-* "not" plus *finis* "end, boundary, border," of unknown prior origin. Something infinite has no boundary or end; it keeps on going forever. To use a related word borrowed from French, it never *finishes*. Although the word *infinity* contains a negative element, most people think of *infinity* as a "positive" noun; we no longer use the separate noun *finity*, which is (or used to be) a real word meaning "finiteness." The English mathematician John Wallis (1616–1703) first introduced the ∞ symbol to represent infinity. In infinitary logic, infinitely long proofs are allowed. [141, 60]

infinitesimal (adjective): from *infinite* (*q.v.*), plus the Latin suffix *-esimus*, traditionally added to the stems of cardinal numbers to convert them to ordinal numbers. For instance, Latin *mille* meant "thousand," while *millesimus* meant "thousandth." In *infinitesimal*, the suffix *-esimus* was used by analogy to convert *infinite* from something unimaginably large to something unimaginably small. Notice that English can get away with saying "infinitely small," which is a contradiction if taken literally. [141, 60]

inflection (noun): from Latin *in*, cognate to and with the same meaning as the identical English word, plus *flectere* "to bend, to curve, to turn," of unknown prior origin. In mathematics, a point of inflection on a curve representing a function is a place where the curve "bends in" in such a way that the concavity changes from up to down or vice versa. [52, 61]

information (noun): from *inform*, whose earliest English meaning, in the 14th century, was "to give form to." During that same century the word also came to mean "give knowledge of a thing to," because the form of something provides knowledge about what the thing is like. Information, then, is knowledge. The word is based on Latin *forma* "contour, figure, shape." The Latin word seems to be a metathesis of the sounds in Greek *morphe* "form, beauty, appearance." Greek *morphe* is of unknown prior origin. The twentieth century, especially the part in which computers have proliferated, has been called the Information Age. If a year has to be chosen as its beginning, it may be 1948, when the American Claude Shannon published two papers about sending and receiving messages. The basic unit of information in the computer world is the bit (*q.v.*). [52, 137]

initial (adjective): from Latin *in*, cognate to and meaning the same as the identical English word, and *it-*, past participial stem of the verb *ire* "to go." The Indo-European root is *ei-* "to go." Your initials are the letters that your names begin (= "go in") with. In differential equations, an initial condition is a set of values that you start with. Similarly, in trigonometry, the initial side of an angle is its "beginning" side, as opposed to its terminal side. [52, 50]

injective (adjective): from Latin *in* "in, into," and *ject-*, past participial stem of *jacere* "to throw." The Indo-European root is *ye-* "to throw." Related borrowings from Latin include *eject* and *adjective* (a word thrown next to a noun). In medicine, when you get an injection, medicine is "thrown into" your body through a needle. In mathematics a mapping of elements from a set *A into* a set *B* is said to be injective. Metaphorically speaking, if we imagine set *B* to be an open box, the relation "throws" elements of *A* into box *B*. Contrast *surjective*. [52, 257]

inner (adjective): a native English word, the comparative degree of *in*, so that *inner* means literally "more inside, further inside." The Indo-European root is *en* "in." When the German Sanskrit scholar

Hermann Günther Grassmann (1809–1877) developed the general algebra of hypercomplex numbers, he realized that more than one type of multiplication is possible. To two of the many possible types he gave the names *inner* and *outer*. The names seem to have been chosen because they are antonyms, rather than for any intrinsic meaning. Grassmann symbolized the inner product of two units with a vertical bar, as in $\epsilon_1|\epsilon_2$. Because the inner product is now represented by the symbol "·" it is often known as the dot product. [52]

inscribe (verb): from Latin *in*, akin to the identical native English word, and *scribere* "to scratch," hence "to write." The Indo-European root is *skribh-*, an extension of *sker-* "to cut, to separate," as found in native English *shear*, *shard*, *short*, and *sharp*, as well as in *scrap* and *scrape*, borrowed from Old Norse. In ancient times geometric figures were scratched on the ground or onto waxed tablets or other physical objects. In geometry, an inscribed figure is "written inside" another figure, most often a circle. Contrast *circumscribe*. [52, 197]

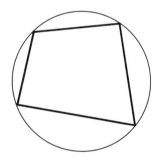

A quadrilateral
inscribed in a circle.

inside (adverb, preposition, noun): a native English compound. The first element is from the Indo-European root *en-* "in." The second element may be from the Indo-European root *se-* "long, late." The inside of a circle is the "side" or region that is in the circle, as opposed to the region outside of it. Textbooks often use the Latin word *interior* rather than native English *inside*. [52, 182]

instantaneous (adjective): Latin *instans*, stem *instant-*, meant literally "standing in or on," from Latin *in* "in, on" and the Indo-European root *sta-* "stand." When you stand on a given spot you literally press down on it, so *instans* developed the extended meanings "present, pressing, urgent." When something is pressing or urgent, it must be dealt with in a short period of time, so our borrowed word *instant* came to mean "a very short amount of time." The related adjective *instantaneous* means "happening in such a short period of time as to be practically immediate." In calculus a derivative is an instantaneous rate of change. [52, 208, 146]

integer (noun): from Latin *in-* "not" and the Indo-European root *tag-* "to touch," so that *integer* means "untouched, intact" and hence "whole." That meaning is evident in *entire*, borrowed from French. In mathematics the set of integers now includes both positive and negative whole numbers; it even includes the number zero, which, contrary to the etymology of integer, can't be touched and has nothing to be whole with. [141, 222]

integral (noun), **integrate** (verb), **integration** (noun): from Latin *integrare* "to make whole," from *integer* "whole." The components are *in-* "not" and the Indo-European root *tag-* "to touch," so that *integer* means literally "untouched, intact." Socially speaking, integration is an attempt to make a nation whole by removing barriers to contact. In calculus, when you integrate you make something whole in the sense of summing up an infinite number of little parts of the whole. The symbol used to represent an integral, $\int$, is a stylized letter S that stands for *sum*, since the integral calculates the entire area of a region by summing up an infinite number of little pieces of it. That version of the letter S was in use in America through the Revolutionary period, as you can see if you look at an original draft of the Declaration of Independence. [141, 222]

integrand (noun): from Latin *integrare* "to integrate" (*q.v.*), with the suffix *-nd-*, which creates a type of passive causative, so that Latin *integrandum* meant "to be integrated." In calculus the integrand is the function that is to be integrated; if x is the variable, then the integrand is whatever appears between the $\int$ sign and the dx at the end of the integral. [141, 222, 139]

intension (noun): from Latin *in* "in(ward)" and *tendere* "to stretch, to direct one's course, to go." The Indo-European root is *ten-* "to stretch," as seen in the English cognate *thin*. When you define a set by intension, you "direct yourself inward" to the criteria that an element must meet to be included in the set. For example, the set of positive even integers may be defined by intension as $\{x \mid x > 0,\ x = 2k$ for some integer $k\}$. Contrast *extension* and *roster*. [52, 226]

intercept (noun): from Latin *inter* "between" and *captus*, past participle of *capere* "to take, to seize." The Indo-European root is *kap-* "to grasp," as seen in the English cognate *have*. When you intercept something, you seize it while it is still between its starting point and its destination. In analytic geometry, an intercept is, figuratively speaking, the place where a curve "seizes "or "takes hold" of one of the coordinate axes. *Intercept* shouldn't be confused with *intersect* (*q.v.*). [52, 86]

interdecile (adjective): from Latin *inter* "between" and *decile*, literally "a tenth." The interdecile range of a frequency distribution is defined as the ninth decile minus the first decile. It includes the eight tenths of the distribution that are between the lowest tenth and the highest tenth. [52, 34]

interest (noun): a Latin compound made up of *inter* "within, inside, between" and *est* "it is," from the verb *esse* "to be." Both Latin *est* and the English cognate *is* are from the Indo-European root *es-* "to be." Latin *interest* means literally "it is in between." Something that is in between you and a goal that you are trying to reach is of importance; in short, it is of interest to you. When someone lends money, it is important to that person that the borrower repay the borrowed amount plus a fee for the use of that money. It is in the lender's best interest (assuming a profit is desired). The fee for the use of borrowed money has come to be known simply as *interest*. [52, 56]

interior (noun, adjective): Latin *inter* meant "within, inside," as seen also in native English *in*. The comparative degree of Latin *inter* was *interior*, literally "more inside." The interior of a closed figure is, rather obviously and somewhat redundantly "more inside" than the outside of the figure. An interior angle of a polygon is inside the polygon, as opposed to an exterior (*q.v.*) angle. [52]

intermediate (adjective): from Latin *inter* "between" and *medius* "which is in the middle," so the compound is redundant. The Indo-European root underlying Latin *medius* is *medhyo-*, found also in native English *mid* and *middle*. One of the most important theorems in calculus is the Intermediate Value Theorem. [52, 127]

interpolate (verb), **interpolation** (noun): from Latin *inter* "between" and *polire* "to smooth, to furbish, to polish, to adorn." The Indo-European root is *pel-* "to thrust, to strike, to drive." Related borrowings from Latin include *expel* and *pulsate*. In the long-ago days

B.C. (before calculators), values of trigonometric, logarithmic, and other special functions were listed in tables. Of course only a finite number of values of each function could be included, and it sometimes happened that the value you wanted to look up fell between two adjacent values in the table. You then had to "smooth over the gap in between" by pretending you were dealing with a linear function and estimating, by the use of a simple proportion, where the desired value fell between the adjacent function values. All the "polishing" that students used to do only made their tables more grimy. [52, 161]

interquartile (adjective): from Latin *inter* "between" and *quartile*, literally "a fourth." The interquartile range covers the middle half of the values in a frequency distribution. It includes the two fourths that are between the lowest fourth and the highest fourth. [52, 108]

intersect (verb), **intersection** (noun): from Latin *inter* "within, in between," and *sectus*, the past participle of *secui* "to cut." The Indo-European root is *sek-* "to cut," as seen in the native English noun *saw*. Two curves intersect when they "cut into" each other; the intersection is the place where that cutting happens. The intersection of two sets is the "place" where the two sets overlap (= cut into each other). The symbol ∩—now sometimes known as a cap (*q.v.*)—is used to represent the intersection of sets; the symbol goes back to Gottfried Wilhelm Leibniz (1646–1716), who also used it to indicate regular multiplication. *Intersect* shouldn't be confused with *intercept* (*q.v.*). [52, 185]

interval (noun): from Latin *inter* "between" and *vallum* "rampart, wall," from *vallus*, "a stake," from the Indo-European root *walso-* "a post." A Latin *vallum* must originally have been made up of poles or similar objects. English *wall* is borrowed from the Latin word. An interval is literally the space "between two walls." For example, the interval on a number line from 2 to 3 includes all the numbers between 2 and 3; metaphorically, the 2 and the 3 are the ramparts that mark the ends of the interval. In the notation [2, 3] that indicates a closed interval, or the notation (2, 3) that represents an open interval, the brackets and parentheses do look a bit like walls at the ends of the interval. [52]

into (preposition, adjective): a native English compound of *in*, from the Indo-European root *en* "in," and *to*, from the Indo-European root *de-*, which had a demonstrative meaning. Doubled-up prepositions are common in Indo-European languages. There are compounds in which the same idea appears twice in a row, as in *underneath*. There are even compounds in which the very same word appears twice in a row, although perhaps in slightly different forms, as in English *off of*. In mathematics a relation which pairs up elements of a set *A* with those of a set *B* is said to be a mapping of *A into B*. Metaphorically speaking, if we imagine set *B* to be an open box, the relation "throws" elements of *A* into box *B*. A relationship which is *into* is also said to be *injective* (*q.v.*). [52, 32]

intrinsic (adjective): from Latin *intra* "on the inside, within," and *secus* "beside, along." The Indo-European root is *sek^w-* "to follow," because something that follows you is often alongside you. The connection between following and being alongside is seen in the common English phrase *to follow along*. In mathematics an intrinsic equation of a curve is based on characteristics such as arc length or curvature that inherently "follow along" with the curve and don't depend on an external coordinate system. [52, 186]

intuitionism (noun): from Latin *intueri* "to look at, to gaze upon," with extended meanings "to look into, to observe, to consider, to contemplate." The prefix is *in* "into," and the Indo-European root of the second element is *teu-* "to pay attention." Related borrowings from Latin are *tutor* and *tuition*. When *intuition* appeared in English around 1500, it retained its etymological meaning of "looking into, contemplation." Later the meaning shifted to what might be called the results of contemplation without the contemplation itself, i.e., a direct understanding of something without reasoning. Intuitionism is a modern philosophy of mathematics that disavows the formalist and logistic manipulation of symbols as abstractions without intrinsic meaning. For intuitionists, mathematical truth is not an objective structure and can never be wholly expressed in symbols. [52]

inverse (noun), **invert** (verb), **invertible** (adjective): from Latin *in*, identical to its English cognate, and *vertere*, with past participle *versus*, "to turn." The Indo-European root is *wer-* "to turn, to bend." Native English cognates include *wrist* and *wrestle*. The inverse of something is that thing turned inside out or upside down. In mathematics the inverse of a function turns that function inside out, so to speak, and gives back whatever value has been put into it. The inverse of a mathematical operation undoes that operation; for example, division undoes multiplication. The multiplicative inverse of a fraction turns the fraction literally upside down. The additive inverse of a quantity is the quantity "turned inside out" in the sense of having its sign changed. A matrix is said to be invertible if it has an inverse. For an explanation of the suffix see *-able*. [52, 251, 69]

involute (noun), **involution** (noun): from Latin *in* "in, on," and *volut-*, past participial stem of *volvere* "to roll." The Indo-European root is *wel-* "to roll, to turn," as seen also in the Germanic *waltz*. To understand what an involute is in mathematics, imagine a string wound (= rolled) up on a given curve; an involute of the given curve is the locus of a point on that string as the string is unrolled while being held taut. In 19th century mathematics books, involution was also the common term for raising a number to a power; the number was figuratively rolled in on itself. For the complementary senses, see *evolute* and *evolution*. In projective geometry an involution is a transformation which when applied twice is equivalent to the identity transformation: a given figure is "rolled [back] in" to where it was at the beginning. [52, 248]

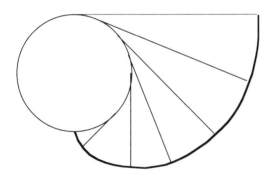

Involute of a circle.

isochrone (noun), **isochronous** (adjective), **isochronal** (adjective): from Greek *isos* "equal" and *khronos* "time," both of unknown prior origin. In mathematics the isochrone is the curve down which a weighted particle will fall with uniform vertical velocity. Jakob Bernoulli (1654–1705) proved that the curve is a semicubical parabola with its cusp pointing upward. [82, 99]

isocline (noun): from Greek *isos* "equal," of unknown prior origin, and *klinein* "to lean," from the Indo-European root *klei-* "to lean." A related native English word is *lid* (older *hlid*), the covering that "leans" over a pot. Related borrowings include

clinic (from the beds on which patients recline) and *climate* (from Greek *klima* "the sloping surface of the earth"). In calculus, points in a direction field of a first-order differential equation that have the same field direction can be connected by a curve called an isocline; all the points "lean the same way." [82, 100]

isogon (noun), **isogonal** (adjective): from Greek *isos* "equal," of unknown prior origin, and *-gon* (*q.v.*) "angle," as seen in native English *knee*, which forms an angle in a leg. An isogon is a polygon with all its angles equal; the most common isogons are rectangles and regular polygons. An isogonal affine transformation is one that doesn't change the size of angles. Isogonal lines pass through the vertex of an angle and are symmetric with respect to the bisector of the angle; in other words, they make equal angles with respect to that bisector. [82, 65]

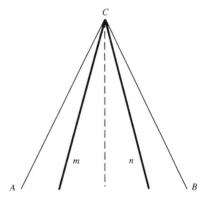

Lines *m* and *n* are isogonal
with respect to angle *ACB*.

isohedral (adjective): from Greek *isos* "equal," of unknown prior origin, and *hedra* "base, seat." The Indo-European root is *sed-* "to sit." As used in the mathematics of tiling, the root *hedr-* refers to a polygon or other closed figure. If all the tiles in a tiling belong to the same transitivity class, the tiling is said to be isohedral. Loosely speaking, all the tiles have "equal bases." [82, 184]

isolate (verb), **isolated** (adjective): via French and Italian, from Latin *insula* "an island," of uncertain origin. When something is isolated, it is figuratively turned into an island; in other words, the thing ends up being separated from other things. In electrical wiring, insulation keeps the current inside a wire and prevents it from leaking out. In mathematics, an isolated point on a graph is all by itself. For example, the point $(0, 0)$ is like a tiny island separated from

the rest of the points on the curve

$$y = \frac{x}{4}\sqrt{x - 2}.$$

To isolate a root of an equation means to find two numbers between which the root is located.

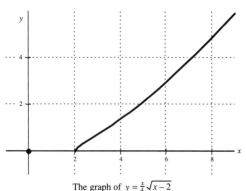

The graph of $y = \frac{x}{4}\sqrt{x-2}$
has an isolated point at the origin.

The three tables are isomorphic:
every E and O can be replaced by a P and N or by a T and B.

isometric (adjective), **isometry** (noun): the first component is from Greek *isos* "equal," of unknown prior origin. The second component is from Greek *metron* "a measure"; the Indo-European root is *me-* "to measure." An isometry is a transformation that leaves figures with the same measurements as before, though usually with different orientation or position. In axonometry (*q.v.*), an isometric projection is one in which all three axes have the same two-dimensional scale when projected onto a flat surface. [82, 126]

isomorphic (adjective), **isomorphism** (noun): from Greek *isos* "equal" and *morph-* "shape, form, beauty, outward appearance." The origin of both Greek components is still unknown. An isomorphism is a one-to-one correspondence between two sets, so that in a figurative sense they have the same shape. [82, 137]

isoperimetric (adjective), **isoperimetry** (noun): from Greek *isos* "equal," of unknown prior origin, and *perimeter* (*q.v.*). In geometry, isoperimetric figures have equal perimeters. In a group of isoperimetric regular polygons, the one with the greatest number of sides has the greatest area. Isoperimetry is the study of isoperimetric figures. [82, 165, 126]

isoptic (adjective, noun): from Greek *isos* "equal" and *optos* "seen, visible." The first component is of unknown prior origin, but *optos* is from the Indo-European root ok^w- "to see." The root is found in native English *eye* and the second part of *window*. Related borrowings from Greek include *autopsy* and

ophthalmology. Related borrowings from Latin include *ocular* and *inoculate* (literally to put in an eye). In mathematics an isoptic of a given curve is the locus of the intersection of two tangents that meet at a constant angle. That angle is always "seen [to be] equal." The word *isogonal* (*q.v.*) might have been more appropriate for the desired meaning, but *isogonal* was already in use with a related but different meaning. See also *orthoptic*. [82, 150]

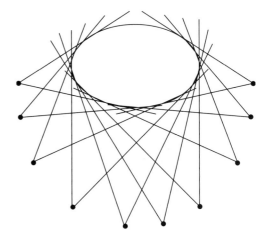

The points of intersection of pairs of tangent lines
making 45° angles lie on an isoptic of the ellipse

isosceles (adjective): from Greek *isos* "equal," of unknown prior origin, and *skelos* "leg." The Indo-

European root *(s)kel-* "curved, bent" is found in *scoliosis* and *colon*, borrowed from Greek. In geometry, an isosceles triangle or trapezoid has two equal legs. It may seem strange that the root means "bent" even though the sides of a triangle and a trapezoid are straight, but each leg is bent relative to the adjoining legs. [82, 195]

isotopic (adjective), **isotopy** (noun): from Greek *isos* "equal" and *topos* "a place, region," both of unknown prior origin. In mathematics an isotopy is an equivalence relation among braids; metaphorically speaking, the two braids "occupy equal regions." [82, 232]

isotoxal (adjective): from Greek *isos* "equal," of unknown prior origin, and *toxa*, a Greek plural noun meaning "bow." The plural was often used because ancient Greek bows were made from two pieces of horn joined in the center. The word may have been borrowed from Iranian or Scythian. In mathematics an isotoxal tiling is one in which each edge (= "bow") can be mapped onto any other edge by one of the symmetries of the tiling. [82]

isotropic (adjective): from Greek *isos* "equal," of unknown prior origin, and *tropos* "a turn, a way." The Indo-European root is *trep-* "to turn." The root appears in words borrowed from Greek such as *entropy* and *tropic* (the Tropic of Cancer, for example, marks the place where the sun "turns back" to the south after the summer solstice). In mathematics, an isotropic plane is an imaginary plane represented by the equation $ax + by + cz = d$, where $a^2 + b^2 + c^2 = 0$. Whichever way you turn, so to speak, the sum of the squares of the coefficients is zero for all such planes. [82]

iterate (verb, noun), **iteration** (noun), **iterative** (adjective): from Latin *iterum* "again," formed on the Indo-European pronominal stem *i-* which is also found in native English *yet* and *yes*. When you perform an iteration, you go through the same routine over and over, using the output of one step as the input of the next step. Each such output is called an iterate. In the most useful type of iteration, successive iterates eventually get as close as desired to a theoretically exact value. [80]

J

jalousie (noun): a French word, also borrowed into English under the form *jealousy*. The French word and the related Italian *gelosia* are from Medieval Latin *zelosus*, from Greek *zelos* "zeal." The Indo-European root is *ya-* "to be aroused." Someone who is zealous diligently pursues a cause; someone who is jealous diligently pursues or watches another person, or keeps that person from contact with others. The type of window known as a jalousie gets its name from the slats that can be turned to keep people outside a house from looking into it. In arithmetic, the jalousie or gelosia method of multiplication uses a jalousie-like grid to keep digits in their proper places. All digit-by-digit multiplications are done first, followed by addition downward and to the left along the diagonals. The jalousie method is very old, and seems to have developed in India. It was in use in China in the 16th century. Were it not for the tedium of drawing the grid each time, the jalousie method might still be the favored one today.

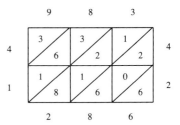

The jalousie method shows that
$$983 \times 42 = 41{,}286.$$

joint (adjective): past participle of French *joindre* "to join." The Indo-European root *yeug-* "to join," is found in native English *yoke*, which joins a pair of oxen. In mathematics, a joint variation is one in which a quantity varies as the product of two others: those two other variables are "joined together" by multiplication. [259]

jump (verb, noun): like *bump* and *thump*, *jump* is presumed to be a word that was invented to sound like what it represents, in this case the sound of feet landing on the ground. The use of the word is attested as early as the 16th century. The meaning of *jump* has now shifted from the sound of feet landing to the

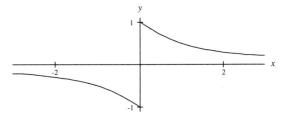

There is a jump at $x = 0$ in the curve $y = \frac{2}{\pi} \arctan\left(\frac{1}{x}\right)$.

motion of the person who is physically active and whose body is going up and down. In mathematics a jump is one type of discontinuity in a function.

K

kampyle (noun): a Greek word meaning "crooked staff." The Indo-European root is *kamp-* "to bend." From the related Italian *gamba* "leg," (which is bent) come our words *gambol* and *gambit*. In mathematics the kampyle (pronounced with three syllables) is a curve that the Greek mathematician Eudoxus studied in connection with the problem of duplicating a cube. The polar equation of the kampyle is $r = a \sec^2 \theta$.

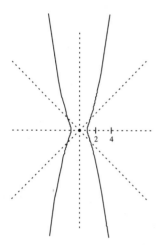

Kampyle of Eudoxus: $r = \sec^2 \theta$.

kappa (noun): a Greek letter, written κ in lower case and K as a capital. The Greek letter is based on Hebrew ⊃, *kaph*, which originally meant "palm." In mathematics κ is often used to represent curvature, because the word begins with a k-sound. The kappa curve, described by the equation $x^4 + x^2y^2 = a^2y^2$, is so named because it resembles the shape of the letter kappa. G. van Gutschoven conceived the curve in 1662.

keratoid (adjective): the main element is from Greek *keras*, stem *kerat-*, "horn," from the Indo-European root *ker-* "head, horn," the latter being the English cognate. The suffix is Greek-derived *-oid* (*q.v.*) "looking like." A keratoid cusp, also known as a cusp of the first kind, is one in which the two branches of the curve lie on opposite sides of the tangent line as they approach the point of tangency.

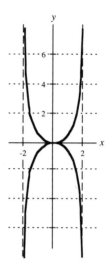

Kappa curve:
$$y^2 = \frac{x^4}{4 - x^2}.$$

An example is $y^2 = x^3$ at the point $(0, 0)$. Such a configuration looks (with some imagination) like an animal's horn. Contrast *ramphoid*. [93, 244]

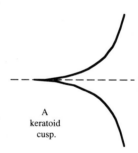

A keratoid cusp.

kernel (noun): from the Indo-European root *greno-*, recognizable in the native English cognate of the same meaning, *corn*. (In England, *corn* is still a generic term referring to grain, rather than to maize, as in America.) A related borrowing from French is *grain*. From the metaphoric sense of *kernel* as "central part, core," comes the mathematical meaning as it relates to homomorphisms. The term is also used in computer programming to refer to that part of a program which is device- or platform-independent. [76]

kilo-, abbreviated *k* (numerical prefix): from Greek *khilioi* "thousand," which is the multiplicative value of the metric prefix. The Indo-European root may have been *gheslo-* "a thousand." The prefixes representing the first three positive powers of ten were chosen from Greek number words by the Paris Academy of Sciences in 1791. In contrast, the prefixes for the first three negative powers of ten were chosen from Latin number words. Due to historical considerations, one anomaly of the current International System of Units is that the kilogram rather than the gram is considered the basic unit of mass. Because multiple numerical prefixes are forbidden, one million kilograms is not a megakilogram but rather a gigagram; numerical prefixes attach to *gram*, and never to *kilogram*. [72]

kissing (adjective): a native English word, from the Indo-European root *kus-* "to kiss." Given a fixed circle, the kissing number is the maximum number of equal circles (6) that can be placed tangent to (= kissing) the given one. In three dimensions, the circles become spheres, and the kissing number is 12; although there is still some room left, it isn't enough for a 13th equal sphere to be squeezed in. Similarly we can calculate the kissing numbers for hyperspheres clustered around an equal hypersphere in *n* dimensions. Compare *osculating*.

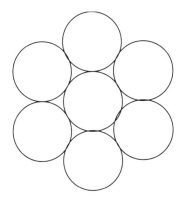

Kissing circles.

kite (noun): a native English word. The flying toy called a kite takes its name from the bird of the same name, which typically has a long, forked tail like that of a man-made kite. The bird, in turn, is

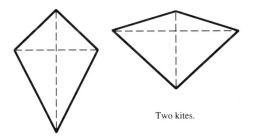

Two kites.

presumed to be named for the sound it makes, as are the cuckoo, whippoorwill, and other birds. In geometry, a kite is a convex quadrilateral with two distinct pairs of adjacent equal sides; contrast that with a parallelogram, in which the equal sides are opposite each other. The geometric kite is named after the characteristic shape of many popular toy kites. Contrast *dart*.

knot (noun): from Old English *cnotta*, from a supposed Indo-European root *gen-* "to compress into a ball." If you pull on a knotted cord, the knot forms a "ball." Related native English words include *knob*, *knuckle*, *knoll*, and *knit*. Mathematically, a knot is a curve in space formed by weaving a string in any manner (including not at all) and then joining the ends. In nontechnical English usage, we don't require that the ends of the string be joined.

kurtosis or **curtosis** (noun): a Greek word meaning "convexity," from *kurtos* "convex," from the Indo-European root *(s)ker-* "to turn, to bend." The convexity (or more commonly the concavity) of a curve at a point is a measure of how the curve bends there. In statistics the kurtosis of a distribution indicates the general form of concentration about the mean, as evidenced by the amount of "bending" in the central part of the graph. For the categories of kurtosis, see *leptokurtic*, *mesokurtic*, and *platykurtic*. [198]

L

lacunary (adjective): from Latin *lacuna* "a ditch, pit, hole, pool, pond," from *lacus* "a hollow thing," and therefore "a basin, tank, tub." The Indo-European root is *laku-* "body of water, lake," as seen in our words *lake* and *lagoon*, borrowed ultimately from Latin. As used in nonmathematical English, *lacuna* emphasizes the "hole" meaning of the Latin word rather than the "pool" meaning. In mathematics, a lacunary function is an analytic function represented by a circle with a natural boundary; that circle represents a "lake" on which the function converges.

lambda (noun): eleventh letter of the Greek alphabet, written Λ as a capital and λ in lower case. It corresponds to the Hebrew letter *lamedh*, ל, meaning "ox goad," and to the sound of our letter *l*. Looking at the Hebrew letter, it is not hard to imagine an ox goad; the Greek letter is a bit further removed in shape from the original object. See further explanation under *aleph*. In calculus λ is used to stand for a Lagrange multiplier because the first letter in *Lagrange* is an *L*.

lamina (noun): a Latin word meaning "a thin piece of metal or wood, a flat plate." The Indo-European root may be *stelǝ-* "to extend." From the Latin word, after many modifications in French, comes *omelette*, which is flat. When a document is laminated, it is encased in a thin, flat plastic covering. A common type of calculus problem involves finding the centroid of a lamina. [210]

large (adjective): a French word, from Latin *largus* "abundant, plentiful, copious, large, much," of unknown prior origin. When English first borrowed *large* from French in the 12th century, it meant "liberal, generous," a meaning still preserved in the borrowed noun *largesse*. By the 13th century *large* meant "wide in range or capacity." By the 14th century it meant "broad," which is the sense that has remained in French. The modern English use of *large* as a synonym of *big* dates back to the 15th century. In statistics the Law of Large Numbers says that the probability that a sample mean will differ from the mean of the population it is drawn from approaches 0 as the sample becomes larger and larger.

last (adjective): the first part of the word is English *late*, from Indo-European *le-* "to let go, slacken." Related borrowing from Latin are *lenient* and *alas* (from *lassus* "weary," which is your condition when you exhaustedly sigh "Alas!"). If you let (another native English cognate) go of your sense of time, you arrive late. The *-st* suffix is the same one that normally appears in English superlatives such as *worst*, *prettiest*, *most*, etc. The word *last* has been contracted from *latst* in the same way that *best* has been contracted from *betst* (compare *better*). By the 16th century, when people no longer recognized that *last* was the superlative of *late*, the new superlative *latest* was created; it is now the only superlative of *late*. In algebra, formulas exist which calculate the last term of a finite arithmetic or geometric sequence.

latent (adjective): from Latin *latens*, stem *latent-*, present participle of *latere* "to lie hidden or concealed." The Indo-European root is *ladh-* "to be hidden," as seen in *Lethe*, the river of forgetfulness in Greek mythology. In mathematics a latent root of the characteristic equation of a matrix is the same as an eigenvalue of the characteristic equation. [146]

lateral (adjective): from Latin *latus*, plural *latera*, "side," of unknown prior origin. In football, a lateral pass is a pass to the side. In geometry, the lateral area

of a cone or cylinder is the "side" area, as opposed to the area of the base(s). [110]

lattice (noun): native English *lath* means a thin, narrow piece of wood or other material used as an underlying framework meant to be covered over with plaster or some other substance. English *lath* cannot be traced farther back than the Germanic languages. French borrowed a Germanic word related to *lath*, and then English borrowed it back later under the form *lattis*. (Re-borrowings like that are common among languages in close or sustained contact). A lattice is essentially a framework. In the Cartesian plane, *lattice* points are points both of whose coordinates are integers. Those integrally-designated points form a sort of framework for the rest of the points in the plane.

latus rectum plural *latera recta* (noun): the Latin noun *latus*, stem *later-*, meant "side" or, in geometry, "chord"; the word is of unknown prior origin. The adjective *rectum*, which is related to English *right*, as in *right* angle, is the past participle of Latin *regere* "to lead in a straight line or in the right direction." The Indo-European root is *reg-* "to move in a straight line." In analytic geometry, the *latus rectum* is the chord of a conic section that passes through the focus and makes a right angle with the axis of the conic. [110, 179]

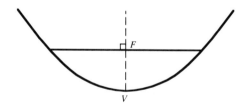

The perpendicular chord through the focus
F is the latus rectum of the parabola.

law (noun): from the Old Norse stem *lag-* "law," literally something that has been set down, from the Indo-European root *legh-* "to lie, to lay." Other native English cognates include *ledge* and *lair*. Two common laws in trigonometry are the law of sines and the law of cosines. What used to be known as the commutative, associative, and distributive laws are now more commonly called properties. [112]

layer (noun): a native English word derived from the verb *lay*, from the Indo-European root *legh-* "to lie, to lay." *Lay* is a causative form of *lie*: to lay something down is to cause it to lie down. English used to have many such pairs, but few remain now. Other examples are *raise* (= cause to rise) and *set* (cause to sit.) A layer is literally something that has been laid down. Because many things are laid down on flat surfaces, layers are frequently horizontal. In geometry, a zone of a sphere is that part of the surface contained between two parallel planes (which are most often drawn horizontally); the solid bounded by the zone and the two parallel planes is called a layer of the sphere. [112]

lb. (noun): abbreviation for *pound*, from Latin *libra*, which meant both "a balance scale" and "a unit of weight used on such a scale." The ultimate origin of the word *libra* is unknown, but variants of a presumed *lithra* "scale" are found in languages around the Mediterranean. Greek *liter*, originally a unit of weight, is believed to be from the same source. Unlike our pound of 16 ounces, the Roman pound was divided into 12 ounces. Because the British monetary unit is called a pound, its abbreviation £ is also drawn from Latin *libra*; furthermore, until recently that division into twelve parts was maintained in a subdivision of the British pound, the shilling, which used to equal twelve pence. Compare *inch* and *ounce*. [120]

leading (adjective): from the native English verb *lead*, from the Indo-European root *leit-* "to go forth." In algebra, given a polynomial of degree *n*, the coefficient of the *n*th-degree term is called the leading coefficient. If the terms of the polynomial are written in descending order, then the leading coefficient metaphorically leads the way for the other coefficients that follow to its right.

leaf, plural *leaves* (noun): a native English word, from the Indo-European root *leup-* "to peel, to break off," because in cool weather leaves "break off" trees. A related borrowing from Germanic is *lodge*, a cabin made from bark that breaks off of tree trunks. An object as common as a leaf turns up surprisingly often in mathematics. Statisticians resort to stem-and-leaf plots (see picture under *stem*). The curve known as the rhodonea (*q.v.*) or rose is often described as having a certain number of leaves. Using the etymologically unrelated Latin word for leaf, *folium* (*q.v.*), mathematicians have studied the bifolium, trifolium, and other folia.

league (noun): from Late Latin *leuca* or *leuga*, possibly of Gaulish origin. A league is a unit of length equal to three miles.

least: see **less**.

left (adjective, adverb): a native English word which originally meant "weak, useless." Because most people are right-handed and have relatively weak left hands, the word came to be applied to the generally weaker left side of the body. Although cognates are found in Dutch and Frisian, the origin of *left* is unknown. This word is unrelated to the past tense of *leave*. In a standard Cartesian coordinate system, smaller values of *x* are found to the left of larger ones. Although the noun *height* refers to how high or low something is, and although there are many other such terms (for example *width*, *length*, *parity*), there is no noun that means "the leftness or rightness" or even just "the leftness" of a point in a coordinate system. Perhaps *leftness* will find its way into the dictionary.

leg (noun): from Old Norse *leggr* "leg," of unknown prior origin. In a right triangle, the two sides that aren't the hypotenuse are known as legs. Because there are two of them, (as opposed to the three sides of a non-right triangle) they are named by analogy with the two legs of the human body. Compare *arm*.

lemma (noun): from the Greek verb *lambanein* "to grasp, to take." The Indo-European root is (*s*)*lag*^{w}- "to seize." An English cognate is *latch*. From Greek comes *epilepsy*, with its seizures. In mathematics a lemma is a theorem that you "grasp hold of" when you proceed to prove another theorem that is usually more complicated or more important.

lemniscate (noun): from Greek *lemniskos* "a woolen ribbon used in fastening a garland for someone's head." The word is derived from the noun *Lemnos*, a Greek island in the Aegean Sea near Turkey, presumably because that type of ribbon originated there. (In a similar way, suede originated in Sweden, damask in Damascus, china in China, etc.) The Latin form of the word, *lemniscus*, is used in anatomy to refer to a bundle of sensory nerve fibers. In mathematics a lemniscate is a closed curve with two loops; a lemniscate looks like a bow tied with ribbon. A common polar equation for the curve is $r^2 = a^2 \cos(2\theta)$. The lemniscate was discovered by Jacques Bernoulli (1654–1705) in 1694. Contrast *bow*.

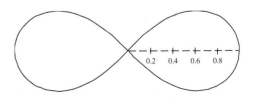

Lemniscate: $r^2 = \cos(2\theta)$.

length (noun): the noun *length* is formed from a variant of the adjective *long* by adding a suffix that in modern English takes the form *-th*. That noun-forming suffix can appear after verbs, as in *birth* (from *to bear*) and *aftermath* (from *to mow*, so that aftermath actually has nothing to do with mathematics, but means "what is left over after a field is mowed"). The suffix *-th* can also appear after adjectives, as in *width* (from *wide*), *breadth* (from *broad*), and *depth* (from *deep*). The noun *length* serves as a generic term that includes both long and short things: the length of something can be a micron or 1000 kilometers. That same generic characteristic is also found in other nouns derived from the "positive" members of opposite pairs: *width* refers to wide and narrow things, *height* refers to high and low things, etc. In mathematics the length of an oval is defined as the maximum value of the breadth (*q.v.*). Contrast *thickness*. [35]

leptokurtic (adjective): the first component is from Greek *lepto-* "slender, fine," from the Indo-European root *lep-* "to peel." A related borrowing used in physics is *lepton* "any of several subatomic particles." The second component is from Greek *kurtos* "convex," from the Indo-European root *(s)ker-* "to turn, to bend." In statistics a distribution is said to be leptokurtic if it is more heavily concentrated around the mean than a normal distribution is. The graph of such a distribution will look "slenderly convex" in the vicinity of the mean. For more information, see *kurtosis*. [198]

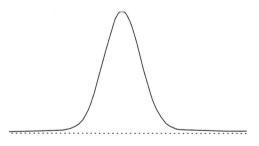

A leptokurtic distribution.

less, least (adjective, adverb): native English words that are, respectively, the comparative and superlative degrees of an adjective that no longer exists on its own. The presumed original adjective was *lais* "small," from the Indo-European root *leis-*, of the same meaning. The predecessor of the modern comparative *less* was something like *lais-iza*, and the predecessor of the modern superlative *least* was something like *lais-ista*. Over time the *s* sound at the end of the original stem merged with the similar sound of the endings (just as many people now pronounce *mirror* as if it were *meer*, in which the two *r*'s have merged) and must have caused confusion of meaning. Because of that confusion, the original adjective ultimately fell out of use. *Less* has now become the *de facto* comparative of the somewhat similar sounding *little*. Although *less* is historically a comparative, it doesn't end in *-er* and therefore doesn't look like a typical comparative; as a result, the modern language has created *lesser*, which is a double comparative meaning literally "more more little." Compare *more*. The fact that *less* and *left* sound similar is a convenient reminder that when two numbers are graphed on a number line, the one farther left is less than the other one. The less-than symbol "$<$" was first used by the British mathematician Thomas Harriot (1560–1621) to represent the relationship "is less than." The word *least* occurs in the least common denominator, the least upper bound, and—last but not least—the method of least squares. [115]

level (noun): Latin *libra* meant a balance scale, as well as the standard weight used in such a scale, the pound (which is why *pound* is abbreviated *lb.*) The diminutive form of *libra* was *libella*, meaning "a level," a device which is used to align something horizontally, because the arms of a balance scale become level when two weights are in balance. Our word *level* comes via French from *libella*. In mathematics a level curve connects all points on a surface that are at the same height; in other words, it connects points that are level, or in the same horizontal plane. [120, 238]

lexicographic (adjective), **lexicographically** (adverb): the first component is from Greek *lexis* "word," from *legein* "to speak." The underlying Indo-European root is *leg-* "to gather": when you speak you gather words together. The second component is from the Greek verb *graphein* "to write," from the Indo-European root *gerbh-* "to scratch." Lexicography is the science of putting a dictionary together. Because dictionaries are arranged in alphabetical order, the phrase *lexicographically ordered* is used by analogy in mathematics to describe a collection of sequences arranged as if the numbers involved were letters of the alphabet. For example, the sequences $\{2, 3, 5, \ldots\}$, $\{2, 3, 8, \ldots\}$, and $\{1, 2, 4, \ldots\}$, are ordered lexicographically. [111, 68, 117]

like (adjective, verb), **likely** (adjective), **likelihood** (noun): from Old Norse *likr* "like." The Germanic root is *lik-* "body, form, like, same." The semantic development is as follows: a body has a characteristic shape, so if two things have similar "bodies" or shapes, then they are alike. When a thing is likely, it looks like the thing is going to happen. When you like something, you approve its form, shape, or appearance. The word *likely* even contains the same root twice, since the adverbial suffix *-ly* is just a reduced version of *like*: if you do something quickly, you do it in a "quick-like" manner; your action has the appearance of being quick. In algebra, like terms have the same form in the sense that they have the same powers of the same variables. Only like terms can (and usually should) be added or subtracted. The Latin borrowing *probability* (*q.v.*) is more commonly used than the English term *likelihood*, at least in its mathematical sense. [117]

limaçon (noun): a French word, from Latin *lamacis* "snail," with the augmentative suffix *-on*. (The little hook under the letter *c*, called a cedilla, indicates that the *c* is pronounced like an *s*). The Indo-European root is *(s)lei-* "slimy." Native English cognates include *slime*, *slippery*, and *slick*. A related borrowing from Latin is *liniment*. In mathematics a limaçon is a curve whose polar equation is of the type

$$r = a + b \begin{Bmatrix} \sin \\ \cos \end{Bmatrix} \theta.$$

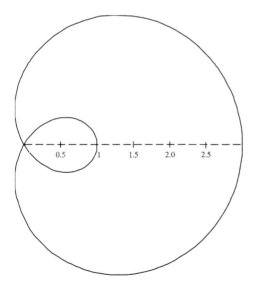

Limaçon: $r = 1 + 2 \cos \theta$

When $a = b$, the curve reduces to a cardioid (*q.v.*). When $b > a$ the curve has an inner loop, and the curve as a whole looks something like a snail's shell with an opening in it. The limaçon is sometimes called Pascal's limaçon, because Étienne Pascal (1588–1640), father of the more famous Blaise, was the first person to study it. G.P. Roberval gave the curve its name in 1650. [151]

limit (noun): from Latin *limes*, stem *limit-*, "boundary between two fields," of unknown prior origin. When a Roman approached the limit of his land, he got closer and closer to its boundary. In mathematics a limit is a value that a function gets closer and closer to under certain conditions. The now-archaic adjective *limitary* meant "having to do with a limit." Its *-ary* ending is believed to have influenced the formation of *boundary* (*q.v.*) from *bound*. [118]

line (noun), **linear** (adjective), **linearity** (noun), **linearly** (adverb), **lineal** (adjective): from Latin *linum* "flax," from the Indo-European root *lino-* "flax." From flax the Romans made *linea*, or "linen thread." Left to its own devices, thread is likely to twist or curl, and so the word *line* could mean what we now call a *curve*. When pulled taut, however, thread can be made to approximate a true (= straight) line. Even now some books refer to any path traced by a point as "a line." The term *straight line* has been used to distinguish a true line from a "curved line." For some people the term *straight line* has become redundant, since a line is now generally taken to mean straight line, but the expression is so much a part of English usage that we tolerate the redundancy. The derived adjective *linear* means "having to do with or being in the shape of a line." Since a first-degree equation in two variables graphs as a straight line, *linear* has come to be a synonym for "first-degree." By extension, a first-degree equation in three variables is also called linear, even though such an equation graphs not as a line but as a plane. That extension to three dimensions makes sense because a plane plays some

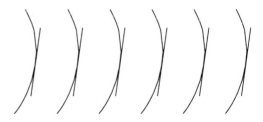

Lineal elements in a direction field.

of the same roles in three dimensions that a line plays in two dimensions. Two quantities x and y are said to be linearly dependent with respect to a field if, for some a and b in the given field, the linear equation $ax + by = 0$ has a solution. Similar to *linear* is *lineal*, a term used in differential equations. Lineal elements are short parallel line segments that typically appear at the same height; they act as tangents to parallel curves in a direction field. [119, 117]

liter or **litre** (noun), abbreviated with lowercase l or uppercase L to avoid confusion with the number 1: from Greek. The original source is a presumed *lithra* "scale," variants of which are found in languages around the Mediterranean. The Greek *litra* was a unit of weight equivalent to 16 ounces. As recently as the 17th century, *lytre* was used in England as a synonym of *pound*. In 1793, after the French Revolution but during some of its many aftershocks, the government of France decided to adopt a new decimal system of weights and measures in order to break with royalist traditions. The litre was chosen as a new unit of capacity equal to a cubic dekameter; the choice of the name was suggested by the *litron*, an obsolete French unit of capacity derived from the ancient Greek term and equal to one-sixteenth of a bushel. A modern liter is about 1.057 quarts. Using a type of additive augmentation, author Mary Blocksma therefore refers to a liter as a "fat quart." (Contrast the multiplicative augmentation in *million* and *thousand*.) Although extremely common in Europe, the liter is not officially part of the International System of Units. [120]

literal (adjective): from Latin *littera* "a letter." The Latin term may have been borrowed from Greek *diphtherai* "writing tablets." If so, the Indo-European root is *deph*- "to stamp," since Greek *diphtherai* were originally made of stamped animal hides. The disease *diphtheria* is from the same source because false membranes that look like animal hide grow in the throat of infected people. In mathematics, literal notation makes use of letters to represent numbers. The adoption of literal notation in Europe during and after the Renaissance helped mathematicians make rapid progress in algebra.

little (adjective): a native English word, from the Indo-European root *leud*- "small." Possibly related is *lout*, originally "a person who bends or bows low," and who therefore looks little.

lituus, plural *litui* (noun): a Latin word. Among the Romans, the *lituus* was the crooked staff carried by an augur (a man who foretold the future by interpreting signs and portents). The word *lituus* was also applied to a kind of curved trumpet, which must have had a similar shape. Later on, Christians, eager to obliterate or appropriate traces of what it considered paganism, transformed the *lituus* into a bishop's crosier. In biology the lituus is a sea creature with a conical shell. In mathematics the lituus is a curve whose shape is reminiscent of the Roman staff and trumpet. Its polar equation is

$$r^2 = \frac{a}{\theta}.$$

The lituus was first studied by Cotes in 1722.

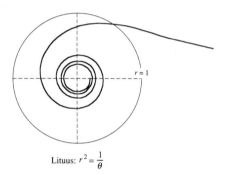

Lituus: $r^2 = \dfrac{1}{\theta}$

The part of the clockwise-
spiraling branch corresponding
to $0 \le \theta \le 10\pi$.

local (adjective); **locate** (verb); **locus**, plural *loci*, (noun): the Latin word *locus* means "place." The Old Latin form was *stlocus*, of unknown prior origin. In geometry, the locus of a point is the set of all places where that point might be found and still fit a certain description. To locate something is to find its proper place. The adjective *local* applies to places close to a give point, as in a local maximum or minimum. The location theorem helps find roots of equations involving continuous functions. [215]

logarithm, often shortened to **log** (noun), **logarithmic** (adjective): from Greek *logos* "reckoning, ratio," and *arithmos* "number." A *logarithm* is literally a "reckoning number." The Indo-European root underlying Greek *logos* is *leg*- "to collect," while that underlying *arithmos* is *ar*- "to fit together," so that the word *logarithm* twice contains the notion of putting things together. An English cognate is *read*, to fit words together to make sense out of them. The Scottish mathematician John Napier (1550–1617) invented logarithms, which he originally called "artificial numbers." By the time he announced his discovery, however, he had adopted the current name. Until the recent advent of electronic calculators and computers, logarithms had been used for hundreds of years as a means of reckoning or calculating, especially when multiplication, division, powers and roots were involved. The shortening of *logarithm* to *log* has given rise to the calculus student's facetious integral

$$\int \frac{d\,\text{cabin}}{\text{cabin}} = \log \text{cabin}.$$

[111, 15]

logic (noun), **logical** (adjective), **logician** (noun), **logicism** (noun): from Late Latin *logica*, from Greek *logike* in the phrase *he logike tekhne* "the art of reasoning." The basic Greek word is *logos* "speech, reasoning, discourse." The Indo-European root is *leg-* "to gather, collect." When you reason, you use logic: you collect your thoughts and gather your statements together to reach a conclusion. A logician is a mathematician who specializes in logic. Logicism is the belief that mathematics is the same thing as symbolic logic. [111]

logistic (adjective): from Greek *logos* "speech, reasoning, discourse," but also "reckoning, ratio, proportion." The Indo-European root is *leg-* "to gather, collect." A logistic curve is defined by the rectangular equation

$$y = \frac{k}{1 + e^{a - bt}}.$$

It takes its name from the fact that as x increases y also increases, and the difference in increments of $\frac{1}{y}$ is proportional to the corresponding difference in $\frac{1}{y}$ itself. A logistic curve is used to model certain types of population growth. [111]

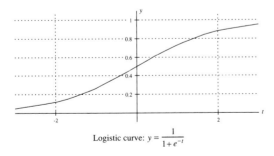

Logistic curve: $y = \dfrac{1}{1 + e^{-t}}$

lognormal (adjective): a combination of *logarithm* and *normal* (qq.v.). A random variable X is said to have a lognormal distribution if $\ln X$ is a normal random variable. [111, 75]

long (adjective), the native English adjective *long* is directly descended from Indo-European and is borrowed neither from Latin *longus* nor from the resulting French *long*. Learned words like *elongation* and *oblong* (qq.v.), however, are borrowed via French from Latin. The Indo-European root underlying *long* is, somewhat surprisingly, *del-*, as found in a learned Greek compound like *dolichocephalic* "having a long head." In pairs of opposite adjectives like *long* and *short*, the "positive" member of the pair also serves as a generic for the entire range of values: the question "how long?" can be answered with a quantity that is very long, such as a light-year, or very short, such as a millimicron. By contrast, the question "how short?" already presumes that the answer will be short. The native English verb *to long for* comes from the adjective *long*: to long for something is to yearn for it as time grows longer and longer. In fact *long* refers to amounts of distance as well as amounts of time. [35]

loop (noun): an English word of uncertain origin. In the 14th century the word referred to a small hole in a wall that a person could shoot through; that's the sense found in *loophole*, which allows you to "shoot your way out" of a contract. During the 14th century the word was also used to describe the part of a string that crossed over itself, perhaps because a loop of string is round, like a hole in a wall. Curves like the folium of Descartes, the limaçon, and the lemniscate possess loops.

low, comparative *lower*, superlative *lowest* (adjective): from Old Norse *lagr* "low," from the Indo-European root *legh-* "to lie, to lay." The etymological connection is still apparent in the phrase "to lie low." A related borrowing from Latin is *litter* (a type of stretcher you lie down on). In mathematics, a low point on a curve is commonly called a minimum. A definite integral possesses a lower and an upper limit of integration. When a fraction is in lowest terms, it has been fully reduced. [112]

loxodrome (noun), **loxodromic** (adjective): the first component is from Greek *loxos* "slanting," of unknown prior origin. The second component is from *dromos* "running," from the Indo-European root *der-* "to run, walk, step." Native English cognates include *tread* and *trap*. In mathematics a *loxodromic* curve "runs slanted" on a surface of revolution in such a way that it cuts every meridian at the same angle. Compare *rhumb*. [38]

lozenge (noun): from Old French, from a presumed Gaulish *lausa* "flat stone." In geometry a lozenge is the same as a rhomb or a diamond, i.e., a non-square rhombus.

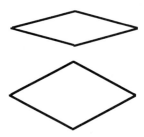

Lozenges.

lucky (adjective): borrowed from a Low Germanic word that was probably a gambling term; of unknown prior origin. Popular wisdom holds that 7 is a lucky number. In number theory the category is broader: it includes all those numbers which are lucky enough to survive the following process of elimination. Start by listing the positive integers: 1, 2, 3, 4, 5, etc. Cross out every second number. After 1, the next surviving integer is 3: go back to the beginning of the list and cross out every third remaining number. After 3, the next surviving integer is now 7: go back to the beginning of the list and cross out every seventh remaining number. The longer you continue the process, the more lucky numbers you'll identify. The lucky numbers apparently have properties in common with the primes. Although common superstition holds that the number 13 is unlucky, mathematically speaking 13 is a lucky number. Contrast *happy*.

lune (noun), **lunar** (adjective): a French word, from Latin *luna* "the moon." The Old Latin form of *luna* is presumed to be *lux-na*, from *lux* "light," from the Indo-European root *leuk-* "light, brightness." Of all objects in the night sky, the moon is the brightest one. An English cognate is *light*, and borrowings from Latin include *luminous* and *luster*. In plane geometry, a lune is a region enclosed by two arcs of circles having the same type of concavity. In solid geometry a lune is a portion of a sphere's surface bounded by two planes passing through a common diameter of the sphere. Some sources claim that the three-dimensional lune is the union of the two semicircular boundaries only, and does not include the interior points on the sphere's surface; those sources refer to the collection of interior points on the sphere's surface as a lunar region rather than a lune. Although two-dimensional lunes were extensively studied in ancient times by people like Hippocrates

of Chios, recent mathematical dictionaries refer primarily to the three-dimensional type of lune. The two-dimensional lune is also known as a crescent. Compare *digon*.

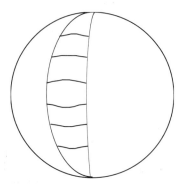

Lune of a sphere.

M

magic (noun): via Latin and Greek, probably from Persian *magus* "member of a priestly caste." The Indo-European root is *magh-* "to be able, to have power," as seen in native English *may*, *might*, and *main*. A related borrowing from Greek is *mechanic*. Magic squares appear frequently in recreational mathematics and in number theory; their "magic" resides in the fact that the sum of every row, column, and diagonal is the same. In a similar fashion, there are magic cubes, magic stars, and various other magic figures. [123]

magnitude (noun): from Latin *magnus* "great," from the Indo-European root *meg-* "great," as found in the English cognate *much*. Related borrowings from Latin are *magistrate* and the month of *May*, named after the "great" Roman goddess Maia; from Sanskrit we have *maharaja* "great king." In mathematics the magnitude of a vector tells how great (= long) the vector is. [129]

main (adjective): a native English word, from the Indo-European root *magh-* "to be able, to have power," as seen also in native English *may* and *might*. A related borrowing from Greek is *machine*. In mathematics the main diagonal of a square matrix is the one that runs from the upper left corner to the lower right corner. [123].

major (adjective): the Latin adjective *magnus* meant "great, mighty" (this last being a native English cognate). The comparative degree was *maior*, literally "greater." The major axis of an ellipse is the greater of the two axes. Two non-diametrically opposed points on the circumference of a circle divide the circumference into two arcs; the greater one is called the major arc. Within a group of people, a majority is any number of people greater than one-half the members. [129]

manifold (noun): a native English compound consisting of *many* and *fold*. The word *many* is from an Indo-European root *menogho-*, of the same meaning, which has produced cognates in other languages but nothing else in English, either directly or through borrowings. *Fold* can be traced back to the Indo-European root *pel-* "to fold." Mathematically, a manifold can be any collection of objects, and in its simplest form is a synonym of *set*. In topology, the meaning of *manifold* is much more "folded up": it is a type of topological space. [160]

mantissa (noun): the Etruscan civilization preceded that of the Romans in northern Italy. In fact the Romans borrowed a lot from the Etruscans, including their alphabet. *Mantissa* is most likely an Etruscan word that meant "a makeweight." A makeweight is an object placed on a scale to attain a desired total; in essence, then, a makeweight is used for subtraction on a balance scale. From the notion of a small object added to a scale, in 16th century English the word *mantissa* developed the meaning "an addition of comparatively small importance." The base-ten logarithm of a number n consists of two parts: the characteristic, which may be any size, is a whole number that indicates the order of magnitude of n; the mantissa, which is always a decimal between 0 and 1, is determined by the digits of n. The characteristic gives n the right order of magnitude; the mantissa is of relatively lesser importance because it merely locates n within a single order of magnitude. (Compare the reversed semantic role inherent in the concept of *significant* digits.) The English mathematician Henry Briggs (1561–1631) introduced the term *mantissa* in 1624.

map (noun, verb), **mapping** (noun): from Medieval Latin *mappa mundi* "map of the world," from *mappa* "sheet," originally "tablecloth" in classical Latin, and *mundus* "world." Other related borrowings include *napkin*, *mop*, and *an apron* (originally *a napron*). In mathematics, if each element x in one set corresponds to a unique element $f(x)$ in another set, there is said to be a mapping from the first set to the second because if the function $y = f(x)$ is

graphed, the resulting points form a kind of map. [124]

margin (noun), **marginal** (adjective): from Latin *margo*, stem *margin-*, "edge, brink, border." The Indo-European root is *merg-* "boundary, border," as seen in native English *mark*. The margin of a page is the white space beyond the printed part of the page. A safety margin is extra space beyond the boundaries that normally mark the beginning of trouble or danger. From that notion of "extra amount beyond" comes the use of margin in finance to represent the difference between the selling price of an item and its cost. Because a margin involves a difference (and hence a rate of change), the term *marginal* has come to designate a derivative.

martingale (noun): the word appears in the French phrase *à la martingale* "in the fashion of the people of Martigue," from the Provençal *martegalo* "person from the town of Martigue, in Provence." Apparently the residents of that town wore pants that fastened in the back, something that people from other towns found ridiculous. The term later came to be applied to a method of betting in which a gambler doubles the stakes after each loss, hoping that an eventual win will more than offset the sum of previous losses. That system is usually ridiculous, however, because a player who suffers sufficiently many consecutive losses will run out of money or hit the bet ceiling of the game. The term martingale has been extended to the study of probability, where it designates a type of stochastic sequence.

mass (noun): from Greek *maza* "kneaded dough." The Indo-European root is *mag-* "to knead, fashion, fit," as seen in native English *make*. A related borrowing from Greek is *magma*, a fiery mass of molten rock. The nontechnical sense of *mass* is "stuff, substance." In physics, mass is defined as a measure of a body's tendency to resist changes in velocity. [122]

match (verb, noun): a native English word based on Old English *gemæcca* "mate, spouse." The Indo-European root is *mag-* "to knead, fashion, fit," as seen in native English *mingle* and *among*. Two people who are matched up and get married fit their lives and personalities together. The term *matchmaker* is a redundancy, etymologically speaking, since both *match* and *make* come from the same root. In statistics, sample pairs are often matched according to various criteria. [122]

material (adjective): from Latin *materia* "material, stuff, wood, matter." The more basic Latin word is *mater* "mother," from the Indo-European root *ma-* "mother." The metaphor explaining the various meanings of Latin *materia* is that "mother Earth" gives birth to material things such as trees. In mathematics a material point (or line or surface) is thought of as having mass, as if it were a physical rather than an abstract object. In logic an if-then statement is sometimes called a material implication; regardless of the abstract form of such an implication, the real-word correctness of the conclusion depends upon the "matter" contained in the antecedent. [121]

mathematics, colloquially shortened to **math** (noun), **mathematical** (adjective), **mathematicize** or **mathematize** (verb), **mathematization** (noun), **mathematician** (noun): Latin *mathematica* was a plural noun, which is why *mathematics* has an *-s* at the end even though we use it as a singular noun. Latin had taken the word from Greek *mathematikos*, which in turn was based on *mathesis*. That word, which was also borrowed into English but is now archaic, meant "mental discipline" or "learning," especially mathematical learning. The Indo-European root is *mendh-* "to learn." Plato believed no one could be considered educated without learning mathematics. A *polymath* is a person who has learned many things, not just mathematics. A mathematician is a person who specializes in mathematics. To mathematicize or mathematize something is to treat it in a mathematical manner. On the other hand, the *-math* in *aftermath* has nothing to do with mathematics, but comes from the native English verb *to mow*, so *aftermath* meant originally "what is left over after a field is mowed." [135]

mathlete (noun): a combination of *mathematics* (*q.v.*) and *athlete*. Ancient Greek *athletes* meant "contestant." The more basic Greek word *athlon* "award, prize" is of unknown prior origin. *Mathlete* is a name commonly applied to secondary school students who compete in mathematics contests. [135]

mathophile (noun), **mathophilia** (noun): from *math*, the colloquial form of *mathematics* (*q.v.*), and Greek *philos* "dear, loving." The Indo-European root may be *bhilo-* "dear, familiar." A mathophile is a person who loves mathematics; not as sinister as necrophilia, mathophilia is the love of mathematics. [135]

matrix, plural *matrices* (noun): the Indo-European root *ma-* "mother" is presumed to derive from the sound that babies make when calling for their mothers or for milk. The addition of the Indo-European

kinship suffix *-ter* resulted in Latin *mater*, which is cognate to English *mother*. When the Latin suffix *-ix*, indicating a female agent, was added to Latin *mater*, the resulting *matrix* meant "a female animal kept for breeding." Referring to people, it meant "a female about to become a mother," in other words "a pregnant woman." From there the word came to be applied to the organ most responsible for making a woman a mother, namely the womb. Metaphorically *matrix* later came to mean anything that generates something else, just as the womb generates a child. In mathematics a matrix is womb-like in that it can be used to generate geometric or algebraic transformations. Also, since the womb surrounds a growing baby, *matrix* came to mean "the material that surrounds the development of something," not necessarily a baby. That also makes sense because when we write a matrix we write an array of numbers surrounded by large brackets or parentheses. In a typical example of semantic inflation, employees of large computer companies have begun referring to any chart or table as a matrix. [121, 236]

matroid (noun): a modern coinage based on Latin *matrix* (see previous entry) and the Greek-derived suffix *-oid* (*q.v.*) "looking like," or more loosely "having something in common with." In a nonsingular square matrix the rows of elements are linearly independent. By analogy, any linearly independent set of subsets of a finite universal set is said to be a matroid. [121, 244]

maximum, plural *maxima* (noun), **maximal** (adjective), **maximize** (verb): the Latin adjective *magnum* meant "great, big." Related borrowings from Latin include *magnify* and *master* (someone who is greater than you). The superlative degree of the adjective *magnum* was *maximum* (for *mag-simum*), meaning "the greatest, the biggest, the most." The maximum is therefore the greatest value something can take on. To maximize something is to cause it to take on its greatest possible value. The Indo-European root is *meg-* "great," as seen in the English cognate *much*. [129]

maze (noun): a native English word whose meaning as a verb in the 13th century was "to stupefy, to daze." The early meaning is more clearly seen in the slightly longer form of the word, *amaze*. Only in the 14th century did *maze* come to mean "a network of paths." The ultimate source of the word is unknown.

mean (noun, adjective): from French *moyen*, in turn from the Latin adjective *medianus*, a derivative of the more basic *medius* "in the middle." The Indo-

European root is *medhyo-* "middle," as found in native English *mid* and *amid*. The arithmetic mean of a group of numbers is a type of "middle" value or average. With regard to proportions, the mean proportional between two positive numbers *a* and *b* is the positive number *m* such that *a* : *m* : : *m* : *b*; the use of the term *mean* stems from the position of the two *m*'s in the middle between the *a* and the *b*. In solid geometry, an ellipsoid typically has three axes of different lengths; the axis that is neither the longest nor the shortest is called the mean axis. Don't confuse *mean* as a mathematical term with the unrelated English *mean* in the sense of "nasty," although it might be considered mean to ask someone all the things that *mean* can mean. [127]

measure (verb, noun): from French *mesure* "a measure," from Latin *mensura* "a measuring," from *mensus*, past participle of the verb *metiri* "to measure." The Indo-European root is *me-* "to measure." The related Greek noun *metron*, also meaning "measure," is the source of the *meter* used in the metric system. A native English cognate is *meal*, originally "a measure, a mark, a sign, or an appointed time," especially "a time for eating"; *meal* is also found in *piecemeal* "a piece at a time." Another English cognate is *month*, a measure of time according to the phases of the *moon*. Historical systems of measurement were a hodge-podge of weird units derived from such strange things as elbows and barleycorns. Only after the French Revolution, when the metric system was first established, was order slowly brought to the chaos of measurement. On the other hand, so many new units and prefixes have been added to the International System that a person is hard-pressed to recognize them all. [126]

medial (adjective): from the Latin adjective *medius* "in the middle." The Indo-European root is *medhyo-* "middle," as seen in native English *mid-* as well as in the borrowed Latin word *medium*. A medial

triangle is one whose vertices are the midpoints of the sides of another triangle. Old arithmetic books taught something called alligation medial, which, in the words of one such book "is when the quantities and prices of several things are given, to find the mean price of the mixture compounded of those things." [127]

median (noun): from the Latin adjective *medianus*, a derivative of the more basic *medius* "in the middle." The Indo-European root is *medhyo-* "middle," as seen in the native English cognates *amid* and *middle*, as well as in borrowings such as *mediocre* and *Mediterranean*. In the world of driving, a median is a strip of land or grass in the middle of two paved lanes. In statistics the median is the value that stands in the middle of a set of data; the median divides the data into two equal groups, one with values less than or equal to the median, and the other with values greater than or equal to the median. In a triangle, a median is a line connecting a vertex to the midpoint of the opposite side. In a trapezoid, a median is a line connecting the midpoints of the two nonparallel sides; it is also known as a midline (*q.v.*). [127]

mediant (noun): from Latin *mediant-*, present participial stem of *mediare* "to be in the middle." The more basic Latin word is *medius* "in the middle." The Indo-European root is *medhyo-* "middle," as seen in the native English cognates *mid* and *middle*, as well as in borrowings from Latin such as *Medieval* and *media*. In mathematics, given positive integers *a*, *b*, *c*, *d*, the *mediant* of the two fractions $\frac{a}{b}$ and $\frac{c}{d}$ is defined to be $\frac{a+c}{b+d}$. The mediant always lies "in the middle" between the values of the two original fractions. Calculating a mediant allows students to "add" fractions in the way so many of them always wanted to. The mediant makes sense in situations where conventional addition of fractions doesn't. For example, if a baseball player has 2 hits out of 4 at-bats in one game, and 1 hit out of 3 at-bats in the next game, the player's mediant of $\frac{2+1}{4+3}$ corresponds to 3 hits in 7 at-bats for both games. [127, 146]

mediation (noun): from Latin *medius* "in the middle." The Indo-European root is *medhyo-* "middle," as seen in native English *mid* and *middle*. If something is cut down the middle, each of the resulting parts is half as big as the original object, so *mediation* means "the process of taking half of something." The term is used in describing a method of multiplying two numbers that involves the repeated doubling (duplation) of one number and the repeated halving (mediation) of the other. The method is sometimes

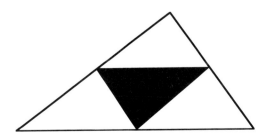

The medial triangle of the outer triangle is shaded.

called the Egyptian algorithm because the ancient Egyptians apparently used it; it has also been called Russian peasant multiplication. [127]

mediator (noun): the first element of this Latin compound is from *mediatus*, past participle of *mediare* "to be in the middle." The Indo-European root is *medhyo-* "middle," as seen in native English *mid* and *middle*. The Latin agental suffix *-or* indicates a male person or thing that performs the indicated action. In nonmathematical English, a mediator is someone who is "in the middle" trying to reconcile quarreling people. In geometry a mediator is a perpendicular bisector of a line segment; it is "in the middle" between the two endpoints of the line segment. [127, 153]

mega-, abbreviated *M* (numerical prefix): from Greek *megas* "great," from the Indo-European root *meg-* "great," as found in native English *much*. Related borrowings from Greek include *megaphone* and *megascopic* as well as the trendy word *megatrend*. In the metric system, the prefix *mega-* multiplies the following unit by one million. The prefix has even appeared with the numerical value of one million in nonscientific English words like *megabucks* and *megadeath*. [129]

member (noun): from Latin *membrum* "limb, member of the body." The Indo-European root is *mems-* "flesh, meat." A related borrowing from Latin is *membrane*. From the literal meaning of *member* as a body part came the figurative sense "part of an organization." We can therefore speak of a member of a club, a member of the cast in a play, etc. In algebra the two "sides" of an equation or inequality are sometimes referred to as members. The elements that make up a set are frequently called the members of the set. For any member x of a set A we write $x \in A$, "x is a member of set A." The $\in$ symbol is really the Greek letter epsilon, which is the first letter of the Greek word *estis* "is." Compare *belong*.

mensuration (noun): from Latin *mensura* "a measuring," from *mensus*, past participle of the verb *metiri* "to measure." The Indo-European root is *me-* "to measure." Mensuration is the act of measuring; it is used particularly of geometric quantities. The term has also been used to refer to that part of mathematics which establishes rules for measuring lengths, areas, volumes, etc. The similar word *menstruation* is related: it comes from Latin *mensis*, "month," literally "the measure of time by the cycles of the moon." In fact English *moon* and *month* are cognates. [126]

meridian (noun): from Latin *meridianus* "having to do with midday," from the compound *meridies* "midday." The first element, *meri-*, was originally *medi-*, from the Indo-European root *medhyo-* "middle," which is an English cognate. The second component is Latin *dies* "day," from the Indo-European root *deiw-* "to shine." A Germanic cognate is *Tiu*, the sky god, found in the first (but not the second) part of English *Tuesday*. Our familiar abbreviations A.M. and P.M. stand for Latin *ante meridiem* and *post meridiem*, literally "before noon" and "after noon." In geodesy, a line of longitude on the earth's surface is known as a meridian because at midday the sun appears to be at its highest point as seen from all points along the meridian. In mathematics the concept of a meridian has been extended from spheres to generalized surfaces of revolution: a meridian is the space curve that results from the intersection of a surface of revolution with a plane containing the axis of revolution. [127]

meromorphic (adjective): the first component is from Greek *meros* "part, division," from the Indo-European root *(s)mer-* "to get a share." A related borrowing from Greek is *polymer*. The second component is from Greek *morphos* "shape, form," of unknown prior origin. In mathematics a function of a complex variable is said to be meromorphic if it is analytic in a given domain except for a finite number of poles. [137]

mesokurtic (adjective): the first component is from Greek *mesos* "middle," from the Indo-European root *medhyo-* "middle." The second component is from Greek *kurtos* "convex," from the Indo-European root *(s)ker-* "to bend, to turn." In statistics a distribution is said to be mesokurtic if it is as heavily concentrated around the mean as a normal distribution is. The category *mesokurtic* is "in the middle" between the more extreme categories *platykurtic* and *leptokurtic*. See more under *kurtosis*. [127, 198]

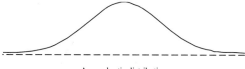

A mesokurtic distribution.

metacompact (adjective): the first element is from Greek *meta* "beside, beyond." The Indo-European root is *me-* "in the middle." The second element is Latin-derived *compact* (*q.v.*). A metacompact space is a type of topological space that involves the notion of point-finiteness. It goes "beyond" the notion of simple compactness. [127, 103, 156]

metamathematics (noun), **metamathematical** (adjective): the first component is from Greek *meta-*, in the sense "after, beside, beyond," with a shift in meaning from the Indo-European root *me-* "middle." The second component is *mathematics* (*q.v.*). If mathematics is taken to be a set of meaningless (= purely formal) statements, as David Hilbert (1862–1943) believed, then metamathematics is a set of meaningful statements about the meaningless statements of mathematics. For example "1 + 2 = 3" is a formal mathematical statement, whereas "It's easy to see that 1 and 2 add up to 3" is a metamathematical statement. In a less formal sense, students acquire lots of metamathematical knowledge: they know, for instance, that the homework problems at the end of a section in their textbook almost invariably involve the topic explained in that section. [127, 135]

meter, also spelled **metre**, (noun), **metric** (adjective, noun), **metrical** (adjective): from Greek *metron* "measure, length." The Indo-European root could be either *me-* "to measure," or *med-* "to take appropriate measures." As first used in English, *meter* referred to poetic rhythms such as pentameter and hexameter. After the Revolution of 1789, the French adopted a new system of weights and measures as a way of breaking with royalist traditions of the past. The meter was chosen as a unit of length equal to one ten-millionth of the distance from the North Pole to the Equator. One meter equals approximately 39.37 inches, which is 3.37 inches longer than a yard. Using a type of additive augmentation, author Mary Blocksma therefore refers to a meter as a "fat yard." In abstract mathematics a metric space is a set of points such that to each pair of points a positive number called a distance (= measurement) is assigned. [126, 128]

method (noun): a Greek compound. The first component is from the Indo-European root *me-* "in the middle of," though, as is common with prepositions, the meaning in Greek was extended to include "between, with, alongside," and ultimately "after, beyond." The second component is *hodos* "a going, a trip, a way," from the Indo-European root *sed-* "to go." A method is literally a way of "going after" something, particularly knowledge. René Descartes' (1596–1650) famous book was entitled *Discours de la méthode* (*Discourse on Method*). Some methods studied in mathematics are the method of least squares and the method of infinite descent. In non-

mathematical English, a methodical person is one who performs a task by systematically following a method. [127, 183]

metrizable (adjective): from the root of Greek *metron* "a measure," plus a Greek verbal suffix *-ize*, plus the Latin suffix *-able* "capable of being done." See more under *-able*. A topological space is said to be metrizable if there exists a transformation between it and a metric (*q.v.*) space. [126, 69]

micro-, abbreviated μ (numerical prefix), **micron** (noun): from Greek *mikros*, a variant of *smikros* "small." The Indo-European root may be *sme-* "to smear." The semantic connection would be that when something is smeared, little pieces of it are spread around. The *Oxford English Dictionary* conjectures that the Greek word may come from the Indo-European root *smelo-*, with the same meaning as the native English cognate *small*. As a nonmathematical prefix, *micro-* refers to small things; it is found in such borrowed words as *microscope* and *microfiche*. In the decimal system, the prefix *micro-* is used with the specific value of one one-millionth of the unit to which it is attached. A micron is one one-millionth of a meter. [199]

midline (noun): a compound of native English *mid* and the French borrowing *line* (*q.v.*) The Indo-European root underlying *mid* is *medhyo-* "middle," which is also an English cognate. A midline is literally "a line in the middle." In geometry the midline of a trapezoid is the line segment connecting the midpoints of the two nonparallel sides. The midline is parallel to the two bases of the trapezoid and is located midway between them. The midline of a trapezoid is also known as a median (*q.v.*). [127, 119]

midpoint (noun): a compound of native English *mid* and the French borrowing *point* (*q.v.*). The Indo-European root underlying *mid* is *medhyo-* "middle," which is also an English cognate. A midpoint is literally "a point in the middle." In geometry we may speak of the midpoint of a line segment or of an arc of a curve. [127, 172]

mile (noun): from Latin *milia passuum* "a thousand paces," the first word coming from *mille* "a thousand"; a Roman mile was originally a thousand steps long. The Roman mile is estimated to have measured 1618 yards, while a modern mile in England and the United States is 1760 yards long. The Indo-European root may have been *gheslo-* "a thousand," as seen also in Greek *kilo-*. [72]

mill (noun): from Latin *millesimus* "one-thous-andth." A mill is an American monetary unit equal to one one-thousandth of a dollar, or one-tenth of a cent. The monetary unit has been in use since the 18th century, although with inflation a mill is worth so little that it is rarely used now. [72]

millennium, plural *millennia* (noun): the first element is from Latin *mille* "a thousand." The Indo-European root may have been *gheslo-* "a thousand," as seen also in Greek *kilo-*. The second element is from Latin *annus* "year," from the Indo-European root *at-* "to go." The original sense must have been that a year is the cycle we go through until the seasons and the night sky are the same again. A similar meaning is found in the Latin-derived *anniversary*, literally "the turning of the year." A millennium is a period of a thousand years. The word was patterned after *biennium* and *triennium*, periods of two and three years, respectively. [72, 17]

milli-, abbreviated *m* (numerical prefix): from Latin *mille* "one thousand," although the prefix is used with the fractional meaning "one one-thousandth." The Indo-European root may have been *gheslo-* "a thousand." In 1791 the Paris Academy of Sciences chose the metric system prefixes for the first three negative powers of ten; all three (*deci-* and *centi-* being the first two) came from the appropriate Latin number words and all three ended in *-i*. In recent years all of the new submultiple prefixes have ended with *-o*, and they are no longer necessarily chosen from Latin. [72]

milliard: see *billion*.

million (numeral): from Old French, which most likely borrowed the term from Italian. The source is ultimately Latin *mille* "thousand," plus the augmentative suffix *-on*, so a million is literally "a big thousand." The word first appeared in English in the 14th century, though the French form is documented a century earlier. Compare the literal meaning of *thousand*. After the initial augmentation via the *-on* suffix, higher powers were formed by replacing the first part of the word with *bi-*, *tri-*, etc. [72, 151]

minimax (noun): a combination of *mini-*, from *minimum*, and *max-*, from *maximum* (*qq.v.*). In calculus a minimax is another name for a saddle point because the point is a minimum on the cross-sectional curve in one plane but a maximum on the cross-sectional curve in a perpendicular plane. [131, 129]

minimum, plural *minima* (noun), **minimize** (verb): *minimum* is a Latin superlative adjective corresponding to the comparative *minor* "less," so the *minimum* is literally the least or smallest value something can have. The Indo-European root is *mei-* "small." A related borrowing from Latin is *minister* (someone who is supposed to make himself less than you and serve you). [131]

minor (adjective): a Latin comparative adjective meaning "less." It is quite similar to the comparative adverb *minus* (*q.v.*). The Indo-European root is *mei-* "small." In law, a minor is a person whose age is less than that required to be considered an adult. In geometry, the minor axis of an ellipse is the lesser of the two axes; similarly, but not quite grammatically, the shortest of the three axes of an ellipsoid is called the minor axis. Two non-diametric points on a circle divide the circumference into two arcs; the smaller one is called the minor arc. [131]

minuend (noun): from Latin *minuere* "to make smaller," with the suffix *-nd-*, which creates a type of passive causative, so that Latin *numerus minu-endus* meant "the number to be made less." In the statement $5 - 3 = 2$, the 5 is the minuend because it is the number that is to be made smaller. The Indo-European root is *mei-* "small." A related borrowing from Greek is *meiosis*, which diminishes (another related borrowing) the number of chromosomes in a cell. [131, 139]

minus (preposition, adjective): a Latin comparative adverb meaning "less," from the Indo-European root *mei-* "small." When we subtract a positive quantity from an amount, the result is less than before. The word *minus* therefore came to represent the operation of subtraction: five minus three makes two. We can also use our native English synonym instead: five less three makes two. During the Renaissance, the word *minus* was commonly abbreviated $\overline{m}$ throughout Europe, and the bar over the *m* is plausibly assumed to have given rise to our minus sign. Our modern plus and minus signs began to be adopted in northern Europe by the 15th century, but southern European countries were slower to accept them. English speakers born before the 1950's are likely to refer to a number like -3 as "minus three" rather than "negative three"; the longer version is usually now taught as the only correct one. The older usage is still fairly entrenched, however, as seen especially in temperatures: $-10°$ is usually read as "minus ten degrees." Some children create (and as adults occasionally continue to use) a verb *to minus*. Non-mathematical English also uses *minus* as a noun, as in a statement like "not having a college degree is a minus." [131]

minute (noun): via Latin *minutus* "little, small, minute," from *minus* "less," from the Indo-European root *mei-* "small." When a degree is divided up into 60 parts, each of the "lesser" parts is known as a minute. In the same way, when an hour is divided up into 60 parts, each of the lesser parts is a minute. A system in which fractions of a degree are expressed as decimals is slowly replacing the traditional system of minutes and seconds. The adjective *minute*, stressed on its second syllable, preserves the original Latin meaning "small." Confusion between the adjective and the noun has sometimes arisen: for example, Chopin's "Minute Waltz" is short, but it can't reasonably be played in one minute. [131]

mirror (noun): from French *miroir* "a mirror," from a presumed Vulgar Latin *miratorium*. The Classical Latin source was *mirari* "to wonder at, to be amazed," from *mirus* "astounding, wonderful, amazing, admirable." The semantic connection is that people stare at amazing things, so the meaning of *mirari* shifted from being amazed to looking at something. The Indo-European root is *smei-* "to laugh, to smile," because something wonderful often makes us smile or laugh. In mathematics, when a curve is reflected in a line or a plane, we describe the transformed curve as a mirror image of the original. The analogy, of course, is with the way objects are reflected in a mirror.

mixed (adjective): from the French adjective *mixte*, from Latin *mixtus*, past participle of *miscere* "to mix, to mingle, to blend." The Indo-European root is *meik-* "to mix." The native English word *mash* may be from the same source. Related borrowings from Latin include *miscellaneous* and *miscegenation* (interracial marriage). From French come *mêlée*, *meddle*, and *medley*, and from Spanish comes *mestizo*. In arithmetic, a mixed number consists of a whole number and a fraction. Similarly, a number like 18.907 is called a mixed decimal because it mixes a whole number and a decimal fraction. In calculus, if $z = f(x, y)$, then

$$\frac{\partial^2 z}{\partial x \, \partial y} \quad and \quad \frac{\partial^2 z}{\partial y \, \partial x}$$

are called the mixed second partial derivatives of the function because the variables with respect to which the derivatives are taken are mixed.

mnemonic (adjective, noun): from Greek *mnemonikos* "mindful," from the Indo-European root *men-* "to think," as seen in native English *mind*. Other related borrowings from Greek are *amnesia* and *amnesty*. From Latin come *memento*, *reminiscent*, and *mental*. A mnemonic is a device that helps you remember something else. In trigonometry, for example, the double-angle formulas $\sin^2 \theta = \frac{1}{2}(1 - \cos 2\theta)$ and $\cos^2 \theta = \frac{1}{2}(1 + \cos 2\theta)$ can be difficult to memorize because the right sides of the equations are identical except for the sign between the two terms. An aural mnemonic to keep the two formulas apart is to note that the sound "ine" in *sine* is the same as the sound "ine" in *mine-us*. In algebra, students sometimes get confused when calculating a slope because they don't remember whether the change in x goes on the top or the bottom of the fraction. A visual mnemonic involves noticing that the letter x has two "legs" and can balance when holding up another letter on top of it, whereas the letter y would fall over if it tried to support another letter on top of it. Students can remember that the slope of a h0riz0ntal line is 0 because each o in *horizontal* resembles the numeral 0. To remember that the value of π through the first seven decimal places is 3.1415926, count the number of letters in each word of the mnemonic "Now I have a great statement to relate." [134]

mode (noun), **modal** (adjective): *mode* is a French word, from Latin *modus* "standard, measure." The Indo-European root is *med-* "to take appropriate measures." Related borrowings include *remedy* (appropriate measures to effect a cure) and *modern* (appropriately up-to-date). French *mode* took on the sense "fashion," which means a current (and fleeting!) standard or manner of dressing. What is fashionable is presumably the thing most popular, most often seen, at a given time. In statistics the mode is the number that occurs most often, or is "most popular," in a set of discrete data. Similarly, a modal class interval is the one with the highest bar in a histogram. [128]

model (noun, verb): from French *modelle*, from a presumed Late Latin *modellus* instead of *modulus*; Latin *modulus* "a small measure," was a diminutive of *modus* "a standard by which something is measured," from the Indo-European root *med-* "to take appropriate measures." A person who acts as a model sets a standard for other people to follow. (Consequently, the currently almost unavoidable term *role model* is redundant; the single word *model* used to be and still should be sufficient.) In mathematics a model of a situation is a pattern that is supposed to represent (= set a standard for) a real situation. From the noun *model* we now have the verb *to model* "to find a model" that does a good job of representing a situation. [128, 238]

modulus, plural *moduli* (noun), **modulo** or **mod** (noun phrase), **modular** (adjective): a Latin diminutive of *modus* "standard, measure," from the Indo-European root *med-* "to take appropriate measures." The root appears in native English *mete*, as in the expression "to mete out justice." In mathematics the modulus is the length or measure of a complex number. In modular arithmetic, the modulus or *mod* is the number being used as a standard against which to measure any other number; for instance in mod 12, or clock time, 15 o'clock is the same as 3 o'clock, because 15 is 3 units beyond the standard measure of 12. In the more formal notation of number theory, the word *modulo* is often used; it is a Latin form of *modulus* that means "in modulus." For example, $3 \equiv 1$ *modulo* 2, or, with further abbreviation, $3 \equiv 1(mod\ 2)$. The word *modulus* may also mean the number by which a logarithm in one base must be multiplied to convert it to the corresponding logarithm in another base; the modulus is the standard factor that brings about the conversion. [128, 238]

modus ponens (noun): Latin *modus* "standard, measure" is from the Indo-European root *med-* "to take appropriate measures." The second word, *ponens*, is the present participle of Latin *ponere* "to put." (See more under *component*.) In logic, *modus ponens* is a standard form of argumentation in which you "put down" the antecedent of an if-then statement and conclude the occurrence of the consequent of that if-then statement. *Modus ponens* is akin to detachment (*q.v.*). [128, 14, 146]

modus tollens (noun): Latin *modus*, which meant "standard, measure," is from the Indo-European root *med-* "to take appropriate measures." The second word, *tollens*, is the present participle of Latin *tollere* "to take away," from the Indo-European root *tela-* "to lift, support." In logic, *modus tollens* is a standard form of argumentation in which you "take away," that is, negate, the consequent of an if-then statement and conclude the negation of the antecedent of that if-then statement. *Modus tollens* is equivalent to using the contrapositive of the original if-then statement. [128, 224, 146]

molding (noun, adjective): via Old French *molde*, from Latin *modulus*, a diminutive of *modus* "measure, standard." The Indo-European root is *med-* "to take appropriate measures"; it appears in native English *mete*, as in the expression "to mete out justice." A mold is a hollow form used to "measure out" objects. A molding is a type of three-dimensional decoration that must originally have been made from a mold; it is typically in strip form and is found along the edge of a wall, a door, etc. By analogy, in mathematics a molding surface is generated by a plane curve whose plane rolls (without slipping) on a cylinder. If the cylinder degenerates to a straight line, a molding surface becomes a surface of revolution. [128, 238]

moment (noun): from Latin *momentum*, shortened from the earlier form *movimentum* "a movement, a motion," from the verb *movere* "to move." The Indo-European root is *meua-* "to push away." Related borrowings from Latin include *motor* and *mob* (a group of people moving recklessly). The Latin noun *momentum* came to refer to a weight or object just big enough to cause movement in a balance scale, hence "a small amount"; that sense is still seen in the way *moment* is used in physics. In nontechnical English, the word *moment* came to refer to a small movement of time rather than mass, hence the current meanings "a brief period of time, an instant." [136]

money (noun): from Old French *moneie* (modern French *monnaie*), from Latin *moneta* "mint, the place where money is coined." In ancient Rome, money happened to be coined in the temple of the goddess Juno, also known as Juno Moneta. The epithet *moneta* "the warner," is from the verb *monere* "to remind, advise, warn." The Indo-European root is *men-* "to think," as seen also in native English *mind*. Juno had come to be called "the warner" after the cackling of her sacred geese warned the Romans that the Gauls were trying to wage a surprise attack on them. As a learned borrowing, the adjective *monetary* remains closer to *moneta* than does *money*. Like old systems of weights and measures, traditional monetary systems were a hodge-podge of unpredictable units, but with the almost universal adoption of the decimal system, there is strong pressure for monetary units to be based on powers of ten. In 1971 even the venerable British system of pounds, shillings and pence was decimalized. [134]

mon(o)- (prefix): from Greek *monos* "alone, only, single." The Indo-European root is *men-* "small, isolated." The many related nonmathematical words incorporating Greek *mono-* include *monastery* (from the *monk* who chooses to be alone), *monocle*, *monogamy*, *mononucleosis*, *monopoly*, *monorail*, *monosyllable*, and *monotone*. Native English *minnow*, a type of small fish, may also be related. As used in mathematics, the prefix *mon(o)-* conveys the notion of oneness. [133]

monic (adjective): from Greek *monos* "alone, only, single," (see previous entry). In algebra a polynomial is said to be monic if all its coefficients are integers and the leading coefficient is 1. The trial-and-error method of factoring trinomials taught in American schools may be acceptable for a monic trinomial, but when it is extended to a non-monic trinomial like $20x^2 - 21x - 54$, the trial-and-error approach may best be described as manic. [133]

monodigit (adjective): from Greek-derived *mon(o)-* "one, single," and *digit* (*qq.v.*). The term *monodigit* doesn't generally refer to a one-digit integer like 4, but rather to an integer in which only one digit has been used, even if repeated. An example of a monodigit number is 444. Compare *repdigit*. [133, 33]

monodromy (noun): the first element is from Greek-derived *mon(o)-* (*q.v.*) "one, single." The second element is from Greek *dromos* "a running," from the Indo-European root *der-* "to run, walk, step." In complex analysis, the monodromy theorem is so named because analytic continuation (= "running") around any closed curve leads back to the original (= one and the same) function element. [133, 38]

monogenic (adjective): the first element is from Greek-derived *mon(o)-* (*q.v.*) "one, single." The second element is from the Indo-European root *gen-* "to give birth," as seen in borrowings such as *gene* and *genesis*. In analysis, monogenic analytic functions can be generated (another cognate) from a given (= one) function element. In other words, the function's rate of variation in any direction is independent of that direction because it is independent of the mode of change of the variable. In geology, the term *monogenic* refers to a rock in which all the parts are of the same nature. In theology, monogeny is the belief that all of mankind sprang from one pair of ancestors. [133, 64]

monogonal (adjective): the first element is from Greek-derived *mon(o)-* (*q.v.*) "one, single." The second element is from the Indo-European root *genu-* "bend, angle" as seen in native English *knee*. In a

monogonal tiling, the figure composed of a vertex and all incident edges is congruent to every other such figure in the tiling. Speaking loosely, only one angular configuration is involved throughout the tiling. [133, 65]

monohedral (adjective): the first element is from Greek-derived *mon(o)-* (*q.v.*) "one, single." The second element is from Greek *hedra* "base, seat." The prehistoric Greek form was *sedra*, in which the Indo-European root *sed-* "to sit" can be recognized. The literal meaning of *monohedral* is "having [only] one base." In mathematics a monohedral tiling involves tiles of only one size and shape (= "base"), though reflections are allowed. Most mathematical words ending in *-hedral* or *-hedron* have to do with shapes in space; *monohedral* is one of the rare words in which the two-dimensionality of the Greek root has been maintained. [133, 184]

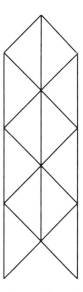

A monohedral
tiling pattern.

monoid (noun): the first element is from Greek-derived *mon(o)-* (*q.v.*) "one, single." The second element is the Greek-derived suffix *-oid* (*q.v.*), literally "looking like," but occasionally used in the looser sense "having something to do with." A monoid is the same as a semigroup. The name may be explained by noting that a monoid fulfills only one part of the definition of a group: a monoid is a set of elements and a binary operation that is closed and associative; however there need be no identity element or inverse elements. [133, 244]

monomial (noun): the first component is from Greek-derived *mon(o)-* (*q.v.*) "one, single." There is disagreement about the origin of the second component. One explanation involves Greek *nomos*, which meant many things: "usage, custom, law, division, portion, part." In that case, a monomial is a mathematical expression consisting of one part. Another explanation involves Latin *nomen*, cognate to the English *name*, so that a monomial is an expression involving one "name," i.e., one term. In biology a monomial is a taxonomic name that consists of just one word. [133, 143, 145]

monomino (noun): the first component is from Greek-derived *mon(o)-* (*q.v.*) "one, single." The rest of the compound is from *domino*. Whereas a standard domino contains two square sections with numbers on them, a monomino contains a single square section with no numbers on it. With higher-order -ominoes such as pentominoes, a player is challenged to combine the pieces to produce simple pictures of certain shapes or recognizable objects. However, because there is just one type of monomino, namely a square, monominoes are unchallenging. See explanation under *polyomino*. [133, 37]

monomorph (noun), **monomorphic** (adjective): the first element is from Greek-derived *mon(o)-* (*q.v.*) "one, single." The second element is from Greek *morph-* "shape, form, beauty, outward appearance," of unknown prior origin. In number theory, an integer that can be expressed in the form $x^2 \pm Dy^2$ in only one way is said to be a monomorph. Contrast *antimorph* and *polymorph*. [133, 137]

monotonic (adjective): the first component is from Greek-derived *mon(o)-* (*q.v.*) "one, single." The second element is from Greek *tonos* "sound, tone." If you listen to a single sound it quickly becomes monotonous. The word *monotonic*, a variant of *monotonous*, stresses the fact that a phenomenon behaves in only one way. Mathematically, if a function increases monotonically, the function either rises or holds steady at every point, but never decreases. The Indo-European root underlying Greek *tonos* is *ten-* "to stretch." A musical tone is created when a stretched string is struck or plucked. A Native English cognate is the adjective *thin*, which describes an object that has been stretched. [133, 226]

month (noun): a native English word based on *moon*. The Indo-European root is *me-* "to measure." In ancient times people often measured time by the moon; a month is therefore one moon-measure. A related

borrowing from Latin is *menstruation*, which measures the female reproductive cycle in what are approximately moon-units. [126]

more, **most** (adjective, adverb): native English words that are, respectively, the comparative and superlative degrees of an adjective that no longer exists on its own except in British dialectal *mo*. The unrelated but conveniently alliterative words *much* and *many* are now taken to be the basic words to which the comparative *more* and the superlative *most* correspond. The Indo-European root is *me-* "big." Compare the development of *less*. In mathematics "more than" is a synonym of "greater than." [125]

morphism (noun): from Greek *morph-* "shape, form, beauty, outward appearance," of unknown prior origin. In abstract mathematics a morphism is one of the two classes of a mathematical category. More specifically, see *automorphism*, *diffeomorphism*, *endomorphism*, *epimorphism*, *hom(e)omorphism*, and *isomorphism*. [137]

multifoil (noun): the first part of the compound is from Latin *multi* "many," from the Indo-European root *mel-* "strong, great." The second component is from Old French, from Latin *folium* "leaf"; aluminum foil, for example, is so named because it is a "leaf" of aluminum. The Indo-European root is *bhel-* "to thrive, to bloom," as seen in *flower*, which

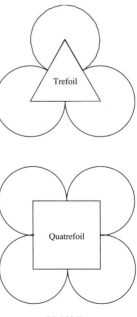

Multifoils.

is borrowed from French, as well as native English *blossom*. In mathematics a multifoil is a plane figure made from adjacent congruent arcs of a circle placed around the vertices of a regular polygon; the figure is made up of many "leaves." Depending on the number of leaves it possesses, a multifoil may be known as a *trefoil*, *quatrefoil*, *cinquefoil*, *hexafoil*, etc. [132, 21]

multigrade (noun): the first part of the compound is from Latin *multi* "many," from the Indo-European root *mel-* "strong, great." The second component is from Latin *gradus* "a step," from the Indo-European root *ghredh-* "to go, to walk." In number theory, an equation like $0^n + 5^n + 6^n + 16^n + 17^n + 22^n = 1^n + 2^n + 10^n + 12^n + 20^n + 21^n$, which is true for more than one value of the exponent n, is called a multigrade. In this example, as n "walks" through "many" consecutive values, namely 1, 2, 3, 4 and 5, the equation remains true. Depending on the number of values of n for which the equation is true, the multigrade may be called by a more specific name: the example cited is a pentagrade. [132, 73]

multigraph (noun): the first part of the compound is from Latin *multi* "many," from the Indo-European root *mel-* "strong, great." The second component is from Greek-derived *graph* (*q.v.*). In graph theory, a multigraph is a graph in which more than one line can join two points. [132, 68]

multimodal (adjective): from Latin *multi-* "many" and *modus* "standard, measure." In statistics a multimodal distribution has several values that appear more prominently than others. If the distribution is continuous, the curve representing the distribution has several relative maxima. [132, 128]

multinomial (noun, adjective): the first part of the compound is from Latin *multi* "many," from the Indo-European root *mel-* "strong, great." One explanation for the second part of the compound involves Greek *nomos*, which meant many things: usage, custom, law, division, portion, part. In that case, a trinomial is a mathematical expression consisting of three parts. A second and more likely correct explanation involves Latin *nomen*, cognate to the English *name*, so that a multinomial is an expression involving many names, i.e., terms. As a noun, the Latin-Greek *multinomial* has now largely been replaced by *polynomial*, in which both roots are Greek. As an adjective, we still use *multinomial* to refer to the coefficients in the expansion of $(x_1 + x_2 + \cdots + x_k)^n$. [132, 143, 145]

multiperfect (adjective): from Latin *multi* "many," from the Indo-European root *mel-* "strong, great," plus *perfect* (*q.v.*). The proper divisors of a perfect number add up to the number. By extension, the proper divisors of a multiperfect number add up to an integral multiple of the number. In 1631 the French mathematician Marin Mersenne (1588–1648) pointed out that the number 120 is a multiperfect number of order 2: its proper divisors add up to 240, which is twice the original number. [132, 165, 41]

multiple (noun, adjective), **multiply** (verb), **multiplication** (noun), **multiplicative** (adjective), **multiplicity** (noun): the first element is from Latin *multi* "many," from the Indo-European root *mel-* "strong, great." Latin *multi* appears in a multitude of borrowed English words such as *multilateral*, *multimedia*, and *multicolored*. The second element is from the Indo-European root *pel-* "to fold." When you multiply two natural numbers (neither of them a 1) the resulting product is "many-fold" compared with either of the starting numbers. The Latin pattern exists in native English as well: we can speak, for example, of a three-fold increase. The use of the symbol $\times$ to represent multiplication goes back to the English mathematician William Oughtred (1574–1660), whereas the use of the "·" is due largely to Gottfried Wilhelm Leibniz (1646–1716). With regard to equations, a root which is repeated n times is said to be of multiplicity n. [132, 160]

multiplicand (noun): from Latin *multiplicare* "to multiply" (*q.v.*), with the suffix *-nd-*, which creates a type of passive causative, so that Latin *numerus multiplicandus* meant "the number to be multiplied." In $A \times B$, the multiplicand is A, because A is to be multiplied by B. Of course, since ordinary multiplication is commutative, $A \times B$ could just as well be rewritten as $B \times A$, in which case B becomes the multiplicand. Contrast *multiplier*. [132, 160, 139]

multiplier (noun): from *multiply* (*q.v.*) and the agental suffix *-er* "the one that does." In $A \times B$, the multiplier is B, because B is doing the multiplying. Of course, since ordinary multiplication is commutative, $A \times B$ could just as well be rewritten as $B \times A$, in which case A becomes the multiplier. Contrast *multiplicand*. [132, 160]

multivariable (adjective), **multivariate** (adjective): the first element is from Latin *multi* "many," from the Indo-European root *mel-* "strong, great." The second element is from *variable* (*q.v.*). Multivariable or multivariate calculus involves functions of more than one variable. [132, 250, 69]

mutually (adverb): from the Latin adverb *mutuo* "in return, reciprocally, mutually." The Indo-European root is *mei-* "to change, go, move." Native English cognates include *mad* (changed from the normal condition of sanity) and the prefix *mis-*, indicating a (negative) change from whatever follows. Related borrowings from Latin include *migrate* and *permeate*. In mathematics, if two sets *A* and *B* are mutually exclusive, *A* excludes all elements of *B*, and in return *B* excludes all elements of *A*. [130, 117]

myria- (numerical prefix): from the Greek stem *muriad-* "myriad, ten thousand," from *murios* "countless." The Indo-European root may be *meu-* "damp," used in the sense "flowing," like the countless waves of the sea. Another hypothesis links the word to Greek *murmex* "ant," because ants swarm in large numbers. A myriad was originally a nonspecific quantity but later it acquired the specific meaning "ten thousand." In the early twentieth century the prefix *myria-* functioned in the metric system with a value of 10,000; it was a logical addition to the sequence *deca-* (10), *hecto-* (100), and *kilo-* (1000). More recently, with the standardization of the International System of Units and its adoption of new prefixes only for every third power of ten (as in *mega-*, *giga-*, *tera-*, etc.), *myria-* has largely fallen out of use. The word *myriad* is still used in nontechnical English with the meaning "a great many."

N

nabla (noun, adjective): the Greek form of a Hebrew word, probably of Phœnecian origin; also known in English as a *nable* or *nebel*. The nabla was an ancient Hebrew stringed instrument believed to have been a type of harp. The Irish mathematician William Rowan Hamilton (1805–1865) named the expression

$$\mathbf{i}\,\frac{\partial}{\partial x} + \mathbf{j}\,\frac{\partial}{\partial y} + \mathbf{k}\,\frac{\partial}{\partial z}$$

the nabla operator because the nabla resembled the ∇ symbol that he used to represent the mathematical operator. Note that the synonymous term *del* also refers to the delta shape of the symbol that represents the operator, rather than to any inherent property of the operator itself. See also *atled*.

nand (conjunction): a combination of *not* and *and* (*qq.v.*). In logic, the nand is the negation of the connective *and*. If *p* and *q* are statements, their nand is frequently symbolized as $\sim (p \wedge q)$. Compare *nonconjunction*; contrast *nor*. [141, 52]

nano-, abbreviated *n* (numerical prefix): from *nana*, a common baby word for a female adult. Greek *nanna* "aunt" had the male counterpart *nannas* "uncle." The slight variant *nannos* came to refer to any old man, especially a small one. The word then came to mean "a dwarf." From that last meaning, the prefix *nano-* has the sense "very tiny." As of 1960, the International System of Units has used *nano-* to mean specifically 10^{-9}, taking advantage of the phonetic resemblance between *nano-* and many Indo-European words for "nine" (*q.v.*). The prefix *nano-*, or *nanno-*, is used in biology to mean "very small," but without any specific numerical value; examples are *nanofossil* and *nanoplankton*.

nappe (noun): a French word meaning "tablecloth," but used figuratively in the phrase *nappe d'eau* "sheet of water" to describe water as it flows over a step or ledge. The Latin original is *mappa* "tablecloth." Related borrowings include *napkin*, *mop*, and *an apron* (originally *a napron*). In mathematics the vertex of a cone separates the cone into two surfaces, each of which is called a nappe of the cone. Textiles are quite at home in mathematics: compare the way we speak of a hyperboloid of two *sheets* (*q.v.*). [124]

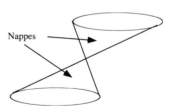

Nappes

narcissistic (adjective): from Greek *narkissos*, a plant with narrow leaves and usually white or yellow flowers. The word may be of Mediterranean origin. It may also be related to or influenced by the Indo-European root *(s)ner-* "to wind, to twist," that is found in Greek *narke* "cramp, numbness," because the narcissus plant is known to have narcotic properties. According to Greek mythology, Narcissus was a young man who was so entranced by his own image in a reflecting pool that he was eventually transformed into a narcissus plant. In mathematics an integer is said to be narcissistic if it can be represented by some form of manipulation of its digits. For example, 153 is narcissistic because $153 = 1^3 + 5^3 + 3^3$. Likewise, $1634 = 1^4 + 6^4 + 3^4 + 4^4$. Figuratively speaking, a narcissistic number is "caught up in itself."

natural (adjective): the Latin adjective *naturalis* is from the noun *natura* "blood relationship, natural affinity, nature," in turn from *(g)natus* "born." The Indo-European root is *gen-* "to give birth." Nature is literally the world as it was in its primeval state, when it was "born." *Natural* refers to that primeval state of things. An English cognate of *natural* is *kin*, people born into the same family. A borrowing from Greek is *gene*, which causes you to be born a certain way. A borrowing from Latin is *native*, a person born in a given place. From Latin via French we have *naive*, referring to the unsophisticated state a person was born in. In mathematics the natural numbers, otherwise known as counting numbers, are the most primeval or basic numbers; they are the ones we are born with (if indeed we are born with a number sense). As the German mathematician Leopold Kronecker (1823–1891) said in a meeting in Berlin in 1886: "God created the whole numbers, and everything else is the work of man." Logarithms to the base *e* arise "naturally" from $\int (dx/x)$, so they are known as natural logarithms. The use of logarithms to any other base requires multiplication or division by a constant, a most "unnatural" complication. The Swiss mathematician Leonhard Euler (1707–1783) popularized use of the letter *e* to represent the base of the natural logarithms. [64]

naught or **nought** (noun): a native English word, a compound of *na*, the forerunner of modern *no*, and *wiht*, modern *whit* "a tiny little bit." *Whit* was an alternative form of what is now the archaic *wight* "a creature, a person." So *nought*, and also its shortened form *not*, originally meant "not [even] a tiny little bit." In somewhat dated American English, *naught* is a synonym for the digit 0. Old folks occasionally say something like "naught seven" when they mean the year 1907. A related word is *naughty*, which originally described a person who was "worth nothing." [141]

necessary (adjective): from Latin *ne* "not" and *cessus*, past participle of *cedere* "to cede, to yield, move back," from the Indo-European root *ked-* "to go, to yield." Related borrowings from Latin include *recession* and, via French, *cease*. *Necessary* means literally "that can't be backed away from." A necessary condition for a given thing to occur is some-

thing that "you can't get away from" if you want that thing to happen; in other words, the thing can never occur without the necessary condition having first occurred. Contrast *sufficient*. [141, 88]

negate (verb), **negation** (noun), **negative** (adjective): from Latin *negatus*, past participle of *negare* "to deny." The Latin verb is based on the adverb *nec* "not," from the Indo-European root *ne* "not." When you negate something you deny it; you literally "say no" to it. When mathematicians began to use negative numbers they had to distinguish them from the familiar numbers that people had always used. The familiar numbers came to be called *positive* (*q.v.*), and the new ones *negative* because many people denied that they had any real meaning. For instance, a negative number couldn't represent the area of a field or the length of a pole or the weight of a stone or the number of people present in a room. In the 15th century numbers less than zero began to be called *negative* or *privative* (= "those which deprive"). Later writers called them *fictitious*, *absurd*, or *defective*, but *negative* eventually prevailed. In logic, negation is one of the basic operations; a lowered tilde, $\sim$, is commonly used to represent negation. That symbol is similar to the familiar minus sign used to represent negative numbers; the minus sign may also be used to indicate logical negation. [141]

neighborhood (noun): the first element of this native English compound is from *neah*, source of the obsolescent *nigh* "near." The second element is from *gebur* "dweller," and is also a compound. Its first component is the prefix *ge-*, from the Indo-European root *kom-* that is so commonly found in Latin *com-* "together with." Its second component is from the Indo-European root *bheu-* that is the source of English *be* and *bower*. A neighbor is literally a "near-dweller." In mathematics, a neighborhood of a point consists of the other points that "dwell" near the given point. [103]

neither (pronoun): a native English compound. The first element is from the Indo-European root *ne-* "not." The second element is *either* (*q.v.*). The meaning of *neither* is therefore "not either one." In logic, the correlated pair *neither . . . nor* is used to indicate the negation of an or-statement. The statement "neither *p* nor *q*" is defined as $\sim (p \lor q)$, which is equivalent to $\sim p \land \sim q$. [141, 103, 109]

nephroid (noun): from Greek *nephros* "kidney" and *-oid* (*q.v.*) "looking like." Related borrowings from

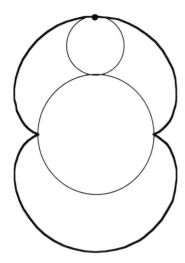

A nephroid is traced by a point
on the small circle rolling
around another circle which
has a radius twice as great.

Greek include *nephritis* and *epinenephrine*. In mathematics a nephroid is a two-cusped epicycloid named for its resemblance to animal kidneys. The nephroid may be represented parametrically by the equations

$$\begin{cases} x = 3\cos t - \cos 3t \\ y = 3\sin t - \sin 3t \end{cases}$$

Christian Huygens (1629–1695) and Ehrenfried Walter Tschirnhausen (1651–1708) discovered the nephroid. [244]

nerve (noun): from Latin *nervus*, "sinew, tendon, nerve, string, bowstring," from the Indo-European root *(s)neəu-*, "tendon, sinew." A related borrowing from Greek is *neuron*. In biology, a nerve is a thread-like bundle of fibers through which impulses pass between the brain or spinal column and other bodily organs. By analogy, a mathematical nerve of a finite family of sets is a type of abstract simplicial complex. For a similar analogy see *skeleton*.

nested (adjective): from *nest*, a native English word. The Indo-European root of the same meaning is the compound *nizdo-*. The first element is from the root *ni-* "down," as found in English *beneath* and *nether*. The second element is from the root "*sed-* "to sit," which is an English derivative found also in *settle*, *saddle*, and even *soot*, which settles on things. The original concept, then, is that a nest is the place where a bird sits down. As early as the 16th century, the word *nest* was used to describe a set of similar

objects of increasing size placed one inside the other. Each object sits inside the next largest one like a bird sitting inside its nest. In a similar way, during mathematical iteration, a certain procedure may be nested inside another identical procedure, and so on. For example, the infinitely nested radical

$$\sqrt{1 + \sqrt{1 + \sqrt{1 + \sqrt{1 + \cdots}}}}$$

converges to approximately 1.618; the exact value is

$$\frac{1 + \sqrt{5}}{2},$$

which is known as the golden mean. [184]

net (noun): a native English word, from the Indo-European root *ned-* "to bind, to tie." In mathematics a net is a two-dimensional fold-up model of a polyhedron. The faces of the polyhedron appear as connected (= bound) polygons in the net. Each face meets another along a common edge. [142]

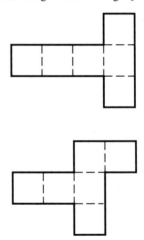

Two possible nets
for a cube

network (noun): a native English compound of *net* (see previous entry) and *work* (*q.v.*). The Indo-European root underlying *work* is *werg-* "to do," which is found in Greek *erg*, now used as a unit of work (prehistoric Greek lost its initial *w*'s). A net is a trap woven (or connected) from fibers and is typically used to catch butterflies, fish, etc. Like the compounds *ironwork*, *handiwork*, and *grillwork*, we also have the compound *network*, meaning anything structured like a net. Mathematically speaking, a network is any system of interconnected points. [142]

nil (adjective): a Latin contraction of *nihil*, a compound of *ne* "not" and *hilum* "a trifle, a little thing," of unknown prior origin; *nil* meant literally "not [even] a little thing," therefore "nothing." In mathematics a line segment of length zero is called a nil segment; it starts and ends at the same point. [141]

nilpotent (adjective): the first component is Latin *nil*, short for *nihil*, a compound of *ne* "not" and *hilum* "a trifle, a little thing," of unknown prior origin. Both *nil* and *nihil* meant literally "not [even] a little thing," therefore "nothing." The second component is from Latin *potens*, stem *potent-* "able, powerful," from the Indo-European root *poti-* "powerful, lord." In mathematics a square matrix is said to be nilpotent if, when raised to some power, the result is the zero matrix. For example, the matrix

$$\begin{bmatrix} 1 & 1 \\ -1 & -1 \end{bmatrix}$$

is nilpotent of order 2 because when it is raised to the second power it yields the zero matrix. [141, 173, 146]

nim (noun): nim is a game, said to be of Chinese origin, that frequently appears in books of recreational mathematics. Typically two players are presented with three piles of matches. Each player in turn can take any number of matches (but at least 1) from any one pile. Depending on the variant being played, the player who takes or leaves the last match is the winner. The game's name is from the German verb *nehmen* "to take," with an imperative singular form *nim*. The game is named after the action involved, taking matches. The Indo-European root *nem-* "to assign, allot, take," appears in native English *nimble* "quick to seize." [143]

nine (numeral): a native English word, from the Indo-European root *newn-*, of the same meaning. When compounded with two forms of the native English word *ten*, the same root has given *nineteen* and *ninety*. From the Latin cardinal number *novem* "nine," English has borrowed *November* (originally the ninth month) and *novena*. From the Latin ordinal number *nonus* "ninth," English has borrowed *nonagenarian*, *noon* (originally the ninth hour of the Roman day), and *nonagon*. From Greek *ennea* "nine" English has borrowed *ennead* "a group or sequence of nine things." See *nano-* for a phonetic but not etymological connection. The number 9, being one less than the base in which we do our calculations, has special properties; the best known is involved in

what has traditionally been called casting out nines. [144]

node (noun), **nodal** (adjective): from Latin *nodus* "a knot." The Indo-European root is either *ned-* "to connect" or *gen-* "to compress, to make compact." In a network (*q.v.*) a node is a place where two paths come together or connect or make a knot, figuratively speaking. The term *node* is also used to describe a place where a curve crosses itself. See more under *acnode* and *crunode*. [142]

noisy (adjective): the noun *noise* is an Old French word meaning "outcry, disturbance, loud dispute." The probable source is Latin *nausea* "discomfort, seasickness," from Greek *nausia*, of the same meaning. The Indo-European root would be *nau-* "boat," as seen in *navy* and *navigate*, borrowed from Latin, as well as *nautical* and *astronaut*, borrowed from Greek. In mathematics a duel is a two-person zero-sum game. A duel is said to be noisy if each player always knows whether or not the opponent has acted. Contrast *silent*.

nominal (adjective): from Latin *nominalis* "having to do with a name," from the Indo-European root *no-men-* "name." A nominal number is one which is used to identify something. An example is the number sewn on a baseball player's shirt or pinned to the clothing of a contestant in a dance contest. As a more common example, in many colleges social security numbers have almost entirely replaced students' names in computer records. Some people might argue that nominal numbers aren't truly "mathematical," but in fact they are akin to the subscripts so commonly used in algebra and calculus; both nominal numbers and subscripts distinguish one individual from another. In finance, a nominal rate of interest is the yearly rate as "named" or stated, in contrast to the effective rate, which takes compounding into account. [145]

nomograph (noun): from Greek *nomos* "law, custom, portion, usage," and *graphein* "to write." The Indo-European root underlying *nomos* is *nem-* "to assign, to take," while that underlying *graph* is *gerbh-* "to scratch." A nomograph is a type of graph showing three lines or curves with scales on them. When a straight line is drawn connecting a number on one scale with a number on another, the result of combining those two numbers using a certain operation is shown where the straight line intersects the third scale. [143, 68]

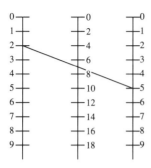

A nomograph for addition.
The slanted line shows that
$2 + 5 = 7$.

nonagon (noun): the first element is from Latin *nonus* "ninth," from the Indo-European root *newn-* "nine." The second element is from Greek *gon-* "angle"; the Indo-European root *genu-* is also found in English *knee*. A nonagon is a polygon with nine angles, and therefore also nine sides. [65, 144]

nonconjunction (noun): the first component is from Latin *non* "not," itself a compound of the Indo-European roots *ne* "not" and *oi-no-* "one," so that *non* originally meant "not [even] one [little bit]." The second part of *nonconjunction* is *conjunction* (*q.v.*) "a joining together." It may sound paradoxical, but in terms of parts of speech, the word *nonconjunction* is a noun that does in fact refer to a conjunction. In logic, a nonconjunction is the negation of the conjunction *and*. If *p* and *q* are statements, their nonconjunction is frequently symbolized as $\sim (p \wedge q)$. Compare *nand*; contrast *nor*. [141, 148, 103, 259]

none (adjective): a (not so) hidden number word, a compound of native English *ne* "not" and *one*, so *none* means "not [even] one." The Indo-European roots are *ne* "not, and *oi-no-* "one." Compare the structure of Latin *non-* as explained under *nonconjunction*. [141, 148]

nonillion (numeral): patterned after *billion* by replacing the *b-* with Latin *non-* "ninth." Since in most countries a billion is the second power of a million, a nonillion is defined as the ninth power of a million, or 10^{54}. In the United States, however, a billion is 10^9, and a nonillion adds seven groups of three zeroes, making a nonillion equal to 10^{30}. [144, 72, 151]

nonresidue (noun): the first component is from Latin *non* "not," itself a compound of the Indo-European roots *ne* "not" and *oi-no-* "one," so that *non* originally meant "not [even] one [little bit]." The second

component is *residue* (*q.v.*). In number theory, if $x^n \equiv c \,(mod\; m)$ has no solution, then *c* is said to be a nonresidue of *m* of the *n*th order. [141, 148, 178, 184]

nonsingular (adjective): the first component is from Latin *non* "not," itself a compound of the Indo-European roots *ne* "not" and *oi-no-* "one," so that *non* originally meant "not [even] one [little bit]." The second component is *singular* (*q.v.*). A determinant which isn't singular is said to be nonsingular. [141, 188, 238]

nontiler (noun): the first component is from Latin *non* "not," itself a compound of the Indo-European roots *ne* "not" and *oi-no-* "one," so that *non* originally meant "not [even] one [little bit]." The second part of *nontiler* is from *tile* (*q.v.*). A nontiler is a tile that cannot by itself tessellate the plane. [141, 148, 209, 238]

noon (noun) from Latin *nona (hora)* "ninth hour [after sunrise]," which put the original "noon" at about 3 o'clock in the afternoon. At that time of day the Catholic church celebrated the logically named *nones*, one of the canonical hours. The ecclesiastical service was later moved to 12:00. Perhaps in anticipation of eating afterwards, people began to use the word *noon* to describe the midday meal that took place following religious services. Finally the word came to refer to midday itself, which we now designate as 12 o'clock. [144]

nor (conjunction): a combination of *not* and *or* (*qq.v.*). In logic, *nor* is the negation of the connective *or*. If *p* and *q* are statements, their nor is frequently symbolized as $\sim (p \vee q)$. Contrast *nand*. [141, 155]

norm (noun), **normal** (adjective, noun), **normalize** (verb): from Latin *norma* "a carpenter's square." A carpenter's square is used for making right angles, and so a normal line is a line perpendicular to something else. In an abstract sense, a carpenter's square is a pattern for perpendicularity, so a norm came to refer to a pattern of behavior or a standard of measurement. A normalized vector has a standard length of 1 unit; more generally, the norm of a vector is its length. Linguists have traditionally attributed Latin *norma* to the Indo-European root *gno-* "to know," as in the English cognates *know* and *can* (= to know how to), but that may not be correct. [75]

not (adverb): a native English word, actually a shortened form of *naught*, (*q.v.*) "nothing." In logic, *not* is a unary operator that is often represented by a minus sign or by a lowered tilde, $\sim$. [141]

notation (noun): from Latin *notare* "to mark, to write down," from *nota* "that by which a thing is known," hence "a mark, a sign, a note." The Indo-European root may be *gno-*, meaning the same as the English cognate *know*: when you mark something, you know what or where it is. Also from that Indo-European root are English *ken* and *can* (= to know how to). Mathematical notation consists of the "marks," that is, the signs and symbols, used to represent the operations and relationships of mathematics. The algebraic notation that was developed in Europe during the Renaissance was superior to any that had gone before, and it aided the rapid advance of mathematics. [75]

November (noun): from Latin *novem* "nine," from the Indo-European root *newn-* "nine." In the Roman calendar, which began in March, November was the ninth month of the year. In spite of its name, it is now the eleventh month of the year because January later became the first month. [144]

novemdecillion or **novendecillion** (numeral): patterned after *billion* by replacing the *bi-* with the root of Latin *novemdecim* (or *novendecim*) "nineteen," a compound of *novem* "nine" and *decem* "ten." Since in most countries a billion is the second power of a million, a novemdecillion was defined as the nineteenth power of a million, or 10^{108}. In the United States, however, a billion is 10^9, and a novemecillion adds seventeen groups of three zeroes, making a novemecillion equal to 10^{60}. [144, 34, 72, 151]

nucleus (noun): a Latin diminutive meaning "kernel, nut," from *nux*, stem *nuc-*, "a nut," especially "a walnut." The Indo-European root may be *ken-* "to compress," since a nut usually grows in a hard shell and is therefore "compressed." From the same root are native English *nut* and *neck*, a narrow "compressed" part of the body. In mathematics a nucleus of an integral equation is the same as a kernel (*q.v.*).

null (adjective), **nullity** (noun): from the Latin adjective *nullus*, composed of *ne* "not" and *ullus* "any." *Ullus*, in turn, is from a presumed *unulus*, the diminutive of *unus*, meaning the same as its English cognate *one*. So Latin *nullus* meant literally "not [even] one little one," therefore "none at all." Compare the structure of English *none*. In the term $\aleph_0$, the subscript is properly read as "null" because *null* is the German word for "zero" (although it was borrowed

from Latin). The null set is the set having "not even one little" member in it; it is represented by the symbol $\varnothing$, which is a stylized zero. A null space is a set of vectors that are transformed by a given linear transformation into the zero vector. A nullity is the dimension of a null space. In logic, a nullity is a statement that a certain type of thing does not exist. Contrast *entity*. [141, 148, 238]

number (noun): from Latin *numerus* "number," which became Old French *nomre*, modern French *nombre*, and English *number*. A related borrowing from Latin is *enumerate*. The Indo-European root may be *nem-* "to apportion, to assign, to allot, to take," from which would have come the notion of a number, something which measures the size of an allotment. Classical Latin *numeri*, the plural of *numerus*, meant "arithmetic" or "astronomy," the sciences that deal with numbers. In algebra the letter *n*, which is the first letter of the word *number*, is frequently used to stand for an unknown or generic number. In the same way that an ordinal number like *sixth* is formed from the cardinal number *six*, we also form the cardinal number "*n*th" from the cardinal number *n*; we speak, for example, of the *n*th term in a sequence. [143]

numeracy (noun), **numerate** (adjective): patterned after *literacy* and *literate*, from Latin *numerus* "number" (*q.v.*). As used by John Allen Paulos in his popular book *Innumeracy*, numeracy might be defined as "an ability to deal comfortably with the fundamental notions of number and chance." Unfortunately, it appears that during recent decades American schoolchildren and young adults have become increasingly innumerate. [143]

numeral (noun), **numeration** (noun): from Latin *numerus* "number" and the adjectival suffix *-alis*. The Indo-European root may be *nem-* "to apportion, to assign, to allot, to take," from which would have come the notion of a number, something which measures the size of an allotment. In mathematics a numeral is any symbol that can represent a number. In informal English, people rarely distinguish between a number and the numeral used to represent it. A system of numeration is a system of counting or numbering things. Western countries now use a variant of the Hindu-Arabic system of numeration, but many other systems of numeration have been used historically and in other parts of the world. One that still somewhat survives in the West is the system of Roman numerals, though with certain regularities that became standard long after the time of the Romans. Asian countries like China and Japan have their own numerals, although even in those countries the Hindu-Arabic numerals have made great inroads. Computer programmers often use a numeration system with a base of two, eight, or sixteen. [143]

numerator (noun): the first element is from Latin *numeratus*, past participle or *numerare* "to number." The more basic word is the Latin masculine noun *numerus* "number" (*q.v.*). The Latin *-or* suffix indicates a masculine person or thing that performs the indicated operation. A numerator is literally "a numberer." In arithmetic, the numerator of a fraction gives the number of parts being considered, where each part is an equal fraction of the whole. Since the numerator tells you how many parts you have, when you add fractions you add only numerators, never denominators (*q.v.*). English writers of the Renaissance sometimes called the numerator the *topterm* or just the *top*. In fact *top* is an excellent choice even today because it is succint and avoids confusion between the similar sounding *numerator* and *denominator*. [143, 153]

numerical (adjective): from Latin *numerus* "number" (*q.v.*). *Numerical* is the "learned" adjective corresponding to the "popular" noun *number*, which English borrowed from French. What is now commonly called absolute value used to be known (and sometimes still is) as numerical value; it is the value of the "number part" of a numeral, without regard to the sign. [143]

O

oblate (noun): the first component is from Latin *ob-* "against, toward," from the Indo-European root *epi-* or *opi-* "against, near, at." The second component is from Latin *latus* "carried," from the Indo-European root *telə-* "to lift, support, weigh." In mathematics the poles of an oblate spheroid are "carried toward" the center a sphere, making the resulting ellipsoid shorter in one dimension than in the other two. An oblate spheroid results from rotating an ellipse about its shorter axis. The earth approximates an oblate spheroid: the diameter through the poles is shorter than a diameter in the plane of the Equator. Contrast *prolate*; compare *oblong*. [53, 224]

oblique (adjective): the first component is Latin *ob-* "against, toward," from the Indo-European root *epi-*

or *opi-* "against, near, at." The second component is of unknown prior origin. Latin *obliquus* meant "side-long, slanting, oblique." In trigonometry an oblique triangle is one which is either acute or obtuse, but not right (in which case the two legs would be perpendicular rather than "slanting.") In a Cartesian coordinate system the axes are usually perpendicular, but they may be oblique. [53]

oblong (adjective): the first component is Latin *ob-* "against, toward," from the Indo-European root *epi-* or *opi-* "against, near, at." The second component is Latin *longus* "long," from the Indo-European root *del-* "long." In mathematics an oblong rectangle is longer in one dimension than in another; in other words, any non-square rectangle is oblong. In the category of figurate numbers, the oblong numbers are 1×2, 2×3, 3×4, etc.; an oblong number of dots can be arranged in an $n \times (n + 1)$ rectangle. [53, 35]

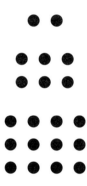

The first three
oblong numbers.

obtuse (adjective): the first component is Latin *ob-* "against, toward," from the Indo-European root *epi-* or *opi-* "against, near, at." The second component is Latin *tusus*, past participle of *tundere* "to strike, to beat." The Indo-European root is *(s)teu-* "to push, to beat." Latin *obtusus* meant "beaten down to the point of being dull." A related borrowing from Latin includes *stupid*, literally "struck dull"; in fact in non-mathematical English we use *obtuse* to refer to a dull or stupid person. In geometry, an obtuse angle, being greater than 90° (but less than 180°), looks "dull" by comparison to an *acute* (*q.v.*) angle. [53, 214]

octadecagon (noun): the first component is from Greek *okto-*, from the Indo-European root *okto(u)-* "eight." The second component is from Greek *deka-*, from the Indo-European root *dekm-* "ten." The ending is from Greek *gon-*, from the Indo-European root *genu-* "angle, knee." An octadecagon is an eighteen-angled (and therefore also eighteen-sided) polygon. [149, 34, 65]

octagon (noun), **octagonal** (adjective): the first component is from Greek *okto*, from the Indo-European root *okto-* "eight." The second component is from Greek *gon-* "angle"; the Indo-European root is *genu-* "knee, angle," as found in English *knee*. An octagon is a polygon with eight angles, and therefore also eight sides. [149, 65]

octahedron, plural *octahedra* (noun), **octahedral** (adjective): from Greek *okto* "eight" and *hedra* (prehistoric Greek *sedra*) "base"; the Indo-European root is *sed-* "to sit," as found in English *sit* and *seat*. An octahedron is an eight-based, or eight-faced polyhedron. A regular octahedron is one of the five regular polyhedra. [149, 184]

octakis- (numerical prefix): a Greek form meaning "eight times," based on *octo-* "eight." The prefix is used as part of the name of a stellated regular solid. It indicates that each original face became eight faces. An octakishexahedron, for example is a polyhedron with eight times six, or 48, faces. [149]

octal (adjective): from Latin *octo*, from the Indo-European *okto-* "eight." Related borrowings include *octave* and *octogenarian*. An octal system of numeration, which uses base eight, is found in computer programming. Octal notation has occasionally been called *octimal* or *octonary* (*qq.v.*). [149]

octant (noun): patterned after *quadrant*, with the Latin root *quadr-* "four" replaced by the Latin root *oct-* "eight." In a two-dimensional rectangular coordinate system the two axes divide the plane into four parts called quadrants. In Latin, an octant was half of a quadrant, a plane region bounded by two lines meeting at a 45° angle. Mathematics has broken with that pattern but established a new one: in a three-dimensional rectangular coordinate system the three mutually perpendicular coordinate planes divide space into eight octants. Following the pattern for numbering quadrants, the first four octants in space are usually numbered counterclockwise about the positive *z*-axis, beginning in the octant in which all three coordinates are positive; the fifth through eighth octants are numbered consecutively beneath the first four. Compare *orthant*. [149, 146]

octic (adjective): from Latin *octo*, from the Indo-European root *okto-* "eight." In group theory, the octic group is a specific group containing eight elements. It may be represented by eight rigid motions (four rotations and four reflections) of a square into itself. [149]

octillion (numeral): patterned after *billion* by replacing the *b-* with Latin *oct-* "eight." Since in most countries a billion is the second power of a million, an octillion was defined as the eighth power of a million, or 10^{48}. In the United States, however, a billion is 10^9, and an octillion adds six groups of three zeroes, making an octillion equal to 10^{27}. [149, 72, 151]

octimal: (adjective): patterned after *decimal*, with the Latin root *dec-* "ten" replaced by the Latin root *oct-* "eight." Related borrowings from Latin include *octane* (a chemical compound containing eight carbon atoms) *octet*, and *octogenarian*. An octimal system of numeration is one which uses base eight. It is found primarily in computer programming. Octimal notation has occasionally been called *octonary*, a form which is patterned after the word *binary*. The most common name nowadays, however, is *octal* (*q.v.*). Compare *decimal*, *hexadecimal*, and *vigesimal*. [149]

October (noun): from Latin *octo* "eight," from the Indo-European root *okto-* "eight." In the Roman calendar, which began in March, October was the eighth month of the year. In spite of its name, it is now the tenth month of the year because January later became the first month. [149]

octodecillion (numeral): patterned after *billion* by replacing the *bi-* with the root of Latin *octodecim* "eighteen," a compound of *octo* "eight" and *decem* "ten." Since in most countries a billion is the second power of a million, an octodecillion was defined as the eighteenth power of a million, or 10^{108}. In the United States, however, a billion is 10^9, and an octodecillion adds sixteen groups of three zeroes, making an octodecillion equal to 10^{57}. [149, 34, 72, 151]

octomino (noun): from Greek *okto* "eight" and all but the first letter of Latin *domino*. For the second component and further explanation see *polyomino*. [149, 37]

octonary (adjective): patterned after *binary* by changing Latin *bi-* "two," to *octo-* eight." An octonary number system is one which uses base eight. It is used primarily by computer programmers, though most often under the name *octal*. Compare *binary*, *denary*, *quinary*, *senary*, *ternary*, and *vicenary*. Also see *octimal*. [149]

octonion (noun): patterned after *quaternion* (*q.v.*) by replacing Latin *quater* "four times," by *octo* "eight." An octonion is a type of eight-dimensional hyper-complex number. [149]

octuple (verb, adjective, noun): from Latin *octuplus* "eightfold," from *octo* "eight" and the Indo-European root *pel-* "to fold." To octuple something is to make it eightfold, i.e., to multiply it by 8. With the advent of in vitro fertilization, whenever the number of children conceived at one time is eight times as great as usual, the babies are called octuplets. An octuple of numbers (a, b, c, d, e, f, g, h) is a set of eight numbers in a given order. Compare *double*, *triple*, *quadruple*, *quintuple*, *sextuple*, and *septuple*. [149, 160]

odd (adjective), **odds** (noun): from Old Norse *oddi*, which referred to pointy or uneven things, including triangles. What distinguishes a triangle from a line is the odd (= third) point "sticking out." For that reason a person who stands out from the norm or who is strange or unusual is called *odd*. An odd sock is one left over after you've paired up your other socks, and so *odd* came to refer to a number that is one greater than a pair, i.e., one greater than an even number. Historically speaking, the ancient Greeks believed that the first odd number is 3, i.e., one more than the first possible pair, 2. They also believed that odd integers are male. In analysis, an odd function gets its name from the fact that if it is represented by a power series, all the exponents are odd integers. The derivative of an odd function is even (*q.v.*), and vice versa. In probability, the odds in favor of some event is defined as the ratio of the number of favorable outcomes to the number of unfavorable outcomes. That usage developed from the sense of *odd* as "uneven," therefore "different," as exemplified in the phrase "at odds with." When you calculate odds, you compare two different groups of things, the favorable ones and the unfavorable ones. The expression "even odds," referring to odds of 1:1, is an etymological (but not mathematical) contradiction. [147]

oddorial (noun): a newly coined term made up of *odd* (*q.v.*) and the end of *factorial*. For a positive integer n, factorial n or n factorial is the product of n and every lesser positive integer down to the number 1. By analogy, for an odd integer d, oddorial d or d oddorial is the product of d and every lesser odd integer down to the number 1. For example, oddorial $9 = 9 \times 7 \times 5 \times 3 \times 1 = 945$. Although the symbol "!" is used to represent factorials, no standard symbol exists for oddorials. The value of an oddorial is always an odd integer, since all the factors are odd. If the term *oddorial* had been invented a century ago, it would probably have been called something like *imparorial*, where the first two components of the word are from Latin *im-* "not" and *par* "even." Oddorials occur in the general term of certain types of infinite series. Compare *primorial*. [147, 153]

ogive (noun): a French architectural term meaning "diagonal rib of a vault." The French word—earlier forms of which were *augive* and *orgive*—may have developed from a presumed Vulgar Latin *obviativa*, based on the past participle of *obviare* "to be in opposition, to resist." The first element is Latin *ob* "against," and the second element is from *via* "way, road." The Indo-European root is *wegh-* "to go, to transport," as seen in native English *way*. In mathematics an ogive, also known as a cumulative frequency curve, is one in which each ordinate is the sum of frequencies in preceding intervals. Figuratively speaking, the values are getting "piled up" like the roof of a vault. In the case of a continuous distribution the ogive curve resembles the ribs of a vaulted ceiling in a church or crypt. [53, 243]

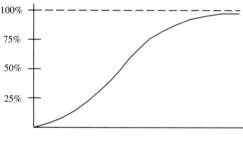

An ogive.

-oid (suffix): from Greek *eidos* "shape, form." A humanoid, for example, is a robot or an extraterrestrial who has more or less the appearance of a human. In mathematics the suffix may refer to a two-dimensional shape, as in *cardioid* "shaped like a heart," or a three-dimensional shape, as in *paraboloid*, a surface whose cross-sections parallel to the axis of the surface have the shape of a parabola. The suffix may also be used metaphorically, as in *centroid*, a point that behaves as if it were a center of mass. The Indo-European root is *weid-* "to see." From the related Latin *videre* "to see," come borrowings such as *video*, *evident*, and *vision*. A native English cognate is *wise*, since a wise person can see (and hence understand) well. Similarly, the English cognate *witty* refers to someone who quickly "sees" what to say in a given situation. [244]

omega (noun): the name of the last letter of the Greek alphabet, written Ω as a capital and ω in lower case. Because there were two kinds of o in ancient Greek, one was called *o-micron* "small o" and the other *o-mega* "big o." The Indo-European root underlying *omega* is *meg-* "great." In the theory of transfinite numbers, ω is the least of the infinite ordinal numbers. The symbol was chosen not because of what its name means, but because the ordinal number ω comes after all the finite natural numbers in the same way that the letter ω comes after all the other letters of the alphabet. [129]

on (preposition): a native English word, from the Indo-European root *an* "on." The same Indo-European root appears in Greek *ana* "on, up, at the rate of." We commonly say that a point lies "on" a line, but we might just as well say that a point lies "in" a line, since a line has no width. Because of the duality of the situation, we could also say that a line lies on a point, though we rarely speak that way. [10]

one (cardinal number): a native English word, from Indo-European *oi-no-* "one." Related native English words are *alone* (= all one), shortened to *lone*; *atone* (= at one, since you are at one with yourself when you atone); *only*; and even the indefinite article *an* (shortened to *a* before consonants), as in *an apple* (= one apple). The pronunciation of *one* as if it had an initial *w-* developed in the 15th century and has now become the standard, although the spelling hasn't changed to reflect the pronunciation. Although *once* and *oneness* reflect the pronunciation with *w-*, the majority of compounds preserve the original pronunciation of *one* without an initial *w-*. In Latin, the Indo-European root *oi-no-* developed into *unus* "one." The root *un-* appears in many borrowings from Latin such as *unilateral* and *unanimous*. Among the ancient Greeks, 1 was not considered a number, but rather the generator of numbers. In modern times, we still continue that tradition when we deal with prime numbers: even though 1 fits the definition of a prime because the only whole numbers that divide it are itself and (redundantly) 1, we don't consider 1 a prime. [148]

onto (preposition, adjective): a native English word, a compound of *on* (q.v.), from the Indo-European root *an* "on" and *to*, from the Indo-European root *de-* with a demonstrative meaning. Doubled-up preposi-

tions are common in Indo-European languages; there are even compounds containing two prepositions of the same meaning, as in *underneath* and *off of*. In mathematics a relationship which maps the elements of a set A into those of a set B is said to be *onto* if every element in B is the image of at least one element in A. Metaphorically speaking, if we imagine set A to lie above set B, the word *onto* indicates that B is completely "covered" by the mapping. An *onto* relationship is also said to be *surjective* (q.v.). [10, 32]

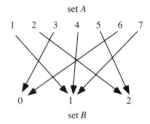

A mapping from set A
onto set B

open (adjective): a native English word. The Indo-European root is *upo*, which meant "under" but also "up from under," and as a result the word developed opposite meanings in different Indo-European languages and sometimes even within the same language. Native English shows the "upward" meaning in *above*, *up* and the related verb *open* (because in order to open something like a box, we have to lift the lid upward; notice that we still often say, somewhat redundantly, *to open up* rather than just *to open*. In mathematics an open interval is one which does not include its endpoints; the interval is open because it doesn't contain a point that closes it at each end. A disk, box, ball, or region may also be open. [240]

operand (noun): the first component is from Latin *operari* "to work, toil, labor" [see next entry]. The second component is from the Latin suffix *-nd-*, which creates a type of passive causative, so that Latin *operandum* meant "[something] to be worked on." In mathematics an operand is a number or a variable that is to be worked on by a function. For example, in $f(x, y)$, x and y are the operands of the function. Compare *argument*. [152, 139]

operate (verb), **operation** (noun), **operator** (noun): from Latin *operari* "to work, toil, labor." The Indo-European root is *op-* "to work, to produce abundantly." Related borrowings from Latin include *opus* and *opera* (actually the plural of Latin *opus*), which are musical works; *office*, *opulent*, and *copious*. The six fundamental operations of arithmetic perform a kind of "work" on numbers: the numbers get added, subtracted, multiplied, divided, raised to powers, or have roots taken. In calculus the "work" of the differential operator is to take derivatives. [152, 153]

opposite (adjective): from Latin *ob*, against, and *positus*, past participle of *ponere* "to put, to place." (See more under *component*.) In geometry, the side of a triangle opposite a given angle is the side that has been "placed against" it, i.e., put across from it. In algebra the opposite of a number is the negative of that number. [53, 14]

optimal (adjective); **optimum**, plural *optima* (adjective, noun); **optimize** (verb); **optimization** (noun): *optimum* is a Latin word meaning "best." The Indo-European root is *op-* "to work, to produce abundantly." From the sense "abundance," the root developed the meaning "wealth, goods." The *-timum* suffix makes a superlative out of the root it is attached to, so *optimum* refers to a situation that yields the most wealth or the greatest good. In other words, the optimum value is the best one. In calculus, optimization problems require finding the "best" value of a function, that is, the maximum or minimum value. [152]

or (conjunction): from Old English *oththe* "or." Under the influence of other words indicating alternatives, such as *either* and *whether*, the original *oththe* developed into *oththerr*. By the 12th century, the two-syllable form *otherr* existed alongside the one-syllable *oththr* before vowels, and the reduced form *orr* before a consonant. The shortest form eventually won out, not only because languages generally move toward shorter forms, but perhaps also as a way of avoiding confusion with the forerunner of the word *other*. In terms of meaning, English *or* is ambiguous. It can be inclusive: this thing or that thing or both things. It can be exclusive: this thing or that thing but not both things. In logic, to avoid ambiguity, the Latin word *aut* is used to indicate the exclusive *or*, which is frequently indicated by the symbol $\veebar$. Latin *vel* is used to indicate the inclusive *or*, which is frequently indicated by the symbol $\vee$. [155]

order (noun, verb): from French *ordre*, from Latin *ordo*, stem *ordin-*, "a straight row." The Indo-European root is *ar-* "to fit together," so that Latin *ordo* developed from the idea of putting things together in a certain way, namely lined up in a row. Latin *ordo* also had the derived meaning "a

rank or class of citizens." That sense is reflected in mathematical usage when we refer to differential equations of first order, second order, etc. In analytic geometry, an ordered pair consists of two numbers in a specified order; the ordered pair indicates the location of a point in a two-dimensional Cartesian or polar coordinate system. The nonmathematical use of *order* to mean "command" follows from the idea that when you carry out someone's instructions you arrange or "order" things in a certain way. Related borrowings include *arthritis*, a disease of the joints (where bones are fitted together), and *adorn* (to fit things together in a decorative way). Also related is native English *read*, to fit words together to make sense out of them. [15]

ordinal (adjective): from Latin *ordo*, stem *ordin-*, "straight row, order." The Indo-European root is *ar-* "to fit together," so that Latin *ordo* developed from the idea of putting things together in a certain way, namely lined up in a row. In arithmetic, the ordinal numbers (*first*, *second*, *third*, *fourth*, etc.) tell what order or position an item occupies in a sequence (= "straight row"). In English there are only two ordinal numbers not derived from the corresponding cardinals: the first is *first* and the second is *second* (*qq.v.*). [15]

ordinary (adjective): from Latin *ordinarius* "pertaining to order, according to the usual order," and therefore "ordinary." The underlying Latin noun is *ordo*, stem *ordin-*, "straight row, order," from the Indo-European root *ar-* "to fit together," so that Latin *ordo* developed from the idea of putting things together in a certain way, namely lined up in a row. In analysis, an ordinary point is one that is "orderly" and "well-behaved"; it acts "according to the usual order" and doesn't cause any trouble, as contrasted with a singular point. [15]

ordinate (noun): from Latin *ordinatus*, past participle of *ordinare* "to order, set in order, arrange, regulate." The more basic Latin word is *ordo*, stem *ordin-* "a straight row, an order." The Indo-European root is *ar-* "to fit together," so that Latin *ordo* developed from the idea of putting things together in a certain way, namely lined up in a row. When you put numbers in order, you put them in a straight row on a number line. With regard to a two-dimensional rectangular coordinate system, the ordinate of a point means the *y*-value of the point. It is the distance measured on the ordered scale of the *y*-axis from the value zero up or down to the point in question. The

German mathematician Gottfried Wilhelm Leibniz (1646–1716) coined the word *ordinate*. The ancient Greeks had used the term *tetagmenos*, a variant of a simpler word meaning "in an orderly manner." Leibniz merely translated the concept from Greek to Latin. [15]

orient (verb), **orientable** (adjective), **orientation** (noun): from Latin *oriens*, stem *orient-*, "the direction where the sun rises," from the present participle of *oriri* "to rise." The Indo-European root is *er-* "to move, to set in motion." The Latin word does not seem to be a cognate of native English *rise*, although the meanings are identical. When you orient yourself, you literally find where the east is, and then where all the other directions are. From the notion of finding a direction comes the use of *orientation* in calculus to mean the direction in which points are traced out on a curve by a vector-valued function or a set of parametric equations. A surface is said to be orientable if it has a definite "direction." For example, the surface of a shoe is oriented for either a right foot or a left foot, but the same shoe can't fit both feet. For an explanation of the suffix see *-able*. [54, 146, 69]

origin (noun): from Latin *origo*, stem *origin-*, "the beginning, the source," from the verb *oriri* "to rise." The Indo-European root is *er-* "to move, to set in motion." The Orient, or East, is the place where the sun rises. In a Cartesian coordinate system, the origin is the place from which all measurements "arise," i.e., the place where the coordinate axes cross. Some students mistakenly pronounce the word as if it were "orgin." [54]

orthant (noun): from Greek *orthos* "straight, upright," hence "perpendicular," from the Indo-European root *wrodh-* "to grow straight, upright." The *-ant* ending is borrowed from *quadrant* and *octant*, after which *orthant* is patterned. In a 2-dimensional rectangular coordinate system, the two mutually perpendicular coordinate axes divide 2-space into four (2^2) regions called quadrants. In a 3-dimensional rectangular coordinate system, the three mutually perpendicular coordinate planes divide up 3-space into eight (2^3) regions called octants. In a 4-dimensional rectangular coordinate system, the four mutually perpendicular coordinate hyperplanes divide 4-space into sixteen (2^4) regions that might be called "hexadecimants," but the terminology begins to grow cumbersome. Beyond three dimensions, instead of using a word

that is etymologically connected to the number of regions involved, mathematicians use the generic term *orthant*, whose etymology refers to the mutually perpendicular axes, planes, and hyperplanes involved. [256, 146]

orthic (adjective): from Greek *orthos* "straight, upright," hence "perpendicular," from the Indo-European root *wrodh-* "to grow straight, upright." An orthic triangle is a triangle whose vertices are the feet of the altitudes (= perpendicular segments) of another triangle. [256]

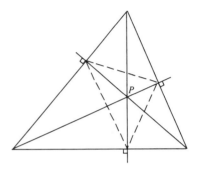

The dashed triangle is an orthic triangle. Point *P* is an orthocenter.

orthocenter (noun): the first component is from Greek *orthos* "straight, upright," hence "perpendicular," from the Indo-European root *wrodh-* "to grow straight, upright"; the second component is *center* (*q.v.*). With regard to a triangle, the place where the three altitudes (which are perpendicular to the sides) meet is called the orthocenter. See picture under *orthic*. The orthocenter is "centered" inside the triangle only when the triangle is acute. Contrast *excenter*, *circumcenter*, and *incenter*. [256, 91]

orthogonal (adjective), **orthogonality** (noun): from Greek *orthogonios* "right-angled." The first component is from Greek *orthos* "straight, upright," hence "perpendicular," from the Indo-European root *wrodh-* "to grow straight, upright"; the second component is from the Indo-European root *genu-* "angle, knee." If two lines are orthogonal, they form a right angle. In 16th- and 17th-century English, the terms *orthogon*, *orthogonium*, and *orthogonion* meant "a right triangle," but only the related adjective *orthogonal* has survived. It is now usually used only for vectors: two vectors are said to be orthogonal if, when the directed segments representing them share an initial point, those segments form a right angle. When vectors aren't involved, we usually use the Latin translation of *orthogonal, rectangular,* or Latin-derived *normal* or *perpendicular.* [256, 65]

orthonormal (adjective): the first component is from Greek *orthos* "straight, upright," hence "perpendicular," from the Indo-European root *wrodh-* "to grow straight, upright"; the second component is *normal* (*q.v.*). A set of vectors is orthonormal if each pair of vectors is orthogonal and if all the vectors involved have been normalized (in other words, the norm or length of each vector is 1 unit). Like the root *ortho-*, the word *normal* may also mean "perpendicular," so it might at first seem as if *orthonormal* should mean "perpendicular perpendicular." As just explained, however, it is the alternate meaning of *normal,* "regularized," that appears in this compound. [256, 75]

orthopole (noun): the first component is from Greek *orthos* "straight, upright," hence "perpendicular," from the Indo-European root *wrodh-* "to grow straight, upright"; the second component is from Greek *polos* "pivot, axis." The Indo-European root is *kwel-*, meaning "to revolve." The meaning of *orthopole* in mathematics may be explained as follows. From each vertex A, B, and C of a triangle, a perpendicular is dropped to a line, meeting the line in A′, B′, and C′, respectively. A perpendicular is then dropped from A′ to side a, from B′ to side b, and from C′ to side c. The point (= pole) at which the second set of perpendiculars meet is the orthopole of the triangle. [256, 107]

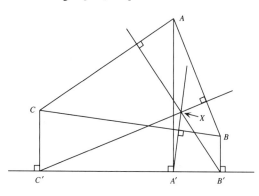

Point X is the orthopole of triangle ABC.

orthoptic (adjective, noun): the first component is from Greek *orthos* "straight, upright," hence "perpendicular," from the Indo-European root *wrodh-* "to grow straight, upright"; the second component is Greek *optos* "seen, visible," from the Indo-European

root *okʷ-* "to see." The root is found in native English *eye* and the second part of *window.* Related borrowings from Greek include *optics* and *autopsy.* In mathematics an isoptic of a given curve is the locus of the intersection of two tangents that meet at a constant angle. If the angle happens to be a right angle, the locus is called an orthoptic because at every intersection the two tangents are "seen to be perpendicular." [256, 150]

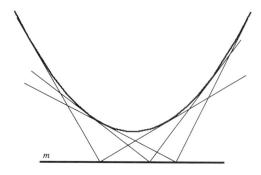

Line *m* is an orthoptic to the parabola.

orthotope (noun): the first component is from Greek *orthos* "straight, upright," hence "perpendicular," from the Indo-European root *wrodh-* "to grow straight, upright"; the second component is from Greek *topos* "place, region," of unknown prior origin. An orthotope is a generalization to *n* dimensions of the 2-dimensional rectangle and the 3-dimensional rectangular solid or cuboid. Each vertex of an orthotope is a place at which *n* mutually orthogonal edges meet. [256, 232]

oscillate (verb), **oscillation** (noun): from Latin *oscillum,* diminutive of *os,* stem *or-* "mouth." The Indo-European root is *os-* "mouth." By extension Latin *os* came to designate the entire face, not just the mouth. An *oscillum* was a little mask representing the face of Bacchus, the god of wine, that the Romans used to hang from trees, especially in vineyards. Since the mask would swing back and forth in windy weather, *to oscillate* came to mean "to move back and forth." Related borrowings from Latin that preserve the "mouth" meaning of the root include *oral* and *orifice.* A borrowing that shows the "swinging" meaning is *oscilloscope.* In analysis an oscillation is the same as a *saltus* (*q.v.*). [154, 238]

osculating (adjective), **osculation** (noun): from Latin *osculum,* diminutive of *os,* stem *or-* "mouth." An *osculum,* literally "a little mouth," was a colloquial Latin way of saying "a kiss." In a kiss, a person's two lips touch each other as well as the person being kissed. From that notion of two things touching, mathematics uses the term *point of osculation* to designate a place on a curve where two branches have a common tangent and each branch continues on both sides of the point. The two branches "kiss" each other, and the curve in the vicinity of the common point looks like a little mouth. An example is the curve $y^2 = x^4(1 - x^2)$ at the point $(0, 0)$. A point of osculation is also known as a tacnode or a double cusp. In a similar fashion, the osculating circle of a curve at a point is a circle in a certain limiting position tangent to ($=$ kissing) the curve at the given point. Compare *kissing*. [154, 238]

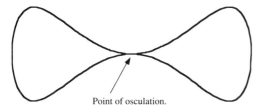

Point of osculation.

osculinflection or **osculinflexion** (noun): a combination of the words *osculation* and *inflection* (*qq.v.*). A point of osculinflection on a curve is one which is both a point of osculation and a point of inflection. [154, 238, 52, 61]

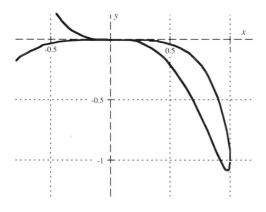

Point of osculinflection at (0,0): $y^2 + 2x^3y + x^7 = 0$.

ounce (noun): from Latin *uncia*, a unit of weight. The Latin word was derived from *unus*, from the Indo-European root *oi-no-* "one." The *uncia* took its name from the fact that it was a unit of a larger unit; specifically, each *uncia* was one of the twelve parts of the Roman pound known as the *libra*. Although an English ounce is still a common unit of weight (as

well as one of fluid volume), it has been redefined to be one-sixteenth (or one thirty-second, in the case of volume) of the next highest unit. The redefinition most likely arose from the process of halving the basic unit, then halving those halves, etc., much the way the cognate *inch* is subdivided. In the Troy system, however, which is used for weighing gold, a pound is still divided into twelve ounces. Compare *inch*. [148]

outcome (noun): a native English compound. The first element is from the Indo-European root *ud-* "up, out." The second element is from the Indo-European root g^wa- or g^wem- "to come." In the study of probability, an outcome is literally the result that "comes out" of an experiment. The only thing unusual in the word *outcome* is the order of the components: for example, the police never say to a criminal "Outcome with your hands up." [237, 77]

outdegree (noun): the first element is from the Indo-European root *ud-* "up, out." The second element is degree (*q.v.*). In graph theory, the outdegree of a point P in a graph is the number ($=$ degree) of arcs in the graph whose initial point is P. The prefix *out-* indicates that each arc goes out from point P. Contrast *indegree*. [237, 32, 73]

outer (adjective): a native English comparative of *out*, so that *outer* means literally "more outside." The Indo-European root is *ud-* "up, out." A related English word is *utmost*, literally "outmost." When the German Sanskrit scholar Hermann Günther Grassmann (1809–1877) developed the general algebra of hypercomplex numbers, he realized that more than one type of multiplication is possible. To two of the many possible types he gave the names *inner* and *outer*. The names seem to have been chosen because they are antonyms rather than for any intrinsic meaning. The outer product is also known as the vector product (since the product is itself a vector) or the cross product (because it is indicated by the cross-like symbol $\times$). [237]

outlier (noun): a native English compound. The first element is from the Indo-European root *ud-* "up, out." The second element is from the Indo-European root *legh-* "to lie." In a statistical scatter diagram, an outlier is a point that lies so far outside the pattern suggested by the other points that it seems to be an anomaly. If statistics had been invented centuries ago, an outlier would probably have been called something like an "abjacent" or "exjacent" point. [237, 112]

outside (adverb, preposition, noun): a native English compound. The first element is from the Indo-European root *ud-* "up, out." The second element may be from the Indo-European root *se-* "long, late." In mathematics a closed plane curve divides the plane into two regions, an inside and an outside. Textbooks often use the Latin-derived *exterior* rather than the simpler native English *outside*. [237, 182]

oval (noun): from Latin *ovum* "an egg," because an oval is egg-shaped. The Indo-European root is plausibly *awi-* "bird," which is where most of the eggs that people see come from. English *egg*, borrowed from Old Norse, is from the same source. Related borrowings from Latin include *ovary*, *avian*, and even *auspicious* (since in ancient times people looked at birds to try to predict the future). In mathematics an oval is a convex closed curve that has an axis of symmetry. As the etymological connection to eggs implies, an oval is generally pointier at one end than at the other. In less strict usage, *oval* may be a synonym of *ellipse*. [20]

over (adverb, preposition): a native English word, from the Indo-European *uper* "over." English *over* is etymologically the same as Greek *hyper-* and Latin *super*. A related borrowing from Latin is *supreme*. Note that in English we can use *over* in its literal, physical sense, as in "somewhere over the rainbow," but we can also use *over* in the metaphorical sense of "more than," as in "I have over a thousand stamps in my collection." In informal English, *over* is used as a synonym of *right*; students often say something like "To get from $(1, 5)$ to $(3, 10)$ you have to go over 2 and up 5." That usage probably developed from phrases like "cross over to the other side" or "over in France," in which *over* takes on a horizontal sense. [239]

overlap (verb, noun): a native English compound. The prefix is *over* (see previous entry). The main part of the word is from Old English *læppa* "flap of a garment," from the Indo-European root *leb-* "hanging loosely." When someone sits down, a loose flap of clothing often falls onto the person's now-horizontal upper legs, so that portion of a sitting body became known as a lap. From the noun came the verb *to lap*, which is what a loose garment does when it covers the lap. In mathematics two sets are said to overlap if they have at least one element in common, i.e., if their intersection is not empty. In a Venn diagram, the circle representing one set "falls onto" a part of the circle representing the other set, and vice versa. [239]

ovoid (noun): the first element is from Latin *ovum* "an egg" (see more under *oval*). The suffix is Greek-derived *-oid* (*q.v.*) "looking like." An ovoid is the three-dimensional counterpart of the two-dimensional oval. It is generated by rotating an oval about its axis of symmetry. Because an ovoid is three-dimensional, it is more like an egg than a two-dimensional oval is. [20, 244]

P

pack (verb), **packing** (noun): a native English word with cognates in the Germanic and Romance languages, but of unknown prior origin. In its oldest English usage, a pack was a bundle or bale. From the noun came the verb *to pack* "to put things into a bundle." When physical objects are put into a (back)pack, they can be pressed together but can't interpenetrate each other. Following that notion, in topology a packing is a set of non-overlapping sets. One mathematical challenge involves packing as many circles of different sizes as possible into a square of a certain size.

pair (noun, verb); **parity** (noun): *pair* is a French word, from Latin *par* "equal, even." The Indo-European root underlying Latin *par* may be *perə-* "to grant, to allot," which is possibly the same as *perə-* "to produce, procure." A native English cognate is *fair*, which describes a situation in which things are equal. French *pair* also appears in an alternate form in *peer* "someone who is your equal." Latin *par* is also used in English as a golf term; it is the number of strokes you must take to equal the average performance on a given hole. In mathematics when you pair things up, you form two equal groups. A pair means two things, with emphasis on the two-ness rather than on the even-ness of the original word. In contrast, a suffixed version of the word, *parity*, emphasizes the evenness (or oddness) of a number, rather than the twoness. As with many words of category—for example length, width, height—the etymological connection is to the "positive" member of the category: *parity* means literally "evenness," even though the word also includes the opposite polarity, oddness. Parity applies not only to integers but also to functions, which may be classified as even or odd (or neither). [167]

pairwise (adverb): a compound of *pair* (see previous entry) and native English *-wise*, from the Indo-

European root *weid-* "to see." How a thing looks or is seen is an indication of its condition or the way in which it is formed, so the English *wise*, as used in the phrases "in no wise," "in this wise," came to mean "manner, way." The word nowadays is rarely used on its own but usually serves as a suffix indicating manner, direction, or topic. The other English word *wise*, literally "able to see (and therefore understand) things," is from the same Indo-European root. In mathematics *pairwise* means "with reference to each pair" that can be formed from a larger group. For example, the numbers 8, 21, and 55 are pairwise relatively prime because each pair of numbers (8 and 21; 8 and 55; 21 and 55) is relatively prime. [167, 244]

palindrome (noun), **palindromic** (adjective): from Greek *palin* "again" and *dromos* "a running." The Indo-European root underlying the first component is $k^w el$- "to revolve," while the Indo-European root underlying the second component is *der-* "to run." In mathematics a palindrome is a numeral like 36763 whose digits read the same forwards or backwards; once you reach the middle of the numeral, the digits "run back" through the same sequence. The term *palindrome* is also applied to words or phrases that read the same in either direction. For example, if blue jeans were found in layers of rock, we could have *mined denim*. [107, 38]

pandiagonal (adjective): the first component is from Greek *pan(t)-* "all," from the Indo-European root *pant-* "all"; the same root is found in *Pandora*, the woman in Greek mythology who was given gifts by all the gods and goddesses. The second component is *diagonal* (*q.v.*). In any magic square, the sum of each row, each column, and each of the two main diagonals is a constant. In a pandiagonal magic square, the columns, the rows, and all the diagonals, including the "broken" ones, add up to the magic constant. A pandiagonal square is also called *panmagic* and *diabolic*. [157, 45, 65]

pandigital (adjective): from Greek *pan(t)-* "all," from the Indo-European root *pant-* "all"; plus *digital* (*q.v.*). A pandigital expression is one which uses all the digits from 0 through 9 (or sometimes from 1 through 9) exactly once. For example, one-half can be written as the pandigital fractions $\frac{9327}{18654}$ and $\frac{6729}{13458}$. The pandigital integer 2,438,195,760 is unusual in that it is divisible by every whole number from 2 through 18. [157, 33]

pangeometry (noun): the first element is from Greek *pan(t)-* "all," from the Indo-European root *pant-*, of the same meaning. The second element is *geometry*, a compound derived from Greek *geo-* "earth" and *metron* "a measure." The first two elements occur in *Pangaea*, a name given by geologists to the presumed "supercontinent" that was originally "all the earth" before separate continents developed. Pangeometry is not, as the word might imply, the measurement of ancient Pangaea, but rather the name that the Russian mathematician Nicholas Lobatchevsky (1796–1856) gave to his non-Euclidean geometry. The prefix *pan-* was intended to show that this new geometry transcended traditional plane geometry and was therefore "all-encompassing." [157, 63, 126]

panmagic (adjective): from Greek *pan(t)-* "all," from the Indo-European root *pant-* "all," plus *magic* (*q.v.*). A panmagic square is a magic square in which all the columns, all the rows, and all the diagonals, including the "broken" ones, add up to the magic constant. A panmagic square is also called *pandiagonal* and *diabolic*. [157, 123]

pantograph (noun): from Greek *pan(t)-* "all," from the Indo-European root *pant-* "all," plus *graph* (*q.v.*). A pantograph is an instrument made up of several linkages. Its purpose is to trace a geometrically similar version of a given figure. Whether the tracing (= graph) is larger or smaller than the original, any

9	16	2	7
6	3	13	12
15	10	8	1
4	5	11	14

A pandiagonal magic square.

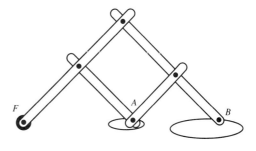

A pantograph anchored at point *F*. The figure that point *A* passes over is enlarged by the writing instrument at *B*.

and all (= *pant-*) figures can be copied. In the days before inexpensive copiers and computer-generated graphics, the pantograph was helpful in making enlargements and reductions. [157, 68]

parabola (noun), **parabolic** (adjective): from Greek *para* "alongside, nearby, right up to," and *-bola*, from verb *ballein* "to cast, to throw." The Indo-European root is *gʷelə-* "to throw, to reach." In mathematics, a parabola is a conic section with eccentricity 1; in other words, the eccentricity is "thrown right up to" the cutoff value of 1 that distinguishes among an ellipse, a parabola, and a hyperbola. Similarly, given a vertical cone centered at the origin of a three-dimensional coordinate system, a parabola results when the angle of the cutting plane is right up to (that is, equal to) the angle between the *xy*-plane and the "edge" of the cone. Since the original meaning of *paraballein* was "put alongside," there developed a subsidiary meaning "to compare." That explains our use of *parables*, stories that compare an easy-to-understand situation with a more complex one. Since parables were originally spoken stories, the Latin verb *parabolare* came to mean "to speak." That accounts for borrowings from French such as *parley*, *parlance*, *parlor*, *parliament*, and even *parole* (release of a prisoner on his word of honor that he will behave). [165, 78]

paraboloid (noun): from *parabola* (see previous entry) and Greek-derived *-oid* (*q.v.*) "looking like." A paraboloid is a three-dimensional figure that looks like the two-dimensional parabola in the sense that each cross-section parallel to the axis of the paraboloid is a parabola. Paraboloids are of two types, either elliptic or hyperbolic, depending on the cross-section that results when the paraboloid is cut perpendicular to its axis. [165, 78, 244]

paracompact (adjective), **paracompactum**, plural *paracompacta* (noun): the first element is from Greek *para* "beside, beyond." The Indo-European root is *per-* "forward, in front of." The second element is Latin-derived *compact* (*q.v.*). A topological space is said to be paracompact if any open covering can be refined to a locally finite open covering. That refinement is conveyed by the prefix *para-*. [165, 103, 156]

paradox (noun), **paradoxical** (adjective): from Greek *para* "alongside," and *doxa* "opinion." The Indo-European root is *dek-* "to take, to accept." Related borrowings from Latin include *docile*, *decorum*, and *dignity*. From Greek *doxa* we also have *orthodox* (= of the right opinion) and *dogma*. A paradox is a situation in which, alongside one opinion or interpretation, there is another, mutually-exclusive one. Probably the most famous paradoxes of antiquity are those of Zeno (c. 450 B.C.). Compare *antinomy*. [165]

paradromic (adjective): the first element is from Greek *para* "alongside," from the Indo-European root *per-* originally "forward, in front," but with many extended meanings. The second element is from Greek *dromos* "a running," from the Indo-European root *der-* "to run." Suppose the end of a strip of paper is twisted a certain number of times and then joined to the opposite end of the strip; if you now use a pair of scissors to cut completely around the strip parallel to and a certain distance from an edge, the resulting rings are said to be paradromic. Before being cut they "ran alongside" each other. The most familiar paradromic rings are the ones that result from cutting a Moebius strip. [165, 38]

parallel (adjective): the first element is from Greek *para* "alongside," from the Indo-European root *per-*, originally "forward, in front," but with many extended meanings. The second element is from Greek *allenon-* "one other." The Indo-European root is *al-* "beyond," as seen in native English *else*. Related borrowings from Latin are *alias*, *alien* and *alibi*. In geometry, parallel lines (or planes) run "alongside" one another, therefore always keeping a constant distance between them. In modern mathematics, parallel lines (or planes) may be considered to meet at a point (or line) infinitely far away. Although the term *parallel* is most often applied to lines or planes, it may also be used to describe curves or surfaces that maintain a constant distance between them. [165, 6]

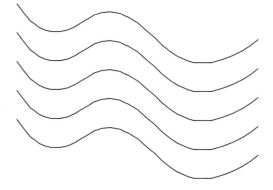

Parallel curves: there is a constant vertical distance between each curve and the curve immediately above or below it.

parallelepiped (noun), **parallelepipedal** (adjective): from Greek *parallel* (*q.v.*) plus *epipedon* "(level) ground," hence "plane." *Epipedon* is composed of *epi* "upon" and the root *ped-* "foot," so that the ground is conceived as being what you put your feet on. A related borrowing from French is *piedmont* (foothills of the mountains); from Greek we have borrowed *chiropodist* and *podiatrist*. In solid geometry a parallelepiped is a solid whose faces lie on three pairs of parallel planes. [165, 6, 53, 158]

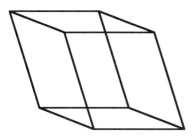

Parallelepiped.

parallelogram (noun): from Greek-derived *parallel* and *gram* (*qq.v.*) "something written," especially a line or letter of the alphabet. A parallelogram is a quadrilateral whose sides lie on two pairs of parallel lines. What we call a parallelogram in the United States is often known in Europe as a rhomboid (*q.v.*). In a parallelogram, two pairs of opposite sides are equal, in contrast to a kite (*q.v.*), in which two pairs of distinct adjacent sides are equal. [165, 6, 68]

parallelotope (noun): from *parallel* (*q.v.*) and Greek *topos* "place, region," of unknown prior origin. In the same way that a 2-dimensional parallelogram is bounded by 2 pairs of parallel lines, and a 3-dimensional parallelepiped is bounded by 3 pairs of parallel planes, a 4-dimensional parallelotope is bounded by 4 pairs of parallel hyperplanes. Compare *polytope*. [165, 6, 232]

parameter (noun), **parametric** (adjective): the first element is from Greek *para* "alongside," from the Indo-European root *per-* "forward, before," with many extended meanings." The second element is from Greek *metron* "a measure," whose Indo-European root is probably *me-* "to measure." In mathematics a parameter is a variable in terms of which two or more other variables are expressed; the parameter is a new "measure" or variable used "alongside" the other variables that are actually of interest. People who want to appear well educated sometimes use big words incorrectly. In recent years people who know little about mathematics have confused *parameter* with *perimeter* and have incorrectly used *parameter* to mean "limit, boundary." Such people should be confined to a room with a very small perimeter and made to study parametric equations. [165, 126]

parenthesis, plural *parentheses* (noun), **parenthesize** (verb): the first element is from Greek *para* "alongside," from the Indo-European root *per-* "forward, before," with many extended meanings. The second element is Greek and Indo-European *en* "in." The third element is from Greek *thesis* "a placing," from the Indo-European root *dhe-* "to set, to put." As early as the 16th century, a parenthetical remark was a subsidiary comment that a person "placed into" a text or talk "alongside" the main points being made. In the 18th century, the symbols used in print to indicate parenthetical comments came to be called parentheses themselves. The symbols commonly used were ( ) and [ ], though now only ( ) are properly called parentheses. Contrast *bracket* and *brace*. The fact that a single parenthetical comment is surrounded by two symbols has caused people to confuse the singular form *parenthesis* with the plural form *parentheses*. Within the realm of mathematics there is a different sort of confusion because the symbol (a, b) may mean the point whose x-value is a and whose y-value is b; or the open interval from a to b; or, for some writers, the vector whose components are a and b; or, in some contexts, the complex number $a + bi$. In the 16th century parentheses seldom appeared in mathematics precisely because they were used in ordinary writing. Eventually parentheses displaced the vinculum (*q.v.*) as the most common grouping symbol because parentheses could be printed on the same line of type as their contents, whereas the vinculum required typesetters to split the line. To parenthesize a quantity is to put parentheses around it. [165, 52, 41]

parity: see *pair*.

part (noun), **partial** (adjective): *part* is a French word, from Latin *pars*, stem *part-*, "part, piece, portion, share." The Indo-European root is *perə-* "to grant, to allot," which is possibly the same as *perə-* "to produce, procure." Related borrowings from Latin include *parcel* and *impart*. Classical geometry assumed that the whole equals the sum of its parts and is greater than any single part; that assumption is challenged by the modern mathematics of infinite sets. In calculus partial derivatives take their name from the fact that part of the time you differentiate with respect to one variable, and part of the time with respect to another. [167]

particular (adjective): from the Latin diminutive *particula* "a small part, a little bit, a particle," a diminutive of *pars*, stem *part-*, "part, piece, portion, share." The Indo-European root is *perǝ-* "to grant, to allot," which is possibly the same as *perǝ-* "to produce, procure." Something particular refers to a very small part of all the members of a group; the part may be so small, in fact, that it consists of a single member. In mathematics a particular solution of a differential equation contains exactly one value for each constant (actually each "variable constant") of integration. [167, 238]

partition (noun): from Latin *partitio*, stem *partition-*, "a sharing, parting, partition." The more basic Latin word is *pars*, stem *part-*, "part, piece, portion, share." The Indo-European root is *perǝ-* "to grant, to allot," which is possibly the same as *perǝ-* "to produce, procure." A partition is a breaking up into parts. In mathematics the integer 5 can be partitioned in seven distinct ways: $1 + 1 + 1 + 1 + 1$ or $1 + 1 + 1 + 2$ or $1 + 2 + 2$ or $1 + 1 + 3$ or $2 + 3$ or $1 + 4$ or 5 itself (which is a kind of non-partitioned partition). In a different context, partition is one of the two interpretations of the arithmetic operation of division. It corresponds to the following type of question: if we start with 30 objects and share them fairly among 5 people how many objects does each person get? The number of groups (5) is fixed, but the number of objects in each group has to be calculated. Contrast *quotition*. [167]

patch (noun): from Old French *pieche*, a dialectical variant of *piece*. The French word is from Medieval Latin *pettia*, in turn from a presumed Gaulish *pettia*. Related words in other Celtic languages show the meanings "quantity, part, share, and piece," though the ultimate source of the word is unknown. In mathematics a surface patch is a "piece" of a surface bounded by a closed curve, as opposed to an infinite surface or a closed surface like an ellipsoid. [170]

pattern (noun): the earliest English usage dates from the 14th century, when the word designated an object that served as a model or specimen. The word was borrowed, with some modification, from Old French *patron*, which has also been borrowed into English unchanged. Because a patron sets an example to be followed, the word *pattern* took on the sense of "model," and later "design." French *patron* was an augmentative of Latin *pater*, stem *patr-*, "father," so that a patron was conceived of as a father figure or "big daddy." The Indo-European root is *pǝter-*, with the same meaning as the native English cognate *father*. Related borrowings from Latin include *perpetrate* (to do something with the authority or effectiveness of a father) and *patrician*. Related borrowings from Greek include *patriot* and *patriarch*. [151]

pearl (noun): from the Indo-European *persna* "heel" came Latin *perna* "a ham together with the leg." By further extension the word came to be applied to a shelled creature shaped like a leg of mutton, i.e., an oyster. The diminutive of *perna*, *pernula*, later came to refer in the early Romance languages to the valuable object produced inside some oysters. In mathematics the family of curves represented by the equation $y^n = k(a - x)^m x^r$, where all exponents are positive integers, is known as the pearls of Sluze. Certain combinations of exponents produce a closed curve that looks somewhat like a pearl (though other combinations of exponents don't produce a closed curve at all). Blaise Pascal (1623–1662) named the family of curves to honor Baron René Française de Sluze, who studied the curves in 1657. [238]

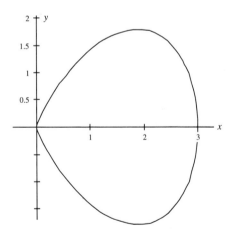

A pearl of Sluze: $y^6 = (3 - x)^3 x^5$.

peck (noun): from Norman French *pek*, of uncertain origin, but possibly related to French *picotin* "a measure of oats." The peck has been used as a unit of dry measure since the 13th century. In the United States Customary System, 1 peck = 8 quarts.

pedal (noun): via French from Italian *pedale*, in turn from Latin *pedalis* "having to do with the foot." The

Indo-European root is *ped-* "foot," as seen in native English *fetter* and *foot* itself. Related borrowings from Latin include *impede* and *expedite*. In nonmathematical English a pedal is a rod or lever meant to be stepped on by a foot. In mathematics a pedal curve is the locus of the foot of a perpendicular dropped from a fixed point to each of a curve's tangent lines. [158]

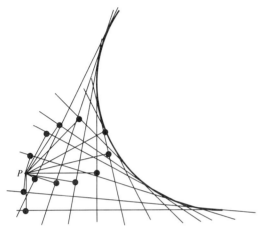

The dots lie on a pedal of the circular arc
with respect to point *P*.

pencil (noun): the earliest use of *pencil* in English was in the 14th century, when the word meant "an artist's paintbrush." The instrument we now call a pencil is so named by analogy with a paintbrush, since both are similar in shape and are used to make marks. Since a paintbrush contains many hairs bound together, the word *pencil* came to be used metaphorically in mathematics to refer to a group of lines, circles, or other shapes, all "bound together" by a certain feature. For example, all the planes passing through a given line comprise a pencil of planes. The shapes in question need not resemble a paintbrush or modern pencil in any way: all the spheres passing through a given circle form a pencil of spheres. The word *pencil* is from Latin *penicillus* "a small brush"; modern penicillin is so named because when looked at in a microscope its small tufts of filaments look like little brushes. *Penicillus* is a diminutive of Latin *penis* "a tail." The word *penis* was used even in Roman times as a euphemism for the male sex organ. What would schoolteachers think if they knew that *pencil* means a little *penis*? The word *pen*, which is so closely related to *pencil* semantically and in physical form, is surprisingly unrelated etymologically; *pen* is from Latin *pinna* "a feather," because early

pens were actually feathers with their tips cut to a sharp point.

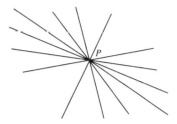

Some of the lines in the pencil
of lines through point *P*.

pentacle (noun): from Medieval Latin *pentaculum*, from Greek *pent-* "five" and a Latin diminutive suffix. A pentacle, literally "a little five[-pointed] thing," is the same as a pentagram (*q.v.*). [164, 238]

pentadecagon (noun): the first component is from Greek *pent-*, from the Indo-European root *penk^we* "five." The second component is from Greek *deka-*, from the Indo-European root *dekm-* "ten." The ending is from Greek *gon-*, from the Indo-European root *genu-* "angle, knee." A pentadecagon is a fifteen-angled (and therefore also fifteen-sided) polygon. A regular pentadecagon is constructible using only a compass and unmarked straightedge. [164, 34, 65]

pentagon (noun), **pentagonal** (adjective): the first component is from Greek *pent-*, from the Indo-European root *penk^we* "five." The second component is from Greek *gon-*, from the Indo-European root *genu-* "angle, knee," as seen in native English *knee*. The Greek root *pent-* is cognate to native English *five* and Latin *quinque*. In geometry a pentagon is a five-angled, and of course also five-sided, polygon. In the past it was occasionally called a *pentangle*, which is a Greek-Latin compound, or a *quinquangle*, which is a Latin-Latin compound. The pentagon was one of the polygons that the ancient Greeks knew how to construct using only a compass and unmarked straightedge. Among the figurate numbers, the pentagonal numbers are the set $\{1, 5, 12, 22, 35 \ldots\}$, because those numbers can be represented by a pentagonal pattern of dots in which each new row contains one dot more than the previous row. See picture under *figurate*. [164, 65]

pentagram (noun): the first component is from Greek *pent-*, from the Indo-European root *penk^we* "five." The second component is from Greek *gramma* "letter, figure." A pentagram is a plane figure that results from extending each adjacent pair of sides of a pentagon until they meet in a point; the resulting figure is a type of five-pointed star. The pentagram has occasionally been called a pentacle, a pentalpha, or a pentangle (*qq.v.*). The ancient Pythagoreans used the pentagram as a symbol of good health. [164, 68]

Pentagram.

pentahedron, plural *pentahedra* (noun), **pentahedral** (adjective): the first component is from Greek *pent-*, from the Indo-European root *penkʷe* "five." The second component is from Greek *hedra* (from prehistoric Greek *sedra*) "base." The Indo-European root *sed-* "to sit," as seen in native English *sit* and *seat*. A pentahedron is a five-based, i.e., five-faced, polyhedron. Only two types of pentahedra are possible: a pyramid with a quadrilateral base, and a "band" of three quadrilaterals bordering two opposing triangles. [164, 184]

pentakis- (numerical prefix): a Greek form meaning "five times," based on *pent-* "five," from the Indo-European root *penkʷe* "five." The prefix is used as part of the name of a stellated regular solid. It indicates that each original face became five faces. A pentakisdodecahedron, for example is a polyhedron with five times twelve, or 60, faces. [164]

pentalpha (noun): the first component is from Greek *pent-*, from the Indo-European root *penkʷe* "five." The second component is from *alpha*, the name of the first letter in the Greek alphabet. A pentalpha is the same as a pentagram (*q.v.*). The name derives from the fact that a pentagram looks as if it contains the capital letter *A* five times. [164, 7]

pentangle (noun): the first component is from Greek *pent-*, from the Indo-European root *penkʷe* "five." The second component is *angle* (*q.v.*). A pentangle is the same as a pentagram (*q.v.*). The name derives from the fact that a pentagram is a type of star with an angle at each of the five vertices. [164, 11, 238]

pentiamond (noun): see *polyiamond*.

pentomino (noun): the first component is from Greek *pent-*, from the Indo-European root *penkʷe* "five." For the second component and further explanation see *polyomino*. [164, 37]

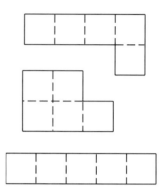

Three pentominoes.

percent (noun), **percentage** (noun): the first component is from Latin *per* "for." The Indo-European root is *per-* "forward, through, in front of," and many other things. The second component is from Latin *centum* "hundred," from the Indo-European root *dekm-tom-*, a form of *dekm-* "ten," because a hundred is ten tens. The word *percent* means literally "for (each) hundred." In older American books the full Latin phrase *per centum* was normally used. Later the abbreviation *per cent.* appeared, and eventually the period after the last letter was dropped. Modern usage allows *percent* as one word or *per cent* as two words. The symbol that commonly represents percent, %, may have originated from the second part of *p c°*, an early Italian abbreviation of *per cento*. By the 17th century the symbol $\frac{o}{o}$ was in common use in Europe to represent a percent. In any case, the current symbol, with its two "zeros," is a convenient reminder that a percent is a fraction whose denominator, 100, also contains two zeros. [165, 34]

percentile (noun): a word based on *percent* (see previous entry). In statistics a percentile is any of the hundred equal parts into which a distribution may be consecutively divided. Compare *fractile*, *quartile*, *quintile*, and *decile*. [165, 34]

perfect (adjective): from Latin *per* "through" and *factus*, past participle of *facere* "to do, to make." Something perfect has been "done through" to completion. (In common usage we say something is done when it is finished.) When something has been perfected, it needs nothing else. A whole number like 9 is a perfect square because nothing needs to be added to make it a square. A number like 6 is said to be perfect because the sum of its proper divisors needs nothing else to make the original number: $1 + 2 + 3 = 6$. The next perfect number is 28. Contrast *deficient*, *abundant*, and *semiperfect*. [165, 41]

perigon (noun): the first element is from Greek *peri* "around," from the Indo-European root *per-* "forward, through, in front of," and many other things. The second element is from Greek *gon-* "angle," from the Indo-European root *genu-* "angle, knee." A perigon is an angle that "goes all the way around," i.e., a 360° angle. Most mathematical words ending in *-gon* refer to polygons; *perigon* doesn't fit that pattern, which may be one reason the term has fallen out of use. Compare *round*. [165, 65]

perimeter (noun): the first element is from Greek *peri* "around," from the Indo-European root *per-* "forward, through, in front of," and many other things. The second component is from Greek *metron* "a measure," from the Indo-European root *me-* "to measure." The word *perimeter* is used ambiguously. The etymology is reflected in the meaning "the distance measured around a closed figure." *Perimeter* may also mean "the points on a polygon or closed curve, as opposed to the points inside such a figure." Although we could speak of the perimeter of a circle, we normally say *circumference* in that case. [165, 126]

period (noun), **periodic** (adjective): the first element is from Greek *peri* "around," from the Indo-European root *per-* "forward, through, in front of," and many other things. The second component is from Greek *hodos* (prehistoric root *sod-*) "way, course." The Indo-European root is *sed-* "to go." Related borrowings from Greek are *episode* and *odometer*. The original meaning of *period* was "a circuit, a revolution." The word was later applied metaphorically to sentences that were considered complete or well-rounded. At the end of a complete or *periodic* sentence the speaker pauses, so the punctuation mark indicating the end of a sentence came to be known as a period. Mathematically speaking, a periodic function is one which "goes around" in the sense of repeating its values at specified intervals. The most common periodic functions are those derived from going around a circle: the sine, cosine, and tangent functions and their reciprocals. The period of a periodic function is defined as the least

amount of time the function takes to complete one cycle. [165, 183]

permissible (adjective): the first component is from Latin *per* "through," from the Indo-European root *per-* "forward, through, in front of," and many other things. The second component is from Latin *missus*, past participle of *mittere*, "to send." The Indo-European root may be *(s)meit-*, "to throw," as seen in the related borrowing *missile*. In mathematics, a permissible value is one that may be "sent through" to a function and not cause problems. For example, in the function $\frac{1}{x}$ the x-value 2 is permissible, but 0 is not. [165, 200, 69]

permute (verb), **permutation** (noun): the first component is from Latin *per*, which can mean "through" or can intensify whatever follows. The Indo-European root is *per-* "forward, through, in front of," and many other things. The second component is from Latin *mutare* "to change," from the Indo-European root *mei-* "to change, go, move." It appears in English *mad*, because an insane person is so changed from normal; it appears as well in Greek *amœba*, which keeps changing its shape. In mathematics, each time the order of a set of objects is changed, the new order is said to be a permutation of the previous order. The number of ways n objects can be permuted r at a time is given by the formula

$$P(n, r) = \frac{n!}{(n - r)!}.$$

[165, 130]

perpendicular (adjective), **perpendicularity** (noun): from Latin *perpendiculum*, a diminutive form meaning "plumb-line." The first component is from Latin *per*, which can mean "through" or can intensify whatever follows. The Indo-European root is *per-* "forward, through, in front of," and many other things. The second component is from Latin *pendere* "to hang." The Indo-European root is *(s)pen-* "to draw, to stretch, to spin." Related borrowings from Latin include *pendant*, *poise*, and *penthouse* (a room "hanging from" or appended to the main living area). If you imagine a plumb-line hanging straight down, it is perpendicular to the ground. Compare *normal*, *orthogonal*, and *rectangular*. [165, 206, 238]

persistence (noun, adjective): the first component is from Latin *per*, which can mean "through" or can intensify whatever follows. The Indo-European root is *per-* "forward, through, in front of," and many other things. The second component is from the present participle of Latin *sistere* "to cause to stand," from the more basic *stare* "to stand." The Indo-European root is *sta-* as seen in English *stand*. When something persists, it "remains standing." In number theory the multiplicative persistence of an integer is the number of times all the digits must be multiplied, and then the digits of the product must be multiplied, etc., until a single digit results. For example, the multiplicative persistence of 39 is 3 because $3 \times 9 = 27$, and then $2 \times 7 = 14$, and finally $1 \times 4 = 4$, which is a single digit. [165, 208, 146]

perspective (noun): the first component is from Latin *per*, which can mean "through" or can intensify whatever follows. The Indo-European root is *per-* "forward, through, in front of," and many other things. The second component is from Latin *spectare* "to look, to watch, to observe." The Indo-European root is *spek-* "to observe." Related borrowings from Latin include *spectacle*, *specimen*, and *prospect*. From French comes *spy*. From Greek, via metathesis, comes the *skeptic* who looks askance at everything, as well as words ending in *-scope*. A perspective is a way of looking at something. In geometry, lines are in perspective if an eye placed at a single point could look down every one of the lines. [165, 205]

peta-, abbreviated *P* (numerical prefix): In the International System of Units, the prefix *peta-* multiplies the unit to which it is attached by 10^{15}, which can be rewritten as $(10^3)^5$, the fifth power of a thousand. *Peta-* is from Greek *pente* "five," with the final *-e* replaced by *-a* for uniformity: all of the numerical prefixes in the International System of Units have two syllables and end in a vowel, which in the case of magnifying prefixes created in recent times is always an *-a*. The *-n-* of the Greek root was dropped to avoid confusion with *pentameter* "a kind of verse with five feet," as well as with other words like *pentagon* in which the prefix refers to five of something rather than to raising something to the fifth power. The prefix *peta-* became a part of the International System of Units in 1975. [164]

petal (noun): from Greek *petalon* "a leaf," from the Indo-European root *petə-* "to spread out"; a typical leaf does indeed spread out as it grows. The fact that a petal is, etymologically and sometimes even botanically speaking, a leaf, explains why some people describe a given rose curve as having a certain number of "petals," while other people refer to the number of "leaves" the curve has. [169]

A rose or rhodonea of 12 petals.

phase (noun): from Greek *phasis* "appearance, phase of the moon," from the Indo-European root *bha-* "to shine." Native English cognates include *beacon* and *beckon*. The meaning of the Greek word shifted from the shining of the moon to its change in shape as time passed. By analogy with the phases of the moon, mathematics now refers to the phase displacement of a sinusoidal function, which is a type of translation.

phi (noun): twenty-first letter of the Greek alphabet, written Φ as a capital and ϕ in lower case, and corresponding approximately to our letter *f*. The Greeks did not borrow the letter from Hebrew or one of the alphabets based on Hebrew, but introduced it themselves. In mathematics ϕ is often used to designate an angle. In spherical coordinates, it is the angle measured from the vertical axis. Some writers use the letter ϕ to designate the Fibonacci [= Phi-bonacci] number,

$$\frac{1 + \sqrt{5}}{2}.$$

An oblique version of the letter ϕ is used to represent the empty set, the set with zero members in it, presumably because the letter looks like a zero with a slash through it. In number theory the function $\phi(n)$, known as Euler's ϕ function, counts the number of positive integers less than and relatively prime to *n*. Compare *indicator*.

pi (noun): the sixteenth letter of the Greek alphabet, written Π as a capital and π in lower case. It corresponds to the Hebrew letter *peh*, **פ**, meaning "mouth." See further explanation under *aleph*. The lower-case letter π was chosen for the most important of mathematical constants because it represents the initial sound of the words *periphery* and *perimeter* (now usually known as the circumference of a circle). In 1647 the English mathematician William Oughtred (1574–1660) used the fraction $\frac{\delta}{\pi}$ to represent the ratio of the diameter of a circle to its periphery. In 1697, David Gregory used the fraction $\frac{\pi}{\rho}$ to represent the ratio of the periphery of a circle to its radius. In 1706, William Jones (1675–1749) became the first English writer to use π by itself to represent the ratio of the circumference of a circle to the diameter. D.E. Smith reports that Euler's adoption of the symbol in 1737 led to its universal acceptance. The upper-case Π is used in mathematics to represent a *product*, again because of the sound of the word's first letter.

pico-, abbreviated *p* (numerical prefix): Spanish *pico* means both "a bird's pointed beak" and, on a larger scale, "the peak of a mountain." Because both a beak and a peak stick out, Spanish *pico* developed the figurative meaning "a small amount 'sticking out' beyond a relatively larger quantity." In the sense of "small amount" the International System of Units in 1960 adopted *pico-* as a prefix that multiplies the unit to which it is attached by 10^{-12}. Spanish *pico* may be from the Celtic *beccus* which, via French *bec*, entered English as *beak*. Another explanation links *pico* to the Indo-European root *(s)peik-* "woodpecker, magpie," both of which are birds with pointy bills. [204]

pictogram or **pictograph** (noun): the first component is from Latin *pictus*, past participle of *pingere* "to paint," from the Indo-European root *peig-* or *peik-* "to cut, to incise." A native English cognate is the tool we call a *file*. Related borrowings from Latin include *pigment* and *picture*. The second component of *pictogram* is from Greek *gramma* "a letter, a character, a figure" from *graphein* "to write." The Indo-European root is *gerbh-* "to scratch," because people originally wrote by scratching letters onto wood, clay, or the ground. Etymologically speaking, the word *pictogram* is redundant because both components have to do with cutting and scraping. In mathematics a pictogram is a figure representing statistical relationships. [159, 68]

pie (noun): a word first attested in English in the year 1199; it may be connected to French *pie*, from Latin *pica* "a magpie." One hypothesis explaining the connection between the bird and the baked dish is that magpies are known to collect all sorts of miscellaneous objects, and pies used to be made with various animal parts, vegetables, grains, and whatever the cook had on hand. The earliest known reference to a fruit pie was in 1590. In statistics, pie charts—which are round, like most pies—are used to show the relationship of parts to the whole. [204]

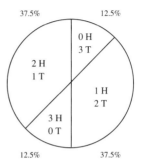

Pie chart representing the outcomes when three fair coins are tossed.

piecewise (adverb): the first element is from Old French *pece*, from Medieval Latin *pettia*, in turn from a presumed Gaulish *pettia*. Related words in other Celtic languages show the meanings "quantity, part, share, piece," though the ultimate source of the word is still unknown. The second element of the word is from the Indo-European root *weid-* "to see." How a thing looks or is seen is an indication of its condition or the way in which it is formed, so the English *wise*, as used in the phrases "in no wise," "in this wise," came to mean "manner, way." The word nowadays is rarely used on its own, but usually serves as a suffix indicating manner, direction, or topic. The other English word *wise*, literally "able to see (and therefore understand) things," is from the same Indo-European root as the suffix of manner. Piecewise means "with reference to each component piece" rather than with regard to the thing as a whole; for example, a curve may be piecewise continuous. [170, 244]

piercing (noun): from French *percer* "to pierce," from Latin *pertusus*, past participle of *pertundere* "to make a hole through"; from *per* "through" and *tundere* "to beat." The Indo-European root is *(s)teu-* "to push, stick, knock, beat." Native English cognates include *steep* and *steeple*. In a three-dimensional coordinate system, a piercing point is a point at which a line intersects one of the coordinate planes. [165, 214]

pigeonhole (noun, verb): the first component is from Old French *pijon* "young bird," ultimately from Latin *pipare* "to chirp, to peep," in imitation of the sounds that birds make. The second component is native English *hole* (*q.v.*). A pigeonhole was originally a small hole that a pigeon would make to use as a nest. By extension a pigeonhole now refers to any small hole or compartment, more often made by people than by birds. By even further extension *pigeonhole* has come to mean "category," with the corresponding verb *to pigeonhole* "to categorize, often quickly and perhaps without basis." In mathematics *pigeonhole* is used in the sense of "compartment, category, slot." The pigeonhole principle says that if more than n objects must be placed in n compartments, then at least one of the compartments will end up containing more than one object. The pigeonhole principle is invoked in many proofs. [89]

pint (noun): from Old French *pinte*, perhaps from Medieval Latin *pin(c)ta* "painted," referring to a painted mark on a container used for measuring. In the U.S. Customary System a pint is a unit of liquid measure as well as a unit of dry measure; in either case, a pint is one-half of the respective quart. [159]

piriform or **pyriform** (adjective): from Latin *pirum* "pear" and *form* (*q.v.*). A piriform curve is pear-shaped. The rectangular equation of the general piriform curve, first studied by G. de Longchamps in 1886, is $b^2y^2 = x^3(a - x)$. The curve often looks more like a fig than a pear, and therefore should perhaps be called *ficuform*, from Latin *ficus* "fig." [137]

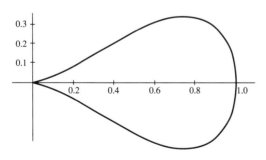

Piriform curve: $y^2 = x^3(1-x)$

pivot (noun, verb): a French word, perhaps akin to Provençal *pivo* (originally *pua*) "tooth of a comb." The word may be related to Latin *pungere* "to prick," from the Indo-European root *peuk-* "to prick." In modern usage, a pivot is a pin or rod about which something else rotates. In linear algebra a pivot is an element of a matrix about which other operations take place. [172]

place (noun): a French word, from a presumed Late Latin *plattus* "flat." Classical Latin had *platea* "open space, broad avenue"; those meanings are still apparent in street names like Park Place and in *plaza*, which English borrowed from Spanish. The Indo-European root is *plat-*, itself an extension of *pelə-* "flat, to spread." Native English cognates include *field*, *floor*, and the noun *flat*, while the adjective *flat* "an apartment" is borrowed from Old Norse. Our decimal system makes use of place value. Compare *exchange*. [162]

plane (noun), **planar** (adjective): from Latin *planus* "flat," which was also used as a noun with the meaning "wide open space, plain." The Indo-European root is *plat-*, itself an extension of *pelə-* "flat, to spread," as seen in native English *flat*. A related borrowing from Latin is *palm*, the flat of your hand. Also related is *Polish*, because Poland was named for its fields. Our adjective *plain*, as in the phrase "plain to see," is also really the same word, because when you are out in a plain you have a clear view all around you. The current spelling of *plane* was introduced in the 17th century to distinguish a mathematical plane from a geographic plain. [162]

planted (adjective): from Latin *planta* "a sprout, shoot, twig, plant." The Indo-European root is *plat-* "to spread"; when a plant grows it spreads out. In graph theory, which abounds with botanical metaphors, a tree is said to be planted if it is rooted (*q.v.*) and if the root is of first degree. [162]

platykurtic (adjective): the first component is from Greek *platus* "flat, broad," from the Indo-European root *plat-* "to spread," as seen in native English *flat*. The second component is from Greek *kurtos* "convex," from the Indo-European root *(s)ker-* "to turn, to bend." In statistics a distribution is said to be platykurtic if it is less heavily concentrated around the mean than a normal distribution is. Compared to the graph of a normal distribution, the graph of a platykurtic distribution will look "flattened" in the vicinity of the mean. For more information, see *kurtosis*. [162]

Platykurtic distribution.

plot (verb, noun): a native English word first attested in the 11th century with the meaning "a small piece of ground"; that meaning is still current. A Middle English variant of the word, *plat*, came to mean "a map of a piece of ground." The change in vowel may have been due to influence from French *plat* "flat." The original word then seems to have been confused with the variant, and *plot* acquired the meaning "a ground-plan, a map." From the noun came the verb *to plot*. The verb *plot* that means "to conspire" is from an unrelated French source. American schools stress the graphing of curves by plotting points rather than by analyzing the functions involved; as a consequence, many students plot many points when they graph even a simple equation like $y = x^2$ that should long since have become familiar.

plus (conjunction, adjective): a Latin word meaning "more," from the Indo-European root *pelə-*, with the same meaning as its English cognate *full*. The semantic connection is that the more you have of something, the fuller you are. The Latin plural of *plus* was *plures*, which is where the word *plural* comes from, because a plural is grammatically "fuller" than a singular. In arithmetic, when you add (indicated by a plus sign) a positive number to something, you end up with more than you had before. The opposite of Latin *plus* is *minus*, meaning "less." We can replace *minus* in "five *minus* three makes two" with the English equivalent and get "five *less* three makes two"; for some reason we can't replace Latin *plus* in "five plus three is eight" with the English equivalent to get "five more three is eight," although we can say "five more than three is eight." In Medieval and Renaissance arithmetic the word *plus* was often written out in full, but sometimes the abbreviations *p*, *P*, and $\bar{p}$ were also used. Our current plus sign, $+$, is presumed to have developed from the "t" of Latin *et* "and." Since Latin *plus* means "more," *plus* came to be used as an adjective with numbers that were more than zero, as in "*plus* three is the opposite of minus three"; Americans born after the 1940's will probably have replaced those terms by "positive three" and "negative three." There is nothing wrong with the older, shorter terminology, even though the longer versions are now taught as the only acceptable ones. [163]

point (noun, verb): a French word meaning "dot, point, period (the punctuation mark)," from Latin *punctus*, past participle of *pungere* "to prick, to puncture." The Indo-European root is *peug-* "to prick." Related borrowings from Latin include *pugilist* and *pugnacious*. When you puncture something, you use a sharp object to make a tiny hole in it. That tiny hole, especially from a distance, looks like a dot, so a point is a dot. Metaphorically speaking, if you're punctual, you arrive right on the dot. In mathematics a point is assumed to be dimensionless, but of course any physical representation of a point must be of some size. A point is often represented in textbooks by the smallest of all printing symbols, the period. In fact printers' type sizes are measured in units called points: the fact that 72 points make an inch tells us the size that printed periods commonly used to be. The verb *to point* developed from the use of the noun *point* to refer to the tapered end of an object like a stick or a pencil. Such objects were and still are used for pointing to things. In mathematics, two vectors of equal length may be distinguished by the direction in which each one points. [172]

pole (noun), **polar** (adjective): from Greek *polos* "pivot, axis." The Indo-European root is *kwel-* "to revolve," as found in native English *wheel*. The axis of rotation of a sphere intersects the sphere in two fixed points. Each of those points came to be known as a pole, as in the North Pole and South Pole of the earth. In the mathematical system of polar coordinates, the pole is the single fixed point from which the polar axis emanates and from which radial distances are measured. [107]

polygon (noun), **polygonal** (adjective): the first component is from Greek *polus* "many," from the Indo-European root *pelə-* "to fill," which is an English cognate. Related borrowings from Greek include *polygamy* and *polyester*. The second component is from Greek *gonia* "angle," from the Indo-European root *genu-* "knee, angle." Although we almost always define a polygon as a figure with many sides, the word actually tells us that the figure has many angles. A polygon with an indeterminate number n of sides is called an n-gon. Polygonal numbers are the same as two-dimensional figurate (*q.v.*) numbers. [163, 65]

polygonometry (noun): the first component is *polygon* (see previous entry). The second component is from Greek *metron* "a measure," from the Indo-European root *me-* "to measure." In the same way that *trigonometry* means literally "the measurement of triangles," the infrequently used word *polygonometry* refers to the measurement of polygons in general. [163, 65, 126]

polyhedron, plural *polyhedra* (noun), **polyhedral** (adjective): the first component is from Greek *polus* "many," from the Indo-European root *pelə-* "to fill," which is an English cognate. Related borrowings from Greek include *polyandry* and *polychrome*. The second component is from Greek *hedra* (originally *sedra*) "base," from the Indo-European root *sed-* "to sit," as seen in native English *sit* and *seat*. In solid geometry a polyhedron is a solid with many bases, i.e., faces. [163, 184]

polyhex (noun): the first component is from Greek *polus* "many," from the Indo-European root *pelə-* "to fill," which is an English cognate. Related borrowings from Greek include *polygyny* and *polyurethane*. The second component is from Greek *hex-* "six," from the Indo-European root *s(w)eks-* "six." Following the concept of a polyomino (*q.v.*), a polyhex of order *n* is a shape made by joining *n* congruent regular hexagons. There are dihexes, trihexes, tetrahexes, pentahexes, hexahexes, etc. Polyhexes occur in connection with tilings of the plane. [163, 220]

Two trihexes.

polyiamond (noun): the first component is from Greek *polus* "many," from the Indo-European root *pelə-* "to fill," which is an English cognate. Related borrowings from Greek include *polysyllabic* and *polytechnic*. The second component is all but the first letter of the word *diamond* (*q.v.*). Following the concept of a polyomino (*q.v.*), a polyiamond of order *n* is a shape made by joining *n* congruent equilateral triangles. The polyiamond of smallest order, 2, is in fact a diamond. Serendipitously, the *di-* of *diamond* happens to look like the *di-* meaning

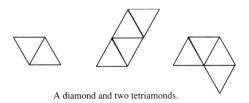

A diamond and two tetriamonds.

"two," as in *digon* and *dihedral*. There are also triamonds, tetriamonds, pentamonds, etc. Polyiamonds are studied in connection with tilings of the plane. [163, 141, 36]

polymorph (noun), **polymorphic** (adjective): the first component is from Greek *polus* "many," from the Indo-European root *pelə-* "to fill," which is an English cognate. Related borrowings from Greek include *polyphony* and *polyglot*. The second component is from Greek *morph-* "shape, form, beauty, outward appearance," of unknown prior origin. In number theory an integer that can be expressed in the form $x^2 \pm Dy^2$ in more than one way is said to be a polymorph. Contrast *antimorph* and *polymorph*. [163, 137]

polynomial (noun, adjective): the first part of the compound is from Greek *polus* "many," from the Indo-European root *pelə-*, as seen in *fill*, the English cognate of the same meaning. One explanation for the second part of the compound involves Greek *nomos*, which meant many things: usage, custom, law, division, portion, part. In that case, a trinomial is a mathematical expression consisting of three parts. A second and more likely correct explanation involves Latin *nomen*, cognate to the English *name*, so that a polynomial is an expression involving many names, i.e., terms. [163, 143, 145]

polyomino (noun): the first component is from Greek *polus* "many," from the Indo-European root *pelə-* "to fill," which is an English cognate. The second component is all but the first letter of Latin *domino*, from Latin *dominus* "master (of the house)." The more basic Latin word is *domus* "house," from the Indo-European root *demə-* "home." In the 16th century, a domino was a priest's winter cloak with hood; the name came from the expression *benedicamus domino* "Let us bless the Lord," which monks used to repeat as a brief prayer. Only in the 19th century did the term apply to the rectangular pieces used in the game of dominoes. One hypothesis to explain the connection between the two dissimilar objects is that the dominoes used in the game are of the same black color as the dominoes worn by monks. The *do-* in *domino* coincidentally but conveniently resembles the *do-* that means "two," as in *double* and *dodecagon*. Whereas a standard domino is made up of two square sections with numbers on each, a polyomino is made up of several square sections without any numbers on them. A player is challenged to combine the pieces in a set of polyominoes to produce certain shapes or simple pictures of recognizable objects. There are 5 distinct tetrominoes, 12 pentominoes, 35 hexominoes, 108 heptominoes, 369 octominoes, etc. Pentominoes have been more popular than any of the other ominoes. The term *polyomino* was coined by Solomon W. Golomb in 1954. [163, 37]

polystar (noun): the first component is from Greek *polus* "many," from the Indo-European root *pelə-* "to fill," which is an English cognate. Related borrowings from Greek include *polycentric* and *polydactyl*. The second component is native English *star* (*q.v.*). The term was coined by Richard L. Francis in 1988. In number theory a star number is a positive integer that fulfills two conditions: the integer has a certain property, and the number of digits in the integer also has the same property. For example, 8 is a star cube because both 8 and 1 (the number of digits required to write eight in base ten) are perfect cubes. A number like 27,000,000 is a double star cube because not only are 27,000,000 and 8, the number of digits in it, perfect cubes, but 8 is itself a star cube. Double star numbers, triple star numbers, etc., are collectively known as polystars. [163, 212]

polytope (noun): the first component is from Greek *polus* "many," from the Indo-European root *pelə-* "to fill," which is an English cognate. A related borrowing from Greek is the polysyllabic word *polysyllabic*). The second element is *topos* "place, region," of unknown prior origin. The word *polytope* was coined by analogy with *polygon*, a 2-dimensional closed figure whose edges are straight lines, and *polyhedron*, a 3-dimensional closed figure whose edges are straight lines: a polytope is a closed figure in 4 or more dimensions; its "faces" are made up of pieces of hyperplanes. A tesseract (*q.v.*) is one of the six regular 4-dimensional polytopes. Compare *parallelotope*. [163, 232]

pons asinorum (noun): a Latin phrase. The first word is *pons*, stem *pont-* "bridge," from the Indo-European root *pent-* "to tread, to go." An English cognate is *find*. The second word is a genitive plural of *asinus* "donkey, ass," probably of Middle Eastern origin. The *pons asinorum*, or bridge of asses, is the colloquial name given by European students to the geometric theorem stating that in an isosceles triangle the angles opposite the two equal sides are equal. The proposition, which can be tricky for beginners to prove, causes some students to make asses of themselves. The reference to a bridge is both literal (the two equal sides of the isosceles triangle make a sort of bridge over the base) and figurative (something

that must be crossed over before going on to more advanced matters).

population (noun): from Latin *populus*, "people," of Etruscan origin. Simply speaking, a population is just a collection of people. In fact English *people* is borrowed from French *peuple*, which developed from Latin *populus*. In statistics a population is a set of items, not necessarily people, defined by one or more common characteristics.

porism (noun): from Greek *porisma*, from *porizein* "to carry, to deduce," from *poros* "a way." The Indo-European root is *per-* "forward, through," with extended meaning "to pass over." Among the ancient Greeks a porism was a proposition that arose as the result of another proposition. In one sense a porism could be a corollary, a theorem that follows directly from another one. In a different sense, a porism was a finding of the conditions that render an existing theorem indeterminate or capable of many solutions. [165]

poset (noun): a combination of the words *partially ordered set* (*qq.v.*) [167, 15, 185]

position (noun): from Latin *positus*, past participle of *ponere* "to put." (See more under *component*.) Related borrowings from Latin are *repose* and *impose*. The position of something is literally the place where it is put. Old arithmetic books taught a topic called position. As explained in the inimitable words of the 1812 American textbook *The Scholar's Arithmetic*, "Position is a rule which, by false or supposed numbers, taken at pleasure, discovers the true one required . . . . [It] is the working with one supposed number, as if it were the true one, to find the true number." Position was a method of solving a simple linear equation by guessing an answer, seeing how well that answer worked, and then adjusting the guess accordingly. [14]

positive (adjective): from Latin *positus*, past participle of *ponere* "to put." (See more under *component*.) Something positive is "put down" so securely that no one can deny it. Before negative numbers were conceived there was no reason to call numbers positive because no other types of numbers were known to exist. When mathematicians began to deal with negative quantities, they needed to distinguish the new numbers from the traditional ones. The numbers that people were accustomed to using came to be known as *positive*, since people had been putting them down in writing and relying on them for thousands of years. The new numbers, whose reality and/or utility people at first denied, came to be called *negative* (*q.v.*). In the 15th century, numbers greater than zero were called *positive* or *affirmative*. Later writers called them *true* or *abundant*, but *positive* eventually prevailed. [14]

postmultiply (verb), **postmultiplication** (noun): from Latin *post* "after" and *multiplication* (*q.v.*). Postmultiplication is not something that you do after you multiply. The prefix *post-* is used in a physical rather than a temporal sense, and since we write from left to right, postmultiplication means multiplying from the right side. In $a \cdot b$, the b postmultiplies the a. Contrast *premultiplication*. Postmultiplication must be distinguished from premultiplication only when multiplication is not commutative, as with matrices, for example. [14, 132, 160]

postulate (noun): from Latin *postulare* "to ask for, to request," from the Indo-European root *prek-* "to ask, entreat." Borrowings from the related Latin *precari* "to entreat" are *pray* and *imprecation* (praying for evil rather than good). In a specific kind of geometry such as Euclidean geometry, postulates are the principles that you request people to adhere to in that system. You have to ask people to use your postulates because postulates by nature cannot be proven, but are starting points for the deductive system that follows.

potency (noun): from Latin *potent-*, the present participial stem of *posse* "to be able." The Indo-European root is *poti-* "powerful; lord." In set theory the potency of a set is the same as the cardinal number of the set. The potency tells how "powerful" (= big) the set is. [173, 146]

pound (noun): from Latin *pondus* "a weight." The Indo-European root *(s)pen-* "to draw, stretch, spin," appears in Latin words having to do with hanging and weight, the connection being that a weight draws or stretches a string straight downward. Related borrowings from Latin include *ponderous*, *pendant*, and *appendix*. From French comes *pensive*, your condition when you weigh things in your mind. From Spanish comes *peso*, now a unit of money but originally a unit of weight. The abbreviation *lb.* (*q.v.*) for *pound* comes from another word entirely, Latin *libra*, which was a Roman pound of twelve ounces. [206]

power (noun): from Old French *poeir*, from Vulgar Latin *potere*, a variant of Classical Latin *posse* "to be able." The Indo-European root is *poti-* "powerful; lord." If you are able to do many things, you are

powerful. A powerful person typically has a large number of possessions (a word derived from *posse*) and a large amount of money. In algebra, when even a relatively small number like 2 is multiplied by itself a number of times the result gets large very quickly; metaphorically speaking, the result is powerful. (Compare *factorial* and the symbol used to represent it for a similar metaphor.) If the term *power* is used precisely, it refers to the result of multiplying a number by itself a certain number of times. Consider $2^3 = 8$, which says that the 3rd power of 2 is 8. The power is 8. In less precise usage, however, 3 is identified as the power, when it is actually the exponent. [173]

precedent (noun): from Latin *præ* "before" and *cedent-*, present participial stem of *cedere* "to go, to yield," from the Indo-European root *ked-* "to go, to yield," Related borrowings from French include *cede*, *proceed*, and *recede*. In nonmathematical English, a precedent is an event that has "gone before" and that has set a pattern for future events. In a mathematical system containing a linear order relation, if $x < y$, then x is said to be a precedent of y because it "goes before" y in the order relation. [165, 88, 146]

precision (noun): from Latin *praecisus* "cut off in front." The prefix is from Latin *præ* "before." The main part of the word is from Latin *cisus*, the past participle of *caedere* "to cut," from the Indo-European root *kaə-id-* "to strike." Related borrowings from French include *concise*, *chisel*, and *scissors*. In mathematics and science, precison refers to the smallest unit of measurement used to express an approximate value. The exact value is "cut off" there. For example, the number 2.38 has a precision of 0.01. [165, 84]

predicate (noun): from Latin *praedicare* "to speak forth, to call out in public, to assert," from *præ* "forth" and *dicere* "to say, to tell." The Indo-European root is *deik-* "to show, pronounce solemnly." In logic, the predicate is what is said or asserted about a subject. Predicate calculus is a type of symbolic logic which takes into account the contents (= predicate) of a statement. [165, 33]

preimage (noun): from Latin *præ-* "in front of, before" and *image* (*q.v.*). In mathematics a function "operates on" a number and produces the image of that number. For example, the tripling function operates on 2 and produces 6 as its image. In the same example, 2 is said to be the preimage of 6 because the 2 must exist or be chosen before its image can be created. [165, 81]

premise or **premiss** (noun): from Latin *præ-* "in front of" and *missus*, past participle of *mittere* "to send, to put." The Indo-European may be *(s)meit(ə)-* "to throw." Related borrowings include *emit*, *mission*, *message*, and *promise*. In logic, the premises of an argument are the statements that are put before the argumentation proper. [165, 200]

premultiply (verb), **premultiplication** (noun): from Latin *præ-* "in front of" and *multiplication (q.v.)*. Premultiplication is not something that you do before you multiply. The prefix *pre-* is used in a physical rather than a temporal sense, and since we write from left to right, premultiplication means multiplying from the left side. In $a \cdot b$, the a premultiplies the b. Contrast *postmultiplication*. Premultiplication must be distinguished from postmultiplication only when multiplication is not commutative, as with matrices, for example. [165, 132, 160]

present (adjective): from Latin *præ-* "before," and *essent-* "being," the present participial stem of *esse* "to be." The Indo-European root is *es-* "to be," as seen in English *is*. A person who is present "is before" you. In finance, a present value is the value something has to have now, before it accrues interest over a period of time and becomes a certain larger amount. [165, 56, 146]

previous (adjective): from Latin *præ-* "before" and *via* "way, road," from the Indo-European root *wegh-* "to go, to transport." In a mathematical sequence, the previous term is the one that "goes before" the current term. [165, 243]

price (noun): from Latin *pretium* "price, worth, value." The Indo-European root is *per-* "to distribute, to traffic in, to sell." Related borrowings from French are *praise* (to give value to something) and *prize*. [165]

primal (noun): from Medieval Latin *primalis* "first, original," based on *prime (q.v.)*, with the suffix *-al* that appears in words like *decimal* "having to do with ten." The primal therapy that was popular in certain circles in the 1970's dealt with the original scream each of us is presumed to have let out when we were born. In number theory a primal is a prime that is not a divisor of the base we're working in. In base 10, for example, 3, 7, 11, and 31 are primals. The only primes that are not primal in base 10 are 2 and 5. The concept of a primal is useful in determining when the decimal expansion (or equivalent in a base other than base 10) of a reduced fraction will terminate: if no primal divides the denominator of a fraction

that is written in lowest terms, then the expansion terminates. [165]

prime (adjective, noun), **primality** (noun): from Latin *primus* "first," from the Indo-European root *per-* "forward, through, in front of"; there are many other extended meanings. Something right in front of you is the first thing you will encounter if you move forward. Something of prime quality is first class. Mathematically speaking, a prime number belongs to the first, most basic category of numbers: a prime cannot be divided by any number other than itself and 1. Geometrically speaking, if a whole number is represented by a corresponding set of physical objects like coins, the number is composite if the coins can be arranged into a rectangle with more than one row; for example, 15 can be represented as three rows of five coins. The coins representing a prime like 7, however, can only be arranged in a minimal rectangle consisting of a single row of seven coins. All the coins representing a prime are in the first (because only) row. The noun *primality* refers to the primeness or non-primeness of a number in the same way that *parity* refers to the oddness or evenness of a number. In calculus the *prime* symbol, as in y', represents the first derivative of a function. That notation is due to the French mathematician Joseph Louis Lagrange (1736–1813). [165]

primitive (adjective): from Latin *primitivus* "first, earliest," from *primitus* "in the first place," from *primus* "first." The Indo-European root is *per-* "forward, through, in front of"; there are many other extended meanings. Something right in front of you is the first thing you will encounter if you move forward. In nonmathematical English the word *primitive*, referring to civilizations in the first stages of development, has taken on the negative connotations "backward, savage." In number theory, however, the etymological meaning is maintained: a primitive kth root of unity is the "earliest" (= smallest positive) of the k kth roots of unity. Similarly, an integer d is said to be a primitive divisor of a function $f(n)$ if and only if d divides $f(n)$ but doesn't divide $f(x)$ for integers x such that $0 < x < n$; in other words, n is the first value of x that allows $f(x)$ to be divided by d. There is nothing "primitive" about such a definition. [165]

primorial (noun): a combination of *prime* and *factorial (qq.v.)*. For a positive integer n, factorial n is the product of n and every lesser positive integer down to the number 1. By analogy, for a prime p, primorial p is the product of p and every lesser prime down to the number 2. For example, primorial $7 = 7 \times 5 \times 3 \times 2 = 210$. The value of a primorial is always an even integer because 2 is always a factor. Although the symbol "!" is used to represent factorials, no standard symbol yet exists for primorials. Euclid used the concept of a primorial in his proof that the number of primes is infinite. The word *primorial* should not be confused with the etymologically related word *primordial* "pertaining to the earliest times or stages." Compare *oddorial*. [165, 153]

principal (adjective, noun), **principle** (noun): from the Latin compound *principium* "beginning." The first element is from Latin *primus* "first," from the Indo-European root *per-* "forward, through, in front of"; there are many other extended meanings. Something right in front of you is the first thing you will encounter if you move forward. The second element is from Latin *capere* "to take, to seize." The Indo-European root is *kap-* "to grasp," as seen in native English *have*, *hawk*, and *heavy*. When you begin something, you take your first steps toward a goal. When you have principles, they are the values you start from and judge things according to; your principles are literally your "first and foremost" beliefs. From that sense comes the use of principal to mean "main, most important," as in the principal square root of a positive number (as opposed to the other root, which is negative). In a somewhat weaker sense, a principle is a rule or law: an example is the pigeonhole principle that is used in various mathematical proofs. In finance, the principal is the amount of money you have at first, before interest makes the amount larger. [165, 86]

prism (noun), **prismatic** (adjective): from Greek *prisma*, "something that has been sawed," from the verb *priein* "to saw," of unknown prior origin. A prism is a polyhedron with two parallel, congruent bases, and sides that are parallelograms. Such an object probably was created in ancient times by sawing pieces off a block of wood. A prism is really a type of cylinder in which the generating curve is closed and consists of line segments. A prismatic surface is more general: the broken line that acts as a generator need not be closed. [174]

prismatoid (noun): the root is from Greek *prisma* "something that has been sawed," from the verb *priein* "to saw," of unknown prior origin. The Greek-derived suffix *-oid* (*q.v.*) means "looking like." A prismatoid looks somewhat like a prism, but the two parallel polygonal bases need not have the same

number of sides. The lateral faces of the prismatoid are triangles, trapezoids or parallelograms. [174, 244]

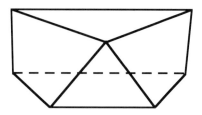

Prismatoid.

prismoid (noun), **prismoidal** (adjective): the root is from Greek *prisma* "something that has been sawed," from the verb *priein* "to saw," of unknown prior origin. The Greek-derived suffix *-oid* (*q.v.*) means "looking like." A prismoid looks somewhat like a prism, and the two polygons acting as parallel bases must still have the same number of sides, but they no longer need be congruent. For some authors, the two bases must be oriented in the same fashion. The vertices of one base are then connected to the corresponding vertices of the other base; in that case, the lateral faces of the prismoid are trapezoids or parallelograms. For other authors, the vertices of one base may be lined up with the edges of the opposite base; each vertex is then connected to the nearest two vertices of the opposite base. The second type of prismoid, all of whose lateral faces are triangles, is also known as an antiprism. In either case, a prismoid is a type of prismatoid. [174, 244]

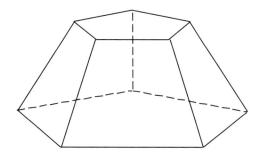

Pentagonal prismoid.

pro- (prefix): a Latin word with many meanings: "for, forward, before, on behalf of, in front of," etc. The Indo-European root is *per-* with a basic meaning "forward, through, toward" but with a great many derivative meanings: "first; around; to lead, to pass over; the young of an animal; to try, to risk; to traffic in, to sell." Of all Indo-European roots, this one has *probably produced* a greater *progeny*, both natively and through borrowing, than any other: the italicized words in this sentence all contain the root. [165]

probability (noun), **probabilistic** (adjective): the Latin adjective *probus* meant "upright, honest," from the Indo-European root *per-* "forward, through," with many other meanings. The derived verb *pro-bare* meant "to try, to test, to judge," especially to test to see how upright or honest someone is. For an explanation of the suffix see *-able*. The Latin compound *probabilis* meant "capable of being made good, able to be proved," but the thing in question had not necessarily already been proved or shown to have occurred. From there came our modern sense of *probable* as "likely, but not certain." The noun *probability* therefore means "likelihood." In mathematics Latin-derived *probability* usually replaces native English *likelihood*. [165, 69]

problem (noun): from the Greek compound *problema*. The first element is *pro-* (*q.v.*) "forward," from the Indo-European root *per-* "forward, through." The second element is from Greek *ballein* "to throw," from the Indo-European root *gʷelə-* "to throw, reach." A related borrowing from Greek is *ballistic*. A problem is a question or a puzzle "thrown out (= forward)" to you that you are expected to try to solve. [165, 78]

produce (verb): from Latin *pro* (*q.v.*) "forward, ahead" and *ductus*, past participle of *ducere* "to lead"; the Indo-European root is *deuk-* "to lead." Native English cognates include the homonyms *teem* and *team*, while some of the many borrowings from Latin are *conducive*, *seduce*, and *abduction*. In geometry, when a side of a triangle is produced, it is "led forward" beyond a vertex of the triangle. Compare *prolong*. In recent usage *produced* has been largely replaced by *extended*. See more under *product*. [165, 40]

product (noun): from Latin *pro* (*q.v.*) "forward, ahead," and *ductus*, past participle of *ducere* "to lead"; the Indo-European root is *deuk-* "to lead." Native English cognates include *tug* and *tow*, while some of the many borrowings from Latin are *conduct*, *induct*, and *duct*. When you produce something, for instance a movie, you "lead it forward" to a successful conclusion. A well-produced movie causes the investors' money to increase. In mathematics a product is the result of multiplying; when

two positive numbers larger than 1 are multiplied, the product "leads you forward" to a result that is bigger than either of the original numbers. Geometrically speaking, the number line must be produced (see previous entry) to accomodate the product. [165, 40]

profit (noun): via French, from Latin *pro* (*q.v.*) "for," plus *facere* "to do, to make," from the Indo-European root *dhe-* "to put." Profit is the money that a business or an investment makes for you. Arithmetically speaking, profit can be calculated as revenue minus expenses. [165, 41]

program (verb, noun), **programming** (noun): the Greek verb *graphein* originally meant "to scratch, to carve," from the Indo-European root of similar meaning, *gerbh-*. Later the Greek verb acquired the meaning "to write" (since writing was carved on tablets or stone). The derived word *gramma* referred to anything used for or in writing, including letters of the alphabet; it also gave rise to the word *grammar*, the set of rules used in writing. Written announcements that were made public were called programs because they were put *pro* (*q.v.*) "before" the public. The word *program* came to refer to the contents of the written announcement, especially if the announcement described a plan of action or a sequence of events. For example, if you go to a classical concert you usually get a program listing the pieces to be played. A computer program gets its name because it is a sequence of steps that the computer has to perform to achieve a certain objective. In mathematics, linear programming is a method of maximizing or minimizing a certain expression subject to a list (= "program") of constraints. [165, 68]

progression (noun): the first element is from Latin *pro-* "forward," from the Indo-European root *per-* "forward," as seen in native English *forth*. The second element is from *gressus*, past participle of *gradi* "to walk, take steps." The Indo-European root is *ghredh-* "go, to walk." When you progress you go forward; because the situation you are heading for is presumed or hoped to be better than the current one, the noun *progress* has come to mean "improvement." In mathematics, a progression is another name for a sequence because the terms of a progression "walk forth" one after the other. [165, 73]

project (verb), **projection** (noun), **projective** (adjective): the prefix is from Latin *pro* (*q.v.*) "forward." The main component is from Latin *iactus*, past participle of *iacere* "to throw," from the Indo-European root *ye-* "to throw." Related borrowings from Latin include *inject* and *subject*. From French comes a *jet* of water, which is thrown into the air. In geometry, when a line segment is projected it is literally "thrown forward." Projective geometry is the study of geometric properties that remain unchanged under projection. [165, 257]

prolate (noun): from Latin *pro* (*q.v.*) "forward" and *latus* "carried." The Indo-European root is *telə-* "to lift." A cycloid is a curve traced by a point on the circumference of a circle that rolls on a straight line. In a prolate cycloid, the distance from the center of the circle to the tracing point has been lengthened (= carried forward); the tracing point of a prolate cycloid is outside of, rather than on, the rolling circle. Whereas a regular cycloid has no loops in it, the tracing point of a prolate cycloid has been "carried forward" far enough that loops are formed. In solid geometry, the poles of a prolate spheroid are "carried forward" away from the center of a sphere, making the resulting ellipsoid longer in one dimension than in the other two. A prolate spheroid results from rotating an ellipse about its longer axis. Contrast *oblate*. [165, 224]

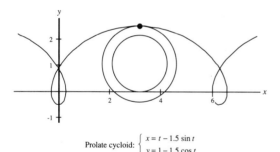

Prolate cycloid: $\begin{cases} x = t - 1.5 \sin t \\ y = 1 - 1.5 \cos t \end{cases}$

prolong (verb): from French *prolonger* "to prolong," from Latin *pro* (*q.v.*) "forward" and *longus* "long." Native English *long* is from the same Indo-European root, *del-* "long." When a line segment is prolonged, it is "pulled" forward and made longer. Compare *produce*. In recent usage, *prolonged* has been largely replaced by *extended*. [165, 35]

proof (noun), **prove** (verb): the Latin adjective *probus* meant "upright, honest," from the Indo-European root *per-* "forward, through," with many other meanings. The derived verb *probare* meant "to try, to test, to judge." One meaning of the verb then came to include the successful result of testing something, so *to prove* meant "to test and find valid." Similarly, if you *approve* of something, you test it

and find it acceptable. In a deductive system like mathematics, a proof tests a hypothesis only in the sense of validating it once and for all. In early 19th century American textbooks, *prove* was used in the etymological sense of "check, verify"; for example, multiplication was "proved" by casting out nines. [165]

proper (adjective): from the Latin compound *proprius* "one's own, special, particular." The first element is *pro* "for," from the Indo-European root *per-* "in front,": when something is intended for you, it is literally or figuratively put in front of you. The second element in the compound is *privus* "particular, one's own." *Privus* is also based on the same Indo-European root *per-* "in front": someone who is far in front of others ends up being alone or "in private." Something proper is your own in the sense that it belongs to you. To behave in a proper way is to act in accord with your true nature; you stay within the boundaries of suitable conduct. That notion of staying within certain bounds is carried over to arithmetic: a positive fraction is said to be proper when the numerator is less than the denominator; in other words, the value of the fraction is less than 1. By analogy, an algebraic fraction is proper when the degree of the numerator is less than the degree of the denominator. A proper divisor of a positive integer is a positive divisor less than the integer itself. [165]

property (noun): via French, from Latin *proprietas* "property." The Latin noun was built on the adjective *proprius*, which in turn was a compound of *pro* "for" and *privus* "particular, one's own." (See *proper*.) A property of a thing is therefore a trait or characteristic that the thing claims for its own. In arithmetic, for example, the commutative property is a characteristic of both addition and multiplication, but not of subtraction or division. [165]

proportion (noun), **proportional** (adjective), **proportionality** (noun): from Latin *proportio*, stem *proportion-*, a compound of *pro* (*q.v.*) "for, according to," and *portio*, stem *portion-*, "share, part, portion." The Indo-European root may be *per-* "to grant, to allot," which is possibly the same as *perə-* "to produce, procure." In mathematics a proportion is a statement of equality between two fractions, as in $\frac{1}{2} = \frac{3}{6}$. The 1 is the same portion of 2 that the 3 is of 6. Also, by a property of proportions, we can read from left to right, instead of from top to bottom, in which case we can say that 1 is the same portion of 3 that 2 is of 6. Latin *proportio* meant the same as modern "proportion," but also took on the extended meanings "analogy" and "symmetry." In a similar way we extend the meaning in English when we say something is well-proportioned, meaning "harmonious, symmetric, esthetically pleasing." [165, 167]

proposition (noun), **propositional** (adjective): from Latin *pro* (*q.v.*) "forward" and *positus*, past participle of *ponere* "to put." (See more under *position*.) A proposition is literally a statement that is "put forward." (Note the not-so-subtle distinction in English between a marriage proposal and a proposition.) In logic, a proposition is a statement that may be true or false. [165, 14]

protasis (noun): the first element of the compound is from Greek *protos* "first." The Indo-European root is *per-*, with a basic meaning "forward, but with many derivative meanings. The second element is from Greek *teinein* "to stretch," from the Indo-European root *ten-* "to stretch." A related English cognate is *thin*, a word that describes the state of something which has been stretched. A protasis is literally "a first stretching" or "a stretching forward." In geometry, the protasis of a proposition is the (necessarily first) part that tells what is to be proved. [165, 226]

prototile (noun): the first element of the compound is from Greek *protos* "first." The Indo-European root is *per-*, with a basic meaning "forward," but with many derivative meanings. The second element of the compound is *tile* (*q.v.*). A prototile is the shape and size common to all the tiles in a monohedral tiling. It is "first" in the sense of being the abstract pattern that all the individual congruent tiles adhere to. [165, 209]

protractor (noun): from Latin *pro* (*q.v.*) "forward" and *tractus*, past participle of *trahere*, meaning the same as its native English cognates *draw* and *drag*. The Indo-European root is *tragh-* "to draw, drag." An Indo-European variant *dhragh-* is the source of native English *draw* as well as *drag*, borrowed from Old Norse. The Latin *-or* suffix indicates a male person or thing that performs an action. In nonmathematical English, a tractor is a machine that draws or pulls something else. A protracted discussion is one which is "drawn forward" for a long time. In geometry, a protractor is an instrument that can "draw [a line] forward" to construct an angle of a certain size. [165, 234, 153]

pseudodiagonal (adjective): the first component is from Greek *pseudes* "false," from *pseudein* "to lie,"

of unknown prior origin. The second element is *diagonal* (*q.v.*). A diagonal matrix is a square matrix in which all the nonzero elements lie on the main diagonal. A pseudodiagonal matrix is "falsely diagonal" because it need not be square. See next entry. [175, 45, 65]

pseudoidentity (adjective): the first component is from Greek *pseudes* "false," from *pseudein* "to lie," of unknown prior origin. The second element is *identity* (*q.v.*). An identity matrix is a square matrix in which every element on the main diagonal is a 1, and every other element is a 0. A pseudoidentity matrix is "falsely" an identity matrix because it need not be square. [175, 80]

$$\begin{bmatrix} 1 & 0 & 0 & 0 & 0 \\ 0 & 1 & 0 & 0 & 0 \\ 0 & 0 & 1 & 0 & 0 \end{bmatrix}$$

A pseudoidentity
matrix

pseudoperfect (adjective): the first component is from Greek *pseudes* "false," from *pseudein* "to lie," of unknown prior origin. The second element is *perfect* (*q.v.*). In number theory, a perfect number is a positive integer like 28 that is equal to the sum of all of its proper divisors; $28 = 1 + 2 + 4 + 7 + 14$. By analogy, a pseudoperfect number is a positive integer like 20 that is equal to the sum of some of its proper divisors; $20 = 1 + 4 + 5 + 10$. Another word for *pseudoperfect* is *semi-perfect* (*q.v.*). [175, 165, 41]

pseudoprime (noun): the first component is from Greek *pseudes* "false," from *pseudein* "to lie," of unknown prior origin. The second element is *prime* (*q.v.*). In number theory, it is proved that if p is a prime, then p divides the quantity $(2^p - 2)$. Although the converse isn't true, Chinese mathematicians believed that it was. With reference to their mistake, the name *pseudoprime* has been given to a composite integer n which divides the quantity $(2^n - 2)$. For example, 341 $(= 11 \times 31)$ divides $(2^{341} - 2)$ even though 341 isn't a prime. The *pseudo-* in *pseudoprime* reflects the fact that a word with *prime* in it is being applied to a non-prime number. Some mathematicians extend the definition of *pseudoprime* to include primes, thereby creating a situation in which primes belong to the category of "false primes." [175, 165]

pseudosphere (noun), **pseudospherical** (adjective): from Greek *pseudes* "false," from *pseudein* "to lie," of unknown prior origin, plus *sphere* (*q.v.*). In mathematics, a pseudosphere is the surface of revolution generated when a tractrix (*q.v.*) is revolved about its asymptote. Whereas a sphere has the same positive curvature at every point, a pseudosphere has the same negative (= "false") total curvature at every point. [175, 207]

pseudotangent (noun): from Greek *pseudes* "false," from *pseudein* "to lie," of unknown prior origin; plus *tangent* (*q.v.*). A true tangent line fulfills two conditions: 1) it touches a curve only once at the point of tangency, and 2) it has the same slope as the curve at that point. A pseudotangent is a false tangent because it fulfills the first property but not the second. For example, the line $y = -x$ is a pseudotangent to the curve $y = \sqrt{x}$ at the origin. A plane may also be a pseudotangent: for instance, the *xy*-plane is a pseudotangent to the half-cone $z = \sqrt{x^2 + y^2}$ at the origin. [175, 222, 146]

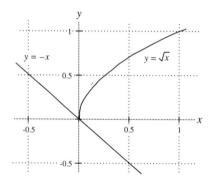

Pseudotangent at the origin.

pseudovertex, plural *pseudovertices* (noun): a newly coined word. The first component is from Greek *pseudes* "false," from *pseudein* "to lie," of unknown prior origin. The second element is *vertex* (*q.v.*). In an ellipse, the vertices lie at opposite ends of the major axis and mark the points where the

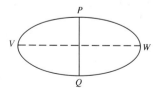

Points V and W are vertices
of the ellipse. Points P and Q
are pseudovertices.

ellipse has the greatest curvature. The two places where the ellipse has the least curvature lie at opposite ends of the minor axis, but those two points have traditionally gone nameless. The term *pseudovertex* has been created to fill the gap in nomenclature. The term may also be applied to the hitherto nameless points at the ends of a hyperbola's conjugate axis. [175, 251]

puncture (verb): from Latin *punctum*, past participle of *pungere* "to prick, to puncture." The Indo-European root is *peug-* "to prick." Latin *punctum* was also used as a noun meaning "the result of pricking something," i.e., "a dot, a point." In nonmathematical English, to puncture something is to make a hole in it. In mathematics a punctured circle is a circle which has had a single *punctum* or point removed from it; likewise a punctured sphere is missing a single point. The usage is curiously negative because an object is being described by the lack of something it normally has; compare a pitted olive, which is an olive that has had its pit taken out, or hulled barley, which is barley that has had its hull removed. English usually works in a positive sense: a dotted *i* has a dot and a crossed *t* has a crossbar. Unless you're familiar with the object in question you don't know whether a given past participle indicates the addition or removal of something. [172]

pure (adjective): from Latin *purus* "clean, free of dirt, unstained, pure." The Indo-European root is *peuə-* "to purify, to cleanse," as found also in borrowings from Latin like *purge* and *expurgate*. In algebra a complex number is of the form $a + bi$. A pure imaginary number is a complex number of the form bi that has been "cleansed of" its real part a. Pure mathematics is the study of mathematical principles for their own sake, as opposed to *applied* (*q.v.*) mathematics.

puzzle (verb, noun): in the 16th century *puzzle* was a verb meaning "embarrass, perplex, bewilder." It may be related to the 14th-century past participle *poselet* "bewildered, confused." It has been conjectured that *puzzle* comes from the verb *to (ap)pose* "to puzzle or confuse by asking a difficult question." If so, the source is Latin *positus* "put": when you puzzle someone, you put a question to the person. It is also possible that *puzzle* and *pose* are from independent sources but became confused because of their similar meaning. [14]

pyramid (noun), **pyramidal** (adjective): from Latin *pyramis*, stem *pyramid-*, in turn taken from Greek *puramid-*, with the same meaning as modern *pyramid* but also with the meaning "cake made out of roasted wheat." One hypothesis is that the meaning corresponding to the Egyptian pyramids was a metaphor based on their resemblance to the cake, but the cake could equally well have been named for the architectural structure. In any case, a pyramid is a polyhedron obtained by connecting each vertex of a polygon to a single point outside the plane of the polygon; the sloping sides of a pyramid are therefore triangles. [176]

pyriform: see *piriform*.

Q

Q. E. D. (clause): an abbreviation of the Latin *quod erat demonstrandum* "which was to be shown." Latin *quod*, like the cognate English *what*, is from the Indo-European root *kwo-* that appears in many relative and interrogative pronouns. Latin *erat* is a past tense of *esse* "to be"; the underlying Indo-European root *es-* is also found in native English *is*. Latin *demonstrandum* is a type of passive causative that means "to be shown," from the infinitive *demonstrare* "to demonstrate, to show." It is from the Indo-European demonstrative *de-* plus the root *men-* "to think." In order to demonstrate the truth of a proposition you must first think through your argument. In traditional geometry books, after the steps of a proof were listed, the abbreviation *Q.E.D.* typically appeared at the end in order to indicate that the proposition in question had now been proved. [109, 56, 32, 134, 139]

Q. E. F. (clause): an abbreviation of the Latin clause *quod erat faciendum* "which was to be done." Latin *quod*, like the cognate English *what*, is from the Indo-European root *kwo-* that appears in many relative and interrogative pronouns. Latin *erat* is a past tense of *esse* "to be"; the underlying Indo-European root *es-* is also found in native English *is*. Latin *faciendum* is a type of passive causative that means "to be done," from the infinitive *facere* "to do." It is from the Indo-European root *dhe-* "to set." In traditional geometry books, after the steps of a construction were carried out, the abbreviation *Q.E.F.* sometimes appeared at the end in order to indicate fulfillment of the original challenge. [109, 56, 41, 139]

quadrangle (noun), **quadrangular** (adjective): from Latin *quadr-* "four," plus *angle* (*q.v.*). A quadrangle is a plane figure made by connecting, in a given order, four points, provided that no three of the points are collinear. If the four points are connected in cyclical order, the quadrangle is a quadrilateral (*q.v.*). In nonmathematical English, a quadrangle (or quad) is a rectangular space surrounded by the facades of three or four buildings, as found typically on a college campus. A quadrangular pyramid is a pyramid whose base has four angles, and therefore also four sides. [108, 11, 238]

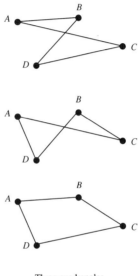

Three quadrangles
determined by the same
four points.

quadrant (noun), **quadrantal** (adjective): the first elemement is from Latin *quadrant-*, present participial stem of *quadrare* "to make square," from *quadrum* "a square." The Indo-European root is *kwetwer-* "four," because a square has four [equal] sides. In a two-dimensional Cartesian coordinate system, the two mutually perpendicular coordinate axes divide the plane into four "squares" or quadrants. Starting with the quadrant in which both coordinates are positive, the four quadrants of a plane are numbered in a counterclockwise direction. The quadrantal angles are 0, $\frac{\pi}{2}$, π, and $\frac{3\pi}{2}$; those angles form the boundaries between adjacent quadrants. Compare *octant* and *orthant*. [108, 146]

quadratic (adjective), **quadric** (adjective): from Latin *quadratum* "square," from the Indo-European root *kwetwer-* "four." To the ancient Romans, the name *square* was literally a description of the figure as "four-sided." The Romans, following the Greek model, conceived of the abstract quantity s^2 as the area of a square of side *s*. That's why something raised to the second power is said to be squared, using the English word, or *quadratic*, using the Latin word. A quadric surface such as an ellipsoid or a cone is one whose rectangular equation is quadratic. The words *quadratic* and *quadric* refer etymologically to the four-sided-ness of a square, but mathematically to the two-dimensionality of a square. [108]

quadratrix (noun): the main component is from Latin *quadratum* "square," from the Indo-European root $k^w etwer$- "four," because a square has four equal sides. The Latin agental suffix *-trix*, when added to a root, meant "the female person or thing that does [whatever the root indicates]." The quadratrix is a curve that can be used to "square the circle," i.e., to construct a line segment of length π. The feminine ending *-trix* was chosen because the Latin word *curva* "curve" was feminine. The rectangular equation of the quadratrix is $y = x \cot(\frac{\pi x}{2a})$. The curve was discovered by the Greek mathematician Hippias of Elis around 430 B.C. [108, 236]

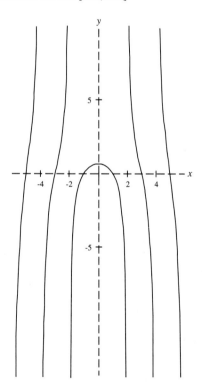

Central portion of quadratrix:

$$y = x \cot\left(\frac{\pi x}{2}\right)$$

quadrature (noun): from Latin *quadratum* "square," from the Indo-European root $k^w etwer$- "four," because a square has four equal sides. Quadrature is the process of finding a square whose area is equal to the area of a given surface. [108]

quadrifolium (noun): the first element is from Latin *quadri-* "four" from the Indo-European root $k^w etwer$- "four." The second element is from Latin *folium* "leaf." The Indo-European root is *bhel-* "to thrive, to bloom." In mathematics the quadrifolium is a symmetric curve with four "leaves." Its polar equation is $r = a \sin(2\theta)$, making it a special case of a rhodonea or rose. If you call the quadrifolium a rose, then the "leaves" might more appropriately be called "petals." Compare the etymologically but not mathematically identical *quatrefoil*. [108, 21]

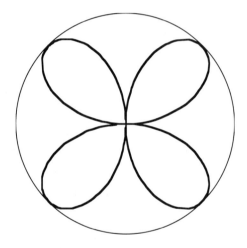

$r = \sin 2\theta$.

Quadrifolium inscribed
in a unit circle.

quadrilateral (noun, adjective): the first element is from Latin *quadri-* "four" from the Indo-European root $k^w etwer$- "four." The second element is from Latin *latus*, stem *later-*, "side," of unknown prior origin. A quadrilateral is a four-sided polygon. The Latin term is a partial translation of Greek *tetragon*, literally "four angles," since a closed figure with four angles also has four sides. Although we use words like *pentagon* and *polygon*, the term *quadrilateral* has completely replaced *tetragon*. [108, 110]

quadrillion (numeral): patterned after *billion* by replacing the *b-* with Latin *quadr-* "four." Since in most countries a billion is the second power of a million, a quadrillion was defined as the fourth power of a million, or 10^{24}. In the United States, however, a billion is 10^9, and a quadrillion adds two group of three zeroes, making a quadrillion equal to 10^{15}. [108, 72, 151]

quadrivium (noun): the first component of the Latin compound is from *quadr-*, from the Indo-European root *$k^w twer$-*, "four." The second component is from Latin *via* "way, path," from the Indo-European root *wegh-* "to go, to transport." The Greeks (and therefore also the Romans) divided knowledge into seven branches, seven being a number of mystic significance to the ancients. The "higher" branch, which consisted of music, astronomy, geometry, and arithmetic, later became known as the *quadrivium*. The three other subjects were collectively known as the *trivium*, which consisted of grammar, rhetoric, and logic. (Note that from a modern point of view logic was divorced from the other mathematical disciplines.) Although the word *quadrivium* existed in Classical Latin with the literal meaning "place where four roads meet," its use to designate the fourfold path to knowledge dates only from the Middle Ages. [108, 243]

quadruple (verb, adjective, noun): from Latin *quadruplus* "four-fold." The first element is from Latin *quadri-* "four," from the Indo-European root *$k^w etwer$-* "four." The second element is from the Indo-European root *pel-*, "to fold." To quadruple something is to make it four-fold, i.e., multiply it by 4. When the number of children born at one time is four times as great as usual, the babies are called quadruplets. A quadruple of numbers (a, b, c, d) is a set of four numbers in a given order. Compare *double*, *triple*, *quintuple*, *sextuple*, *septuple*, and *octuple*. [108, 160]

quantic (noun): from Latin *quantus* "how much, how great," from the Indo-European root *$k^w o$-* that appears in many relative and interrogative pronouns such as native English *why* and *which*. In mathematics, a quantic is a homogeneous algebraic polynomial in one or several variables. The degree of each term in a quantic is the same: in $3x^2 - 8xy - 5y^2$, all terms are of the second degree. In a quantic, each term answers the question "how great is the power of the expression as a whole?" The *-ic* suffix was chosen so that *quantic* would be the general term representing the sequence *quadratic*, *cubic*, *quartic*, *quintic*, *sextic*, ... A quantic may be further characterized as binary, ternary, quaternary, etc., depending on the number of variables in each term. [109]

quantify (verb), **quantifier** (noun): the first component is from Latin *quantus* "how much, how great." The Indo-European root *$k^w o$-* appears in many relative and interrogative pronouns such as native English *who* and *what*. The ending *-fy* is a much reduced version of Latin *facere* "to do, to make"; the underlying Indo-European root *dhe-* is apparent in native English *do*. To quantify something is literally "to make something into a quantity," that is, to give it a numerical value. In logic, a quantifier limits the scope (= "quantity") of a proposition. The two most common quantifiers are the universal and existential quantifiers. [109, 41]

quantity (noun): from Latin *quantus* "how much, how great," from the Indo-European root *$k^w o$-* that appears in many relative and interrogative pronouns such as native English *when* and *where*. In mathematics, a quantity is any expression concerned with value; a quantity answers the question "how much?" [109]

quart (noun): via French, from Latin *quartus* "fourth," from the Indo-European root *$k^w etwer$-* "four." Since the 14th century, a quart has been a unit of liquid measure equal to a quarter, that is a fourth, of a gallon; compare the way alcoholic beverages used to be sold in fifths, which were fifths of a gallon. The term *quart* was later used as a unit of dry measure equal to 67.2 cubic inches. With that extension the original fractional meaning of the word was lost: the dry quart isn't a fourth of any other unit, but is actually one-eighth of a peck. [108]

quarter (noun): from Latin *quartus* "fourth," from the Indo-European root *$k^w etwer$-* "four." A quarter of something is one-fourth of it. In the United States a quarter is one-fourth of a dollar. Compare the way a quart is one-fourth of a gallon. [108]

quartic (adjective): from the Latin ordinal number *quartus* "fourth," from the Indo-European root *$k^w etwer$-* "four." In algebra an equation of the fourth degree is called quartic; it is also sometimes called biquadratic, since a quadratic equation is of the second degree. Compare *quintic* and *sextic*. [108]

quartile (noun): from Latin *quartus* "fourth," from the Indo-European root *$k^w etwer$-* "four." In statistics a quartile is any of the four equal parts into which a distribution may be consecutively divided. Compare *fractile*, *quintile*, *decile*, and *percentile*. [108]

quasi-perfect (adjective): the first word is Latin *quasi* "as if, just as." The first element of that compound is Latin *quam* "as, than," from the Indo-European root $k^w o$- that appears in many relative and interrogative pronouns such as native English *who* and *what*. The second element is Latin *si* "so," ultimately from the Indo-European reflexive root *s(w)e*- that appears in native English *so* and *such*. In English, *quasi* is now used as a prefix meaning "almost." The second word in *quasi-perfect* is *perfect* (*q.v.*). In mathematics a perfect number is one whose proper divisors, when added up, equal the original number; for example, $6 = 1 + 2 + 3$. A quasi-perfect number is "almost" perfect: its proper divisors add up to one more than the number itself. Although the term *quasi-perfect* exists, no quasi-perfect number has ever been found, and it is not known if one exists. Contrast *deficient* and *abundant*. [109, 219, 165, 41]

quaternary (adjective): from Latin *quater* "four times," from the Indo-European root $k^w etwer$- "four." The word *quaternary* is patterned after the words *binary* and *ternary*. In algebra, a quaternary quantic (*q.v.*) is a type of polynomial in which each term contains four variables. [108]

quaternion (noun): from Latin *quater* "four times," from the Indo-European root $k^w etwer$- "four." A quaternion is a type of number symbolized by $x_0 + x_1 i + x_2 j + x_3 k$. It is so named because of its four terms. Quaternions were invented in 1843 by the Irish mathematician William Rowan Hamilton (1805–1865). [108]

quatrefoil (noun): the first component is from French *quatre* "four," from Latin *quattuor*, and the Indo-European root $k^w etwer$- "four." For the second component and further explanation see *multifoil*. [108, 21]

quattuordecillion (numeral): patterned after *billion* by replacing the *bi*- with the root of Latin *quattuordecim* "fourteen," a compound of *quattuor* "four" and *decem* "ten." Since in most countries a billion is the second power of a million, a quattuordecillion was defined as the fourteenth power of a million, or 10^{84}. In the United States, however, a billion is 10^9, and a quattuordecillion adds twelve groups of three zeroes, making a quattuordecillion equal to 10^{45}. [108, 34, 72, 151]

queue (noun), **queuing** (adjective): a French word meaning "tail," from Latin *cauda*, of the same meaning but of unknown prior origin. People queuing up in single file for a bus form a kind of tail. In mathematics, queuing theory examines things like the

frequency with which people join the end of a line at a supermarket checkout. A queue is distinguished from a stack because in a queue, the first come is the first served.

quinary (adjective): from the Latin ordinal number *quintus* "fifth," from the Indo-European root $penk^w e$- "five." A quinary number system is one which uses base five. A Chinese or Japanese abacus combines the decimal and quinary systems. Compare *binary*, *denary*, *octonary*, *senary*, *ternary*, and *vicenary*. [164]

quindecillion (numeral): patterned after *billion* by replacing the *bi*- with the root of Latin *quindecim* "fifteen," a compound of *quinque* "five" and *decem* "ten." Since in most countries a billion is the second power of a million, a quindecillion was defined as the fifteenth power of a million, or 10^{90}. In the United States, however, a billion is 10^9, and a quindecillion adds thirteen groups of three zeroes, making a quindecillion equal to 10^{48}. [164, 34, 72, 151]

quintal (noun): in spite of its use as a unit of mass equaling 100,000, or 10^5, grams, the word *quintal* is not derived from Latin *quintus* "fifth," or words like *quintic* and *quintuple*. The word is actually from Latin *centenarius* "having to do with one hundred," from *centum* "one hundred" (*q.v.*). Latin *centenarius* became Late Greek *kentenarion*, from which it was borrowed by the Arabs in the form *qintar*. The Arabic word later reentered Medieval Latin as *quintale*, eventually spreading to Old French and Middle English, where it had the meaning "hundredweight." (Curiously, although an American hundredweight weighed 100 pounds, at varying times a British hundredweight weighed 100, 112, or 120 pounds.) A quintal is, in fact, 100 kilograms in the metric system. [34]

quintic (adjective): from the Latin ordinal number *quintus* "fifth," from the Indo-European root $penk^w e$- "five." In the Roman calendar, which began in March, the old name for the fifth month was *Quintilis*. With some political maneuvering, it was later changed to what we now call *July*, for Julius Cæsar. In mathematics an equation of fifth degree is called a quintic equation. Compare *quadratic*, *quartic* and *sextic*. [164]

quintile (noun): from the Latin ordinal number *quintus* "fifth," from the Indo-European root $penk^w e$- "five." In statistics a quintile is any of the five equal parts into which a distribution may be consecutively divided. Compare *fractile*, *quartile*, *decile*, and *percentile*. [164]

quintillion (numeral): patterned after *billion* by replacing the *b–* with Latin *quint-* "fifth." Since in most countries a billion is the second power of a million, a quintillion was defined as the fifth power of a million, or 10^{30}. In the United States, however, a billion is 10^9, and a quintillion adds three groups of three zeroes, making a quintillion equal to 10^{18}. [164, 72, 151]

quintuple (verb, adjective, noun): from Late Latin *quintuplus* "five-fold." The first element is from the Latin ordinal number *quintus* "fifth," from the Indo-European root *penk^{w}e-* "five." The second element is from the Indo-European root *pel-* "to fold." To quintuple something is to make it five-fold, i.e., multiply it by 5. When the number of children born at once is five times as great as usual, the babies are called quintuplets. A quintuple of numbers (a, b, c, d, e) is a set of five numbers in a given order. Compare *double*, *triple*, *quadruple*, *sextuple*, *septuple*, and *octuple*. [164, 160]

quotient (noun): from Latin *quotiens* "how often, how many times," from the more basic *quot* "how many." Both forms are from the Indo-European root k^wo- that appears in many relative and interrogative pronouns such as native English *who* and *which*. In arithmetic, when two numbers are divided the quotient tells how many times the divisor goes into the dividend. As alternatives to the technical term *quotient*, people have historically used more general terms such as *answer* or *result*. [109, 146]

quotition (noun): from Latin *quot*; "how many," from the Indo-European root k^wo- that appears in many relative and interrogative pronouns such as native English *how* and *what*. The arithmetic operation of division can be interpreted in two ways. One interpretation, quotition, corresponds to the following type of question: if we start with 30 objects, how many groups of 5 can we make from them? Group size is fixed (5), but the number of groups has to be calculated. Contrast *partition*. [109]

R

radian (noun): An invented word based on *radius* (*q.v.*); in a circle, a radian is the size of a central angle subtended by an arc equal in length to the radius of the circle. The earliest use of *radian* recorded in the *Oxford English Dictionary* dates back only as far as 1879. Radian measure has also been called

circular measure, and on rare occasions π measure and natural measure. [177]

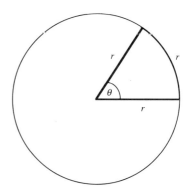

The size of angle θ is 1 radian.

radical (noun): from Latin *radix*, stem *radic-* "root," from the Indo-European root *wrad-* "branch, root." An English cognate of *radix* is *root*, borrowed from Old Norse. A radical reformer wants to get to the root of social problems, often by means considered extreme; as a consequence, any extremist may now be called a radical. In different words, a radical wants to eradicate society's ills by pulling them up the roots. In algebra the radical symbol, $\sqrt[n]{\ }$, indicates that the nth root of a quantity is being taken. The symbol itself is abstracted from the first letter of Latin *radix*. For quite some time the symbol was used without a horizontal bar at the top. The current version of the symbol is actually a fusion of the stylized letter *r* and the vinculum (*q.v.*) that was later used when a sum of terms was to have its root extracted. Some students mistakenly think a radical always involves a square root, probably because a square root is almost always indicated by a radical without an explicit index. [254]

radicand (noun): from Late Latin *radicare* "to take root," from Latin *radix*, stem *radic-* "root," from the Indo-European root *wrad-* "branch, root." The suffix *-nd-* turns the Latin verb into a type of passive causative, so that Latin *numerus radicandus* meant "a number that is to have its root taken." The radicand is the quantity that appears inside the radical. [254, 139]

radius, plural *radii* (noun); **radial** (adjective), **radiate** (verb): a Latin word of unknown prior origin meaning "staff, rod." In geometry the radius of a circle looks like a little rod connecting the center of the circle to the circle itself. Even in Classical Latin the word had its mathematical sense. Modern *radio* is borrowed from the same Latin word; radio waves radiate out from a center like the radii of a circle. Although many people think a radius can be found only in a circle or a sphere, analytic geometry deals with the focal radii of non-circular ellipses as well. In polar coordinates a segment beginning at the pole and ending on a curve is also frequently called a radius, and the distance from the pole to a given point on a curve is called a radial distance. A ray (*q.v.*) that begins at a given point is said to radiate from that point. [177]

radix, plural *radices* (noun): a Latin word meaning "root," from the Indo-European root *wrad*- "branch, root." Latin *radix* developed into French *radis*, which is where our word *radish* (a root vegetable) comes from. In mathematics a radix is the base (= root) of a system of logarithms or of a system of numeration. A radix fraction is analogous to a decimal fraction but the base of a radix fraction can be any appropriate number, not necessarily ten. [254]

ramphoid (adjective): the main element is from Greek *ramphos* "the crooked beak of birds, especially birds of prey," of unknown prior origin. The suffix is Greek-derived *-oid* (*q.v.*) "looking like." A ramphoid cusp, also known as a cusp of the second kind, is one in which both branches of the curve lie on the same side of the tangent line as they approach the point of tangency. An example is $y = x^2 \pm x^{5/2}$ at the point $(0, 0)$. Such a configuration looks (with some imagination) like the beak of a bird of prey. Contrast *keratoid*. [244]

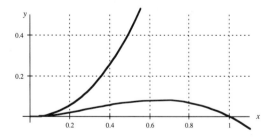

Ramphoid cusp at the origin: $y = x^2 \pm x^{\frac{5}{2}}$

random (adjective): from Old French *randir* "to run impetuously, to gallop." From the verb came the adverb *randon* "haphazardly, running off in one direction or another." The French word may be from Frankish *rant* "running," which would make it akin to English *run*. Another explanation connects *random* to an assumed Germanic *randa* "edge of a

shield," perhaps because when a man was in battle he struck wildly at enemies all around him with the edge of his shield. In statistics a sample of a population is said to be random if each member in the population has an equal chance of being chosen. To get such a sample you might have to "run around wildly." [180]

range (noun): an Old French word which was borrowed from a presumed Frankish *hring*, akin to English *ring*. The Indo-European root is *(s)ker*- "to turn, to bend." Originally the meaning of *range* was the same as that of *rank*. When a group of people is large enough, it can be divided up into many ranks, that is, rows, which range from the first row to the last one. The meaning of *range* shifted from the rows themselves to the extent of the rows; the sense "extent" is seen in a phrase like "home on the range." In mathematics the range of a function is the extent or set of all the values that the function can ever take on. Compare *codomain*. Contrast *domain*. [198]

rank (noun): from Old French *ranc*, which was borrowed from the Germanic predecessor of English *ring*. The Indo-European root is *(s)ker*- "to turn, to bend." In Old French the word *renc* referred to an assembly of people sitting in a circle. When there are many people assembled, some begin sitting in curved rows behind the innermost circle. The meaning of *rank* shifted from the circularity of the rows to the number or relative position of the rows; eventually the rows no longer needed to be circular, either. The English *rank and file* really means "rows and columns." In mathematics, one classification of a matrix is its rank, which is a cardinal number corresponding to the number of rows and columns in the largest nonzero determinant that appears in the matrix. [198]

rare (adjective): a French word, from Latin *rarus* "thin, loose in texture, not close, not thick." The Indo-European root is *erə*- "to separate." From the original sense "loose in texture," English *rare* has taken on the extended meanings "scarce, uncommon, unusual, valuable." As used in mathematics, the word retains its original meaning: a rare set is a set which is nowhere dense. [55]

rate (noun): an Old French word, from Medieval Latin *rata* in the phrase *pro rata parte* "according to a calculated part." The phrase *pro rata* is still used today both in its Latin form and as the Anglicized verb *to prorate*. For instance, if your car battery is guaranteed for 24 months but fails after 8 months, the store or battery company gives you credit for two-thirds of the battery's price, since the unexpired 16-month portion of your guarantee is in a ratio of 2 : 3 with the full 24-month term of the guarantee. The Latin word *rata* is the past participle of the verb *reri*, originally "to think, suppose, imagine, deem," with a derived sense "to reckon, calculate." The Indo-European root is plausibly *ar-* "to fit together," so that the Latin verb *reri* developed from the notion first of fitting ideas together, and later of fitting numbers together. Related borrowings from Latin include *artist*, a person who fits shapes together, and *ratio*, a "fitting together" of two numbers to compare them by division. When units are compared by division, the quotient is expressed in units like miles per hour or percent per year, so we began speaking of a rate of speed, a rate of interest, etc. [15]

ratio (noun), **rational** (adjective), **rationalize** (verb): Latin *ratio*, stem *ration-*, had many meanings: "thinking, reasoning, reckoning, relating." The closely related word *ratus* meant "reckoned, calculated." A ratio is a reckoning of the relationship of one thing to another. Mathematically speaking, it is the relationship gotten by dividing two things. A rational number is one that can be expressed as the quotient (= ratio) of two integers. The verb *rationalize* is formed from *rational* by adding the suffix *-ize*, of Greek origin, meaning "to make." In algebra when you rationalize a quantity you make it rational. Textbooks often ask students to rationalize the denominator of a fraction, i.e., to find an equivalent fraction that has no radical in the denominator. Before calculators were in common use, it was easier to find a decimal approximation for a fraction when the radical was on top. Nowadays, however, it is hard to rationalize many textbooks' slavish insistence on rationalizing all denominators. The Indo-European root underlying Latin *ratio* is plausibly *ar-* "to fit together," so that the Latin *ratio* developed from the idea of fitting numbers together in the sense of comparing them. Related borrowings include *arthritis*, a disease of the joints (where bones are fitted together), and *adorn* (to fit things together in a decorative way). Also related is native English *read*, to fit words together to make sense out of them. [15]

ray (noun): from Old French *rai*, from Latin *radius* "staff, rod," of unknown prior origin. In ancient Latin *radius* also meant "a ray of light," which is the meaning preserved in the Old French *rai* and English *ray*. It's easy to imagine a ray of light breaking

through the clouds and looking to people a lot like a staff or rod. In geometry a ray is a line bounded at one end but infinite at the other; in just that way, a ray of light emanates from a finite source but travels away from it (theoretically) forever. Old French *rai* has been replaced by modern French *rayon*, a word that has been borrowed into English as the name of a shiny (= radiant) fabric. [177]

reach (verb), **reachable** (adjective), **reachability** (noun): *reach* is a native English word. The Indo-European root is *reig-* "to reach, stretch out." In graph theory, a reachability matrix consists of 0's and 1's. The value of each entry n_{ij} is 1 if point P_j is reachable from point P_i, and 0 otherwise. For an explanation of the suffix see *-able*. [69]

real (adjective): the Latin word *res* meant "thing" (so a re-public is literally the "public thing"). The adjective corresponding to *res* was *realis*, which meant "thing-like." Something real was originally like a thing or an object: you could see or touch it. Something unreal was something fantastic, imaginary, nonexistent. In mathematics a real number is distinguished from an imaginary number (*q.v.*). Algebra textbooks typically ask students to rewrite a fraction like $\frac{3}{2-i}$ as $\frac{3(2+i)}{5}$ by multiplying the top and bottom of the original fraction by the complex conjugate of the bottom; if making a denominator rational is called rationalizing, then making a denominator real should be called "realizing." The verb *realize* does exist and does mean "to make real," as in "he realized his dream of finding a cure for tuberculosis," but few people realize that *realize* can be given a mathematical meaning.

reason (noun): from French *raison*, from Latin *ratio*, stem *ration-*, "thinking, reasoning, reckoning, relating." The Indo-European root is plausibly *ar-* "to fit together," so that Latin *ratio* in the sense of "relating, reasoning," developed from the concept of fitting ideas together. The related adjective *rational* means "capable of thinking, reasoning." In traditional high school geometry a proof is written in two columns: each statement on the left is followed by a corresponding reason on the right. [15]

reciprocal (noun), **reciprocity** (noun): from the opposite Latin roots *re-* "back(wards)" and *pro* "forwards." Latin *re-* is of uncertain origin, but *pro* is from the Indo-European root *per-* "forward, through, in front of," and many other things. Something reciprocal literally goes back and forth. A reciprocating engine, for example, is one in which a piston moves back and forth in a cylindrical shaft. In algebra, when you take the reciprocal of a fraction, the numerator and denominator go back and forth in the sense of switching places. Algebraically speaking, the reciprocal is the minus first power of something. Only two other powers are indicated by a special word: *squared* for the second power and *cubed* for the third power. The law of quadratic reciprocity plays an important role in number theory. [178, 165]

rectangle (noun), **rectangular** (adjective): from Latin *rectangulus* "right-angled," a literal translation of Greek *orthogonios* (see *orthogonal*). The first component is from Latin *rectus* "straight, upright, perpendicular," from the Indo-European root *reg-* "to move in a straight line"; the second component is *angle* (*q.v.*). In older English usage, *rectangle* meant "a right angle." In modern usage, a rectangle is a quadrilateral in which every adjacent pair of sides is perpendicular; in other words, all four angles are right angles. [179, 11, 238]

rectify (verb), **rectifiable** (adjective), **rectification** (noun): the first component is from Latin *rectus* "straight, upright, perpendicular," from the Indo-European root *reg-* "to move in a straight line." The second component is, via French, from Latin *facere* "to make," from the Indo-European root *dhe-* "to set, to put." When a section of a curve is rectifiable, it is "capable of being made straight or straightened out": in other words, its exact length can be found. For an explanation of the suffix see *-able*. In the calculus of space curves and surfaces, a rectifying plane is the same as a binormal plane because it makes a right angle with each of the other two planes involved. [179, 41, 69]

rectilinear (adjective): the first component is from Latin *rectus* "straight, upright, perpendicular," from the Indo-European root *reg-* "to move in a straight line." The second component is from Latin *linea* "a line" (*q.v.*). Rectilinear motion is motion in a straight line. A rectilinear figure is made up entirely of straight line segments. [179, 119]

recurring (adjective), **recursion** (noun), **recursive** (adjective): from Latin *re-* "back, again," of uncertain prior origin, and *curs-*, past participial stem of *currere* "to run," from the Indo-European root *kers-* "to run." Related borrowings from Latin and French include *excursion*, *concourse*, and *occur*. An infinite continued fraction is called recurring if it is periodic: the sequence of quotients keeps "running back by." Recursion is a technique in which the result of

a mathematical process is "run back through" the process again to be further refined. For a similar but not etymologically related term, see *iteration*. [178, 95]

reduce (verb), **reduction** (noun), **reducible** (adjective): from Latin *re-* "back, again," of uncertain prior origin, and *ducere* "to lead," from the Indo-European root *deuk-* "to lead." The same root appears in native English *tie*: when you tie your shoe, you "lead" the laces into a knot. In arithmetic, when you reduce a fraction you "lead it back" to lowest terms. Some people speak of "reducing" an improper fraction like $\frac{5}{3}$ to the mixed number $1\frac{2}{3}$, but no reduction has taken place there. What we now call reduction used to be known as *abbreviation* or *depression*, and the word *reduction* sometimes meant converting a group of fractions to a common denominator. For an explanation of the suffix in *reducible* see *-able*. [178, 40, 69]

reductio ad absurdum (phrase): the first Latin word is from *re-* "back (again)," of uncertain prior origin, and *ducere* "to lead," from the Indo-European root *deuk-* "to lead." Latin *ad* "to," is cognate to native English *at*. Latin *absurdum* has the same meaning as the English borrowing *absurd*. It appears to be a compound of *ab* "away from" and *surdus* "deaf." Before the age of standardized sign languages for the deaf, deaf people couldn't easily communicate with hearing people. The Latin compound *absurdus* therefore meant "senseless," and by extension "not reachable by reason," therefore "absurd." The Indo-European roots are *apo-* "away" and *swer-* "buzz, whisper." In mathematics if you are trying to prove that a proposition is true, you may use the type of argumentation known as *reductio ad absurdum*: you assume the opposite of what you want to prove, and then show that that "takes you back to" something absurd, so that the original proposition must be true. [178, 40, 1, 14, 221]

redundant (adjective): from Latin *redundant-*, present participial stem of *redundare* "to flow back, to overflow." The prefix is *red-* (usually *re-* before a vowel) "back, again," of uncertain prior origin. The main component is Latin *unda* "a wave, water," from the Indo-European root *wed-* "water, wet" (both of which are English cognates). In nonmathematical English a phrase like "positive improvement" is redundant because the "positiveness" inherent in an improvement "flows back again" with the unnecessary addition of the adjective *positive*. In mathematics an equation having all the solutions of another equation, and other solutions as well, is called redundant. The two most common ways in which a redundant equation arises are: (1) multiplying both sides of an equation by a variable expression that can take on the value zero; (2) raising both sides of an equation to the same power, especially the second power. Compare *extraneous*. [178, 242, 146]

reentrant or **re-entrant** (adjective): the first component is from French and Latin *re-* "back, again," of uncertain prior origin. The second component is the present participle of French *entrer* "to go in," ultimately from Latin *intra* "within," from the Indo-European root *en-* "in." In geometry a reentrant angle is a vertex angle of a polygon that is greater than 180°. A polygon with a reentrant angle is necessarily concave because the involved vertex "reenters" the polygon. [178, 52, 146]

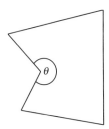

Angle *θ* is a
reentrant angle.

reference (noun): via French, from a Latin compound. The first element is from Latin *re-* "back, again," of uncertain prior origin. The second element is from Latin *ferrent-*, present participial stem of *ferre* "to carry, to bring"; the Indo-European root is *bher-* "to carry," as seen in native English *to bear*. In trigonometry you are "carried back" to a reference angle when you want to find a trigonometric function of an angle greater than 90°. The reference angle is most commonly the difference between the given angle and either 180° or 360°. For example, $\cos 120° = -\cos 60°$; the reference angle is 60°. [178, 23, 146]

reflect (verb), **reflection** (noun), **reflex** (adjective): from Latin *re-* "back" and *flectere* "to bend," both of unknown prior origin. In geometry, when a point is reflected in (or about) an axis, the point is "bent back" to a symmetric position on the opposite side of the axis. Also in geometry, a reflex angle is "bent back" more than 180°. A reflex polygon is one in

which a side is "bent back" so far that it crosses another side. [178, 61]

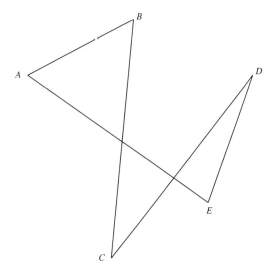

Reflex polygon *ABCDE*.

region (noun): from Latin *regio*, stem *region-*, "a making straight, a direction." The Indo-European root is *reg-* "to move in a straight line." The original linear sense was later extended in Latin, and *regio* came to mean "a boundary line." Because a boundary surrounds a country or a piece of land, *regio* finally acquired the meanings "area, province." Although both the one- and two-dimensional meanings existed simultaneously in Classical Latin, we have not borrowed the linear sense into English. On the other hand, we allow a region to be three-dimensional. [179]

regression (noun): from Latin *re-* "back again," of uncertain prior origin, and *gressus*, past participle of *gradi* "to walk, take steps." The Indo-European root is *ghredh-* "to go, to walk." Related borrowings from Latin include *congress* and *ingredient*. In statistics, when there is a regression to the mean, certain values "go back" toward the mean. [178, 73]

regula falsi (noun phrase): Latin *regula* meant "a ruler, a rule," from the Indo-European root *reg-* "to move in a straight line." Latin *falsus* is the past participle of *fallere* "to deceive, trick, dupe, cheat, lie under oath," of unknown prior origin. The *regula falsi* is usually translated into English as the "rule of false [position]." Popular for centuries, it was an iterative method of converging to the true root of an equation based on a pair of approximate roots. [179, 238, 59]

regular (adjective): the Indo-European root *reg-* meant "to move in a straight line." Figuratively speaking, a person who leads his people straight ahead is a leader or ruler. Consequently Latin *rex* meant "king," as seen in our borrowed word *regal*. The diminutive of *rex* was *regulus* or *regula*, from which we get *regular*. A good king lays down laws and set standards for people to live by; in such a kingdom everything is orderly, that is, well regulated. In geometry a regular polygon is the most orderly type because it has all its sides equal and all its angles equal. [179, 238]

regulus, plural *reguli* (noun): the Indo-European root *reg-* meant "to move in a straight line." That meaning appears in borrowed words like *regulate* and *ruler*. The *-ulus* ending is a Latin diminutive, so a regulus is literally "something small that regulates." In analytic geometry, each of the two distinct families of lines that generates a hyperboloid of one sheet is known as a regulus. The word *regulus* existed in Classical Latin with a pejorative, nonmathematical meaning: "a petty king or chieftain." [179, 238]

relagraph (noun): a combination of *relation* and *graph* (*qq.v.*). In graph theory, a relagraph is a pictorial representation of a relation. The term is used to avoid confusion with the term *graph* as used in analytic geometry and other branches of mathematics. [178, 224, 68]

relation (noun): from Latin *re-* "back, again," of uncertain prior origin, and *latus* "carried," from the Indo-European root *telə-* "to lift, support, weigh." That root is found in *toll*, borrowed from Greek, and *tolerate* and *extol*, borrowed from Latin. Among all possible people, your relations are the ones that you are "carried back to" along blood lines. In a mathematical relation, one thing is figuratively carried back to another. Mathematically, a relation is a set of ordered pairs (x, y) in which the value y is "carried back" or connected to the matching x. Contrast the more restrictive kind of relation known as a *function*. [178, 224]

relative (adjective), **relatively** (adverb): from Latin *re-* "back, again," of uncertain prior origin, and *latus* "carried," from the Indo-European root *telə-* "to lift, support, weigh." The root is found in *retaliate* (to carry hostilities back to your enemy), borrowed from Latin, and possibly *Atlas* (who carried the world on his shoulders), borrowed from Greek. When something is relative, it is "carried back" to, i.e., compared to, certain standards or values. In calculus a relative maximum is a maximum when compared to nearby values, but not necessarily to values farther away. In number theory, the integers 10 and 21 are said to be relatively prime because, although neither number by itself is a prime, insofar at the integers relate to each other they have no common divisor except the number 1. [178, 224, 117]

reliable (adjective), **reliability** (noun): via French, from Latin *re-* "back, again," of uncertain prior origin, and *ligare* "to bind, to tie." The Indo-European root is *leig-* "to bind." A related borrowing from Latin is *league*, a group of people who are bound together for a certain purpose. When something is reliable it can be "tied back" to something true or trustworthy. For an explanation of the suffix see *-able*. In mathematics reliability is a measure of the precision of a statistic. [178, 113, 69]

remainder (noun): from *remain*, from Old French *remanoir*. The original source is Latin *remanere*, a compound of *re-*, an intensifying prefix of unknown prior origin, and *manere* "to stay, to remain, to endure." Related borrowings include *mansion* (a fancy house you stay in) and the subsequent French *ménage*, a household. In modern usage the term *difference* applies to the result of subtraction in general, whereas a *remainder* is what remains after two numbers are subtracted at any stage of the larger process of long division. In older books *remainder* could be used as a synonym of *difference*. A typical question from the 1845 *Intellectual Algebra*, published in New York, is: "If one fourth of x be taken from x, what will express the remainder?" Even earlier writers sometimes referred to the remainder as the *residuum*. [178]

removable (adjective): from French *remouvoir*, from Latin *re-* "back," of uncertain prior origin, and *movere* "to move." The Indo-European root is *meuə-* "to push away." Related borrowings from Latin include *commotion* and *emotion*. In mathematics a discontinuity is said to be removable at a point if the value of the function can be "moved back" to that point (as opposed to the function's being defined differently or not at all). For an explanation of the suffix see *-able*. [178, 136, 69]

repdigit (noun): short for *repeated digit* (*qq.v.*). In number theory a repdigit is an integer made up entirely of the same digit. Two examples of repdigits are 555 and 99999. Note that unless the number of digits is 1, a repdigit is not a digit. [178, 168]

repeat (verb): from French *répéter*, from Latin *re-* "again," of uncertain prior origin, and *petere* "to head for, to seek, to attack." The Indo-European root is *pet-* "to rush, to fly," as seen in native English *feather*. In arithmetic, a repeating decimal eventually "heads back" to the same digit(s). Every rational number can be expressed as a repeating decimal (provided that 0 is allowed to be a final repeating digit.) A repeating decimal is also known as a circulating decimal. [178, 168]

repetend (noun): from Latin *repetere* "to repeat (*q.v.*), with the suffix *-nd-*, which creates a type of passive causative, so that Latin *repetend* means "which is to be repeated." In arithmetic, when an integer is divided by a larger integer, the digits of the quotient eventually begin repeating (if 0 is allowed to be a final repeating digit). The repetend is the set of digits that keeps repeating. A bar is usually written over a representative part of the repetend to distinguish it from the non-repeating digits of the quotient. [178, 168, 139]

replace (verb), **replacement** (noun): from Latin *re-* "back again," of uncertain prior origin, and *place* (*q.v.*). In combinatorics, when you choose with replacement you put each chosen object back in its original place before you choose again. A replacement set is a set of values that can be substituted for a given variable. [178, 162]

reptile (noun): from *repeat* and *tile* (*qq.v.*). As used by Solomon W. Golomb, who invented polyominoes, a reptile is a polygon that can be wholly subdivided into smaller congruent polygons that are geometrically similar to the original. [178, 168, 209]

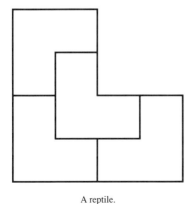

A reptile.

repunit (noun): short for *repeated unit* (*qq.v.*). The term was coined by Albert Beiler. In number the-

ory a repunit is an integer made up entirely of 1's. The repunit 1111, for example, is the product of 11 (another repunit) and 101. [178, 168, 148]

residual (adjective, noun), **residue** (noun): via French, from Latin *residuus* "left behind, remaining," from *residere* "to settle down, to sit back, to remain sitting." The first component is from Latin *re-* "back, again," of uncertain prior origin. The second component is from the Indo-European root *sed-*, with the same meaning as the English cognates *sit* and *seat*. When you reside somewhere, you "sit" there. A residue is something that "remains sitting" or is left behind. In number theory, 7 is the residue of 73 (mod 11) because 7 is left behind when 66—which is the multiple of 11 closest to but not greater than 73—is subtracted from 73. In statistics a residual is what is left when a predicted amount is subtracted from an actual amount. The residual tells "how far back" the actual value "sits" from the predicted value. [178, 184]

resolvent (adjective): from Latin *re-*, used as an intensifier, of uncertain prior origin, and *solvent-*, present participial stem of *solvere* "to free, to loosen." The Indo-European root *leu-* "to cut, to set free," is found in native English *lo(o)se*. In non-mathematical English, something resolvent helps "loosen up" a situation by separating a substance into its components. In algebra a resolvent cubic is a third-degree equation used to help find (= "free up") the roots of a fourth-degree equation. [178, 116, 146]

respectively (adverb): from Latin *re-* "back again," of uncertain prior origin, and *spectare* "to look." The Indo-European root is *spek-* "to observe," as found in Germanic *spy*. In mathematics we often encounter a statement like "7 and 8 are, respectively, prime and composite." The use of *respectively* indicates that the second pair of words, *prime* and *composite*, appears in the same order as the numbers with which they are being matched up, 7 and 8. If there is doubt, we can "look back" to see what the original order was. [178, 205, 117]

result (verb, noun), **resultant** (noun): via French, from Latin *re-* "back, again," of uncertain prior origin, and *saltare*, a popular version of *salire* "to jump, to leap." The Indo-European root is *sel-* "to jump." Related borrowings from French include *assail* and *somersault*. Figuratively speaking, a result is what jumps back at you after you perform a certain operation or calculation. If two vectors are added, the resultant is represented by the diagonal of the parallelogram that "leaps back out" from (= is determined by) the two vectors. [178, 187, 146]

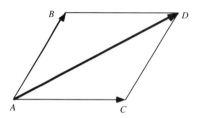

$\overrightarrow{AD}$ is the resultant when
vectors $\overrightarrow{AB}$ and $\overrightarrow{AC}$ are added.

retinend (noun): from Latin *retinendum* "to be retained." The prefix is *re-* "back," of uncertain prior origin. The main part of the word is from Latin *tenere* "to hold," from the Indo-European root *ten-* "to stretch." To retain something is literally to hold it back. The Latin suffix *-nd-* creates a type of passive causative: retinends are things which are to be retained. In symbolic logic the retinends are the parts of the premises which are to be retained in the conclusion. Lewis Carrol provides this example: All cats understand French; some chickens are cats; [therefore] some chickens understand French. The retinends are "some chickens" and "understand French." [178, 226, 139]

retract (noun): the prefix is from Latin *re-* "back," of uncertain prior origin. The main part of the word is from Latin *tractus*, the past participle of *trahere* "to draw, drag, haul." The Indo-European root is *tragh-* "to draw, drag." Related borrowings from Latin include *tractor* and *contract*. A retract is a type of subset of a topological space; there exists a continuous function that maps the topological space onto the retract in such a way that the identity map of the retract has a continuous extension to the topological space. Metaphorically speaking, the identity mapping draws each element back to its original value. [178, 234]

revenue (noun): from French *re-* "back, again," of uncertain prior origin, and *venue*, past participle of *venir* "to come." The Indo-European root is $g^w a$- or $g^w em$- "to come," as seen in the English cognate of the same meaning. In finance, revenue is money that "comes back" to a company; in other words, revenue is in*come*. [178, 77]

reverse (verb), **reversion** (noun): from Latin *re-* "back, again," of uncertain prior origin, and *versus*, past participle of *vertere* "to turn." The Indo-

European root is *wer-* "to turn, to bend." When something reverses, it turns back again. In mathematics, if the parametric equations $x = \cos^2 t$, $y = \sin^2 t$, represent the position of a point at time t, then the point moves along the line segment between $(1, 0)$ and $(0, 1)$, reversing direction each time it reaches either endpoint. Reversion of a series involves expressing x as a series in y, given y as a series in x. [178, 251]

revolve (verb), **revolution** (noun): from Latin *re-*, an intensifying prefix, of uncertain prior origin, and *volvere* "to roll." The Indo-European root is *wel-* "to turn, to roll." A native English cognate is *wallow*, and a related borrowing from German is *waltz*. Socially speaking, a revolution is a "rolling over" of the established order. In mathematics an axis of revolution is a line about which a curve "turns"; the surface generated is called a surface of revolution. [178, 248]

rho (noun): the seventeenth letter of the Greek alphabet, written P as a capital and ρ in lower case. It corresponds to the Hebrew letter *resh*, ר, meaning "head." See further explanation under *aleph*. In spherical coordinates, ρ is used to stand for the radial distance from the origin to another point. In some books it is similarly used to represent the radial distance in two-dimensional polar coordinates. In statistics ρ represents a coefficient of correlation.

rhodonea (noun), **rose** (noun): from a root of unknown origin, *wrod-*, meaning "rose." It is found with that meaning in Greek *rhodo-*, as in *rhodo-*

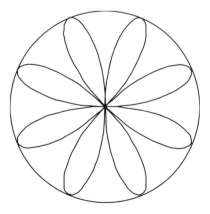

$r = \sin 4\theta.$

Rhodonea, or rose, of eight petals
inscribed in a unit circle.

dendron (= rose tree).It is also found in the less recognizable Persian word *gulab* "rose water," from which we get the sweet drink known as a *julep*. In mathematics the name *rhodonea* or *rose* is used for the family of curves with polar equation $r = \sin n\theta$ or $r = \cos n\theta$, where n is a positive integer greater than 1. If n is odd, the rose has n "petals" and occupies 25% of the area of the circumscribed circle; if n is even, the rose has $2n$ "petals" and occupies 50% of the area of the circumscribed circle. The Italian mathematician Guido Grandi (1671–1742) studied the rhodonea in 1723 and gave it its name. [255]

rhomb (noun): from Greek *rhombos*, known among anthropologists as a bull-roarer, a small object rapidly swung about on a cord in order to make a noise. The Indo-European root is *wer-* "to turn, to bend." Apparently the Greek *rhombos* was similar to what we now call a rhombus in geometry, a quadrilateral with all sides equal. Because a rhombus may be a square, the variant form *rhomb* is used to indicate a non-square rhombus. Compare *diamond* and *lozenge*. [251]

rhombicosidodecahedron, plural *rhombicosidodecahedra* (noun): a compound made up of *rhombus*, *icosahedron* and *dodecahedron* (*qq.v.*). A rhombicosidodecahedron is a semiregular polyhedron. There are two distinct rhombicosidodecahedra, one bearing the unmodified name and the other called the great rhombicosidodecahedron. Like the rhombus, the icosahedron and the dodecahedron, all of the great rhombicosidodecahedron's 62 faces are either [square] rhombi (30), equilateral triangles (20) or regular pentagons (12). [251, 253, 34, 48, 184]

rhombicuboctahedron (noun), plural *rhombicuboctahedra*: a compound made up of *rhombic* (in the sense of *orthorhombic*) and *cuboctahedron* (*qq.v.*). A rhombicuboctahedron is a semiregular polyhedron. There are two distinct rhombicuboctahedra, one bearing the unmodified name and the other called the great rhombicuboctahedron. Like the cube, octahedron, and cuboctahedron, all of the "smaller" rhombicuboctahedron's 26 faces are either squares (18) or equilateral triangles (8). It is [ortho]rhombic in the sense that alternate square faces are mutually orthogonal. [251, 96, 149, 184]

rhombohedron (noun), plural *rhombohedra*: a compound of *rhombus*, (*q.v.*) and Greek-derived *-hedron* "base of a convex solid." A rhombohedron is a prism each face of which is a rhombus or, more generally, a parallelogram. [251, 184]

rhomboid (noun): a compound of *rhombus* and *-oid* (*qq.v.*) "looking like." A rhomboid looks like a rhombus in the sense that both figures are parallelograms, but whereas adjacent sides of a rhombus are equal, adjacent sides of a rhomboid must be unequal. [244, 251]

rhombus (noun), **rhombic** (adjective): *rhombus* is a Latin word borrowed from Greek *rhombos*. The Indo-European root is *wer-* "to turn, to bend." A native English cognate is *wrap*. A *rhombos* in Greek was what is known among anthropologists as a bullroarer, a small object rapidly swung about on a cord in order to make a noise. Such objects were used in religious ceremonies by many cultures, not just the ancient Greeks. Apparently the shape of the Greek *rhombos* was akin to what we now call a rhombus: a parallelogram with all sides equal. Contrast *rhomb* and *lozenge*. [251]

rhumb (adjective): from French *rumb*, probably borrowed from Dutch *ruim* "space," akin to English *room*. The Indo-European root is *reuə-* "to open; space." Related borrowings from Latin include *rural* and *rustic*. A rhumb line is a spiral on the surface of the earth that crosses the meridians at a constant angle. Because a rhumb line is "oblique," the word has been confused with the word *rhombus*; that accounts for the *-h-* in the current spelling that was not originally in the word. Compare *loxodrome*.

right (adjective): a native English word, from the Indo-European *reg-* "to move in a straight line." When a weight is attached to the end of a string, the string hangs straight down and forms a line that makes a right angle with the ground. When you act right (or correctly, to use a Latin cognate), you walk a straight line, morally speaking. In geometry, a right angle is a 90° angle. A right cylinder is one whose axis makes a right angle with the base. The Indo-European *reg-* also developed into words like Latin *rex* "king" and English *rich*, that indicate power or might. For that reason, the hand that is stronger in most people became known as the right hand. In a standard rectangular (another cognate) coordinate system, the x-coordinate indicates how far right or left a point is located. A three-dimensional rectangular coordinate system may be right-handed or left-handed. In algebra a distributive property may likewise be right-handed or left-handed. Although the noun *height* refers to how high or low something is, and although there are many other such terms (for example *width*, *length*, *parity*), there is no noun that means "the rightness or leftness" or even just "the rightness" of a point in a coordinate system. We understand *rightness*, but the word isn't yet in the dictionary in its directional sense. [179]

rigid (adjective): from Latin *rigidus* "stiff, hard." The Indo-European root may be *reig-* "to reach, stretch out." Classical geometry permits only rigid transformations of a figure such as rotation and reflection, whereas modern mathematics, particularly topology, allows a figure to be deformed in many ways.

ring (noun): the Indo-European root *(s)ker-* meant "to turn, to bend." From there comes Old English *hring*, as well as many similar forms in the other Germanic languages. Modern English has lost the initial *h-* (as is also the case with modern *lid* and *loaf*). Mathematically speaking, a ring is a set that has two operations, called addition and multiplication, and that fulfills certain conditions. Small finite examples of rings may be set up using modular arithmetic, with the numbers arranged around a circle. Perhaps for that reason the mathematical structure in question was named a ring. In solid geometry a torus (*q.v.*) is a type of ring. [198]

rise (verb, noun): a native English word with cognates in other Germanic languages but of unknown prior origin. In analytic geometry the term *rise* is used to indicate a change in *y*-values; it appears in the phrase "rise over run" that defines the slope of a straight line.

robust (adjective): via Latin *robus* "red oak," from the Indo-European root *reudh* "red," which is an English cognate, as are *rust* and *ruddy*. Because an oak tree is strong, *robust* means literally "strong as an oak." A related borrowing from Latin is *corroborate*, metaphorically "to add the strength of an oak," therefore "to confirm." In statistics an inference procedure is said to be robust if the associated probability calculations are not affected by violations of the given assumptions.

root (noun): from Old Norse, from the Indo-European root *wrad-* "branch, root." A plant can't exist without its roots, so metaphorically speaking *root* means "basis" or "source." When you say that the cube root of 8 is 2, for example, 2 is the basic number that has to be raised up, like a plant from its roots, until it becomes 8. Also, when we solve an equation, we are looking for the solutions that we know or assume to be "rooted" in the equation; each solution is called a root of the equation. In graph theory, which abounds with botanical metaphors, a

graph is said to be rooted if one of its points, known as the root, is distinguished from all the other points. [254]

rose: see *rhodonea*.

rosette (noun): a French compound made up of *rose* and the diminutive ending *-ette*, so a rosette is "a little rose," or "something derived from a rose." French *rose* is from a root of unknown origin, *wrod-* "rose." In architecture, a rosette is a circular design that resembles a [sometimes highly stylized] rose. In mathematics a rosette is a curve with *n* petals that partially overlap and an *n*-fold point at the origin. In the mathematics of tilings, a rosette is a "round" figure composed of polygons that looks like a stylized rose. [255, 57]

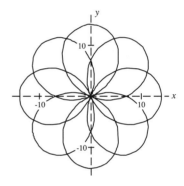

Rosette: $\begin{cases} x = 7\cos t + 7\cos 7t \\ y = 7\sin t - 7\sin 7t \end{cases}$

roster (noun): from Dutch *rooster* "a gridiron, a metal grid made up of parallel rods of metal on which meat is placed to be roasted." The word is of Germanic origin. A list of names written on successive lines of ruled paper came to be known as a roster by analogy with the parallel bars of a gridiron. By further analogy, a list of the members of a set is called a roster. Compare *extension*. Contrast *intension*.

rotate (verb), **rotation** (noun): from Latin *rotatus*, past participle of *rotare* "to turn around, to whirl about," from *rota* "a wheel." The Indo-European root is *ret-* "to run, to roll." Related borrowings include *roll*, from French, and *rodeo*, from Spanish. In mathematics a rotation is one type of rigid transformation. [181]

rotund (adjective, verb): from Latin *rotundus*, from the verb *rotare* "to turn around, to whirl about," from *rota* "a wheel." The Indo-European root is *ret-* "to run, to roll." A related borrowing from Latin is *ro-*

tary. In topology a rotund space is the same as a strictly convex space. [181]

roulette (noun): a French word, from *rouler* "to roll" and the French diminutive ending *-ette*. *Rouler* is from Old French *roueller*, from an assumed Vulgar Latin diminutive *rotella*, from Classical Latin *rota* "a wheel." The word *roulette* is therefore really a double diminutive meaning "a little little wheel." The game of chance known as roulette gets its name from the fact that a little ball rolls around on a turning wheel. In mathematics a roulette is the path of a point attached to the plane of a curve which rolls on a fixed curve. Compare *glissette*. [181, 238, 57]

As the ellipse rolls smoothly on the other curve, the point at its vertex traces a roulette.

round (adjective, verb): via French *rond*, from Latin *rotundus*, from the verb *rotare* "to turn around, to whirl about," from *rota* "a wheel." The Indo-European root is *ret-* "to run, to roll." Related borrowings from Latin include *rotary* and *prune*, from a presumed *prorotundiare* "to cut round in front." That sense of trimming is preserved in the expression "to round off" a decimal: we "cut off" digits beyond a certain place. In geometry a round angle is an angle of 360° (compare *perigon*). [181]

row (noun): a native English word, from the Indo-European root *rei-* "to scratch, to tear, to cut." A related borrowing from French is *river*. In a matrix, a determinant, or a magic square, the rows go across, like rivers cutting across the land, as opposed to the columns, which go up and down.

rule (noun), **ruler** (noun), **ruling** (noun): the Indo-European root *reg-* meant "to move in a straight line." The Latin diminutive based on that root, *regula*, meant "a straight stick, a bar, a pattern." The "pattern" sense appears in *rule*. The "straight line" sense appears in *ruling*, the momentary position of a moving straight line that generates a surface. Indo-European *reg-* also developed the meaning "leader,

king," because a king leads people in the right direction. Our word *ruler* reflects both the "straight line" meaning and the "leader" meaning. There have been many rules in mathematics. One that is still taught but that is no longer given a special name is the Rule of Three; it involves finding the fourth member of a proportion when the other three are known. It occurs in an anonymous student rhyme: "Multiplication is vexation; / Division is as Bad; / The Rule of Three perplexes me, / And Fractions drive me mad." [179, 238]

run (verb, noun): a native English word, from the Indo-European root *rei-* "to flow, to run." In algebra the term *run* is used to indicate a change in x-values: the slope of a straight line is sometimes defined as "rise over run." In the days of slide rules, the little piece containing the hairline was known as the runner because it ran up and down the slide rule. [180]

S

saddle (noun): a native English word, from the Indo-European root *sed-* "to sit," as found in the English cognates *sit* and *seat*. A saddle is a seat that sits on a horse, and you sit on the saddle. In mathematics a saddle point is a point on a surface $z = f(x, y)$ such that $\frac{\partial z}{\partial x}$ and $\frac{\partial z}{\partial y}$ are both zero yet the point is neither a relative maximum nor a relative minimum. In the vicinity of a saddle point, the surface looks a typical saddle: a vertical cross-section from front to back is concave up, while a vertical cross-section from side to side is concave down. [184]

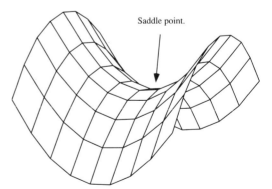

Saddle point.

salient (noun): from Latin *salient-* "jumping," present participial stem of *salire* "to bound, to jump, to leap." The Indo-European root is *sel-* "to jump." Related borrowings from French are *som-*

ersault and *sally (forth)*. In mathematics a salient point is one at which two branches of a curve meet with different tangents but do not cross. Functions of the type $y = |f(x)|$ typically have salient points wherever $f(x) = 0$ has a single root; as the curve passes through such a point it "jumps back up." Compare *corner* and contrast *cusp*. [187, 146]

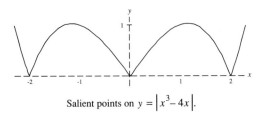

Salient points on $y = \left| x^3 - 4x \right|$.

salinon (noun): a Greek word meaning "saltcellar," which is a small dish in which salt is kept. The Indo-European root is *sal-* "salt," which is a native English cognate. Ancient Greek saltcellars must have resembled the salinon, which is an enclosed region bounded entirely by four semicircular arcs. Archimedes showed that the area of a salinon equals the area of the circle which is tangent to the salinon at its highest and lowest points. Compare *arbelos*.

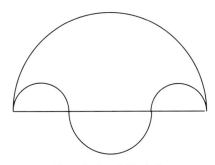

The region bounded by the four
semicircles is a salinon.

saltus (noun): a noun made from the past participle of Latin *salire* "to bound, to jump, to leap." The Indo-European root is *sel-* "to jump." A related borrowing from French is *sautée* (something made to "jump around" in a frying pan). In mathematics the saltus of a function is the difference between the least upper bound and the greatest lower bound at a certain point. The saltus is a measure of how much the function "jumps" at the point in question. Another word for *saltus* is *oscillation*. [187]

same (adjective): a disguised number word, from the Indo-European root *sem-* "one," with extended meaning "one and the same." If two numbers are the same, then we are really dealing with just one value. If two tangents are drawn to the same circle, then only one circle is involved. One of Euclid's axioms was: two things equal to the same thing are equal to each other. [188]

sample (noun): a shortened form of Anglo-Norman French *assample*, from Old French *essample*, from Latin *exemplum* "an example." The Latin compound is made up of *ex* "out of" and *emptus*, past participle of *emere* "to take." The Indo-European root is *em-* "to take, to distribute." If you have a group of things that are of the same type or follow a common pattern, a sample is one of those things that you have "taken out" to represent the group. Although *sample* was already used in English in the 13th century, the full form of the word, *example*, didn't appear till the 14th century. In statistics a sample is a representative subset of a population. [51, 49]

sandwich (noun): the first element of the compound is from the Indo-European root *bhes-* "to rub," because sand is what is "rubbed off" of rocks. The second element is a variant of the now obsolete native English *wick* "dwelling," which survives only in place names such as Greenwich and Norwich. The Indo-European root is *weik-* "clan." Related borrowings from Latin include *vicinity* and *village*. The connection between all of the above and the popular food consisting of two slices of bread with something between them is as follows. A certain British politician named John Montagu (1718–1792) was the Fourth Earl of Sandwich, a small borough close to the coast of southeast England; the place had most likely been named for its sandy appearance. The Earl of Sandwich was quite a gambler, and the sandwich was supposedly created for him so that he could eat without having to take a break from his playing. In calculus the sandwich theorem says that if two distinct sequences approach a common limit as more and more terms are considered, and if a third sequence is "sandwiched" between the other two, then the sequence in the middle also approaches the common limit. [247]

satisfy (verb): from French *satisfaire* "to satisfy." The first element is from Latin *satis* "enough," from the Indo-European root *sa-* "to satisfy." Surprisingly related is native English *sad* which originally meant "*sated*." The second element is from a much reduced form of Latin *facere* "to do, to make," from the Indo-European root *dhe-* "to set, to put," as seen in native English *do*. In mathematics a value is said to satisfy an equation (or inequality) if, when substituted into the equation (or inequality), the value "is enough to make" it true. [41]

scale (noun), **scalar** (adjective, noun): from Latin *scala* "a ladder," from the Indo-European root *skand-* "to climb." The meaning of the Latin word was later extended to a set of regular marks (like the steps on the ladder) used as a standard. That sense of "a standard for measuring" appears in phrases like "on a grand scale" as well as in the name of the object we weigh ourselves with, a scale. In mathematics the word *scale* has been used to mean the base of a number system; as recently as the middle of the 20th century, mathematics books would say that the Babylonians wrote their numbers "on a scale" of sixty. In the realm of vectors, multiplication by a constant changes only the scale (that is, the length) of a vector, but leaves its orientation unchanged. As a result, a constant has come to be known as a scalar, in contrast to a vector, which has both scale (= length) and direction. [194]

scalene (adjective): from the Indo-European root *skel-* "to cut." Greek *skalenos* originally meant "stirred up, hoed up." When a piece of ground is stirred up, the surface becomes "uneven," which was a later meaning of *skalenos*. A scalene triangle is uneven in the sense that all three sides are of different lengths. The scalene muscles on each side of a person's neck are named for their triangular appearance. A scalene cone or cylinder is one whose axis is not perpendicular to its base; opposite elements make "uneven" angles with the base. [196]

scalenohedron (adjective): from *scalene* and *-hedron* (*qq.v.*). A scalenohedron is a polyhedron each of whose faces is a polygon with all sides of different lengths; most often each face is a scalene triangle. [196, 184]

scatter (verb, noun), **scattergram** (noun): *scatter* is probably a variant of *shatter*, with *sh-* replaced by *sk-* under Scandinavian influence. Neither form is attested farther back than the year 1137. The linguist Calvert Watkins traces the origin of the word to an Indo-European root *sked-* "to split, scatter," itself an extended form of the root *sek-* "to cut." In mathematics a scattergram, short for *scatter diagram*, shows the frequencies with which joint values of variables are observed. [185, 68]

scope (noun): from Italian *scopo* "aim, purpose," from Greek *skopos* "a mark for shooting at," from the verb *skopein* "aim at, observe, examine." The

Greek words developed by metathesis from the Indo-European root *spek-* "to observe." Related borrowings from Greek include *microscope*, *periscope*, and *telescope*, all of which convey the sense "observe, look." In mathematics the scope of a symbol is, metaphorically speaking "how far it looks," i.e., "the extent to which it operates." For example, in the expression -3^2, the scope of the squaring is one and only one factor leftward, so that the -1 represented by the minus sign does not get squared. In the expression $\sin 2x$, "by rights" the sine function should apply only to the first factor to its right, but a generally-agreed-upon exception allows the sine function to operate on the quantity $(2x)$ even though no parentheses are used. Mistakes often occur because students don't understand the scope of a symbol. [205]

score (noun, verb): borrowed from Old Norse *skor* "a notch, a tally mark," from the Indo-European root *sker-* "to cut." Old Norse *skor* already had the extended meaning "twenty," presumably because tally marks were grouped in 20's. Compare the way Americans make four vertical tally marks and then cross them with a diagonal tally mark to form a group of five. Abraham Lincoln's "fourscore and seven years" is more picturesque than "eighty-seven years." [197]

secant (noun): from Latin *secans*, stem *secant-*, "cutting," the present participle of *secare* "to cut." The Indo-European root is *sek-* "to cut," as seen in native English *saw*. In geometry, a secant line "cuts" a circle into two parts (as opposed to a tangent line, which just "touches" a circle). In trigonometry the secant function gets its name from the length of a secant segment drawn to a unit circle and passing through the center of the circle. See additional picture under *exsecant*. [185, 146]

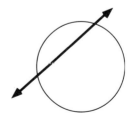

A secant line cuts
across a circle.

second (adjective, noun), **secondary** (adjective): from Latin *secundus*, literally "the one following," hence "the next one, the second one." The

Indo-European root is *sek^w*- "to follow." Related borrowings from Latin include *persecute* and *subsequent*. When a degree or hour is divided into sixty parts the first time, each part is known as a *minute* (*q.v.*). When the minute is further divided up, that is, when the degree or hour is divided a second time— each of the resulting parts is known as a *second*. In a sequence (another cognate), the second term is the one that follows the first term. The ordinal number *second* is one of only two ordinals that don't come from the corresponding cardinals in the way that *third* comes from *three* and *fifth* comes from *five*: the other anomaly is *first* (*q.v.*), which clearly doesn't come from *one*. [186]

section (noun): from Latin *sectio*, stem *section-*, "a cutting," from *sectus*, past participle of *secui* "to cut." The Indo-European root is *sek-* "to cut," as seen in native English *saw* and *scythe*. The same root is also found in the Latin borrowing *insect*, a creature whose body is "cut" into well-defined sections. A section of something is literally a piece cut off of or out of something larger. In mathematics the conic sections are obtained when a plane cuts a cone at various angles. [185]

sector (noun): from Latin *sectus*, past participle of *secui* "to cut," The Indo-European root is *sek-* "to cut," as seen in native English *saw* and *scythe*. The Latin *-or* suffix indicates a male person or thing that performs an action, so a sector is literally "a thing that cuts." A sector of a circle is a piece of the circle "cut out" by two radii. The sector itself "cuts into" the circle. A sector of a sphere is a portion of the surface bounded by a circle (but not a great circle). [185, 153]

Two radii cut a circle
into two sectors.

segment (noun): from Latin *segmentum* "a cutting, a piece cut off," from the Indo-European root *sek-* "to cut." The word *secant* comes from the same root. In plane geometry a segment of a circle is either of the two regions into which a secant "cuts" a circle; the larger region is called the major segment, and the smaller one the minor segment. In solid geometry a segment of a sphere consists of the part of the sphere "cut off" between parallel planes, as well as the region contained inside the sphere and between the planes. [185]

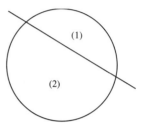

A secant line cuts a
circle and its interior
into two segments.

semi- (prefix): from Latin *semi-* "half." The Indo-European root is *semi-* "half." In prehistoric Greek, initial *s-* became *h-*, so the corresponding Greek cognate is *hemi-*. The only English cognate from this root appears in the first part of the archaic expression *sand-blind*, which has nothing to do with sand, but means "half blind." [189]

semicircle (noun), **semicircular** (adjective): from Latin *semi-* "half," plus *circle* (*qq.v.*). A semicircle is half a circle. Compare *hemicycle*. [189, 198]

semicubical (adjective): from Latin *semi-* "half," plus *cubical* (*qq.v.*) in its algebraic sense "of third degree." In analytic geometry, the semicubical parabola is the curve whose rectangular equation is $y = kx^3$. The name *semicubical* comes from the fact that if we solve for *y*, the equation becomes $y = \pm cx^{3/2}$, and the exponent of *x* is half of three. In spite of its name, the semicubical parabola is not a parabola. [189, 96]

semiellipse (noun): from Latin *semi-* "half," plus *ellipse* (*qq.v.*). A semiellipse is half of an ellipse, usually either of the two halves into which the major or minor axis divides an ellipse. [189, 52, 114]

semigroup (noun): from Latin *semi-* "half," plus *group* (*qq.v.*). The prefix *semi-* is used here in the loose sense of "partial, incomplete," rather than the literal "half." In mathematics a semigroup has closure and associativity, but lacks the other properties of a group. [189, 66]

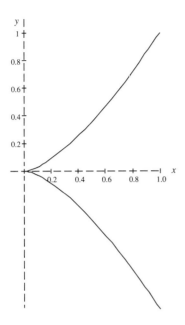

Semicubical parabola: $y^2 = x^3$.

semilogarithmic (adjective): from Latin *semi-* "half," plus *logarithmic* (*qq.v.*). Semilogarithmic graph paper has two perpendicular axes: on one axis the scale is uniform but on the other axis the scale is logarithmic. In other words, half of the two axes use a logarithmic scale. [189, 111, 15]

semimajor (adjective): from Latin *semi-* "half," plus *major* (*qq.v.*) "greater." In analytic geometry the semimajor axis of an ellipse or an ellipsoid is half of the figure's longest axis. Contrast *semiminor* and *semimean*. [189, 129]

semimean (adjective): from Latin *semi-* "half," plus *mean* (*qq.v.*), the literal meaning of which is "middle." A related borrowing from Latin is *medium*. In analytic geometry an ellipsoid has three axes. The semimean axis is half of the figure's "medium" (= neither the longest nor the shortest) axis. Contrast *semimajor* and *semiminor*. [189, 127]

semiminor (adjective): from Latin *semi-* "half," plus *minor* (*qq.v.*) "smaller." In analytic geometry the semiminor axis of an ellipse or an ellipsoid is half of the figure's shortest axis. Contrast *semimajor* and *semimean*. [189, 131]

semiperfect (adjective): from Latin *semi-* "half," plus *perfect* (*qq.v.*). The prefix *semi-* is used here in the loose sense of "partial, incomplete," rather than the literal "half." In number theory, a perfect number

is a positive integer like 28 that is equal to the sum of all of its proper divisors; $28 = 1 + 2 + 4 + 7 + 14$. By analogy, a semiperfect number is a positive integer like 20 that is equal to the sum of some of its proper divisors; $20 = 1 + 4 + 5 + 10$. A synonym of *semiperfect* is *pseudoperfect*. [189, 165, 41]

semiperimeter (noun): from Latin *semi-* "half," plus *perimeter* (*qq.v.*). The Swiss mathematician Leonhard Euler (1707–1783) introduced the use of the letter *s* to stand for the semiperimeter of a triangle. Using the semiperimeter rather than the full perimeter is useful in finding a formula for the area of a triangle: if *a*, *b*, and *c* are the lengths of the sides of the triangle, then $A = \sqrt{s(s-a)(s-b)(s-c)}$. This is known as Heron's formula, named after Heron of Alexandria, who is believed to have lived in the first century A.D. [189, 165, 126]

semiprime (adjective, noun): from Latin *semi-* "half," plus *prime* (*qq.v.*). In number theory a semiprime is a positive integer like 35 that has exactly two positive integral factors other than itself and 1, in this case 5 and 7. In other words, a semiprime has exactly four factors in all; a true prime has only two factors (itself and 1), which is half as many factors as a semiprime has. [189, 165]

semiregular (adjective): the first element is from Latin *semi-*, literally "half," but used here in the sense of "partial, incomplete, imperfect." The second element is *regular* (*q.v.*). In solid geometry all the faces of a regular polyhedron are congruent regular polygons and all the polyhedral angles are equal. The category of semiregular polyhedra is more general: the faces are still regular polygons, but not necessarily all with the same number of sides. The 13 semiregular polyhedra include the 5 regular polyhedra. [189, 179, 238]

senary (adjective): from Latin *senarius* "having to do with six," from *seni* "six each," from *sex* "six." The Indo-European root is *s(w)eks* "six." A senary number system uses a based of six. Compare *binary*, *denary*, *octonary*, *quinary*, *ternary*, and *vicenary*. [220]

sense (noun): via French, from Latin *sensus* "the faculty of feeling, sensation," from the verb *sentire* "to feel." From physical sensations, the word also took on meanings relating to the mental interpretation of physical stimuli; as a result, we speak of the physical senses and also of common sense and being sensible. The Indo-European root may be *sent-* "to head for," as seen in native English *send*. In mathematics the sense of an inequality is the direction of the inequality, either greater than or less than. [190]

sentence (noun), **sentential** (adjective): from Latin *sententia*, most literally "an opinion, a sentiment," from the verb *sentire* "to discern by the senses: to see, feel, hear, etc." The Indo-European root may be *sent-* "to head for," as seen in native English *send*. Latin *sententia* developed the extended meanings "a way of thinking, a determination, a decision," and finally "a sentence." In nonmathematical English a sentence is any complete thought or utterance. In mathematics, however, a sentence is a statement that is either true or false; in algebra, that amounts to an equality or an inequality. A sentence may be closed, as in the false sentence "$2 > 3$," or open, as in "$x^2 = 4$," which is true for $x = 2$ or $x = -2$, but false for all other values. The type of symbolic logic commonly encountered in "potpourri" mathematics courses is known as sentential calculus. [190]

separate (verb, adjective), **separable** (adjective): from Latin *separare* "to put apart, separate, divide." The first element is from Latin *se-* "apart," from the Indo-European third-person pronoun *s(w)e-* as seen in native English *self*; your self is apart from the rest of the world. The second element is from Latin *parare* "to set, put, get ready," from the Indo-European root *perə-* "to produce, procure," possibly the same root as *perə-* "to grant, allot." In geometry, a point separates a line into two rays (though the point may or may not be considered part of either ray). In calculus, separation of variables is one approach to solving differential equations. A separable space is a topological space which contains a countable set of points which is dense in that space. For an explanation of the suffix see *-able*. [219, 165, 69]

separatrix (noun): the feminine version of Latin masculine *separator* "a thing or person that separates," from *separare* "to separate" (see previous entry). A separatrix is any mark that separates digits into groups. In the United States, for example, a comma has traditionally been used to separate large whole numbers into groups of three digits, as in 9,785,246; in Europe a space serves the same purpose. Our decimal point is another type of separatrix: it separates a whole number from a decimal fraction. In Europe a comma is generally used for that purpose. [219, 165, 236]

September (noun): from Latin *septem* "seven," from the Indo-European root *septm* "seven." In the Roman calendar, which began in March, September was the seventh month of the year. In spite of its name, it is now the ninth month because January later became the first month of the year. [191]

septemdecillion or **septendecillion** (numeral): patterned after *billion* by replacing the *bi-* with the root of Latin *septendecim* "seventeen," a compound of *septem* "seven" and *decem* "ten." Since in most countries a billion is the second power of a million, a septemdecillion was defined as the seventeenth power of a million, or 10^{102}. In the United States, however, a billion is 10^9, and a septemdecillion adds fifteen groups of three zeroes, making a septemdecillion equal to 10^{54}. [191, 34, 72, 151]

septillion (numeral): patterned after *billion* by replacing the *b-* with Latin *sept-* "seven." Since in most countries a billion is the second power of a million, a septillion was defined as the seventh power of a million, or 10^{42}. In the United States, however, a billion is 10^9, and a septillion adds five groups of three zeroes, making a septillion equal to 10^{24}. [191, 72, 151]

septuple (verb, adjective, noun): from Latin *septuplus*, "sevenfold." The first component is from Latin *septem* "seven" from the Indo-European root *septm*, of the same meaning. The second component is from the Indo-European root *pel-* "to fold." To septuple something is to make it seven-fold, i.e., to multiply it by 7. With the advent of in vitro fertilization, whenever the number of children conceived at one time is seven times as great as usual, the babies are called septuplets. A septuple of numbers (a, b, c, d, e, f, g) is a set of seven numbers in a given order. Compare *double, triple, quadruple, quintuple, sextuple,* and *octuple*. [191, 160]

sequence (noun), **sequential** (adjective): from Latin *sequent-*, the present participial stem of *sequi* "to follow," from the Indo-European root of the same meaning, *sek^w-*. A sequence is a group of terms that have been put in a row, with one following another in a specific order. In older terminology a sequence was sometimes called a series (*q.v.*). [186, 146]

series, both singular and plural (noun): a Latin word meaning "row, succession, chain, series," from the verb *serere* "to join, connect, arrange, put in a row." The Indo-European root is *ser-* "to line up." The related Latin word *sermon* refers to a group of words put in a row. In mathematics a series is the sum of the terms of a sequence. In older usage, *series* sometimes meant what we would now call a sequence (*q.v.*); for example, the Fibonacci "series" is actually a sequence.

serpentine (adjective): from Latin *serpens*, stem *serpent-*, "a creeping animal," especially "a snake, a serpent." The Indo-European root is *serp-* "to crawl, to creep," as found in Greek-derived *herpetology* "the study of snakes" and *herpes* (otherwise known as shingles), a disease in which there are eruptions along a snake-like nerve path. In mathematics the curve whose rectangular equation is $x^2y + b^2y - a^2x = 0$ is called the serpentine curve because of its snake-like appearance. In 1701 Isaac Newton (1642–1727) studied and named the serpentine curve; it is a projection of the horopter (*q.v.*).

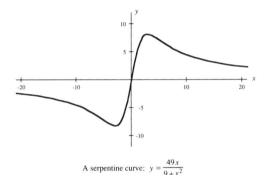

A serpentine curve: $y = \dfrac{49x}{9 + x^2}$

set (noun): from Old French *sette*, from Latin *secta* "a sect." The Indo-European root is *sec-* "to cut," so a sect was a group "cut off from" the mainstream. The evolved French form *sette* was less specific; it referred to any group of people or things gathered together. That meaning of *set* became confused with, and soon merged with, the meaning of the native English homonym *set* "to put, to place." People began to think of a set as any collection of things that have been "set down" together for a certain purpose. [185]

seven (numeral): native English, from the Indo-European root *septm*, of the same meaning. When compounded with two forms of the native English word *ten*, the same root gave *seventeen* and *seventy*. From Latin *septem* English has borrowed *Septentrion* (the seven stars of the Big Dipper), *septet* (a group of seven singers), and similar words. From related Greek *hepta* English has borrowed *hebdomad* (a period of seven days), *heptad* (a group of seven), and mathematical terms such as *heptagon* and *heptahedron*. [191]

sexagesimal (adjective): from Latin *sexagesimus* "sixtieth," from *sexaginta* "sixty," a compound of *sex* "six" and *decem* "ten." The suffix *-esimus* indicates an ordinal as opposed to a cardinal number. In a sexagesimal number system, each unit is one-sixtieth of the next greatest unit. The Babylonians and other ancient peoples did at least some of their calculating in a sexagesimal system. Following them, we still divide a circle into six times sixty degrees; each degree (as well as each hour) is divided into sixty minutes, each of which is further subdivided into sixty seconds. [220, 34]

sexdecillion (numeral): patterned after *billion* by replacing the *bi-* with the root of Latin *se(x)decim* "sixteen," a compound of *sex* "six" and *decem* "ten." Since in most countries a billion is the second power of a million, a sexdecillion was defined as the sixteenth power of a million, or 10^{96}. In the United States, however, a billion is 10^9, and a sexdecillion adds fourteen groups of three zeroes, making a sexdecillion equal to 10^{51}. [220, 34, 72, 151]

sextic (adjective): from the Latin ordinal number *sextus* "sixth," from the Indo-European root *s(w)eks* "six." In the Roman calendar, which began in March, the old name for the sixth month was *Sextilis*. With some political maneuvering, it was later changed to what we now call *August*, for Augustus Cæsar. In mathematics, a sextic equation is one of the sixth degree. Compare *quartic* and *quintic*. [220]

sextillion (numeral): patterned after *billion* by replacing the *b-* with Latin *sext-* "sixth." Since in most countries a billion is the second power of a million, a sextillion was defined as the sixth power of a million, or 10^{36}. In the United States, however, a billion is 10^9, and a sextillion adds four groups of three zeroes, making a sextillion equal to 10^{21}. [220, 72, 151]

sextuple (verb, adjective, noun): from Latin *sextuplus*, "sixfold." The first component is from Latin *sex* "six" from the Indo-European root *s(w)eks*, of the same meaning. The second component is from the Indo-European root *pel-* "to fold." To sextuple something is to make it six-fold, i.e., to multiply it by 6. On very rare occasions in times gone by, and especially now with the advent of in vitro fertilization, whenever the number of children conceived at one time is six times as great as usual, the babies are called sextuplets. A sextuple of numbers (a, b, c, d, e, f) is a set of six numbers in a given order. Compare *double*, *triple*, *quadruple*, *quintuple*, *septuple*, and *octuple*. [220, 160]

shape (noun): a native English word, from the Indo-European root *(s)kep-* "to cut, scrape, hack." Other

English cognates include *shave* and *shabby*. A related borrowing from Old Norse is *scab*. The Old English predecessor of *shape* meant "creation, creature, form": when something is created, it is shaped or given form. In early English, the meaning "to create" predominated; in fact *shape* even served as a euphemism for the female sexual organs. In modern English, the meaning "form, appearance, condition" predominates, although we can still say something like "George Washington shaped a new nation." In geometry, two figures that have the same shape are said to be similar; two figures that have the same shape and size are said to be congruent.

sheaf, plural *sheaves* (noun): a native English word, from the Indo-European root *skeup-* "cluster, tuft." A sheaf was originally a bundle of cut and bound stalks. By extension, a sheaf came to mean a collection of bound objects of any sort, not necessarily having anything to do with agriculture. In mathematics the term has been abstracted away from physical objects: a sheaf is a collection of all the planes passing through a given point. The physical metaphor is even carried over to the naming of the given point, which is called the *carrier* (*q.v.*) of the sheaf. A mathematical sheaf is also known as a bundle (*q.v.*).

shear (verb, noun): a native English word meaning "to cut with a scissor-like instrument." The Indo-European root is *sker-* "to cut." In physics, the word *shear* is used to describe a certain kind of cutting or tearing in which a body is deformed by forces that cause parts of the body to slide in parallel but opposite directions. Similarly, wind-shear is believed responsible for some airplane crashes. In mathematics a shear is a type of transformation patterned after the one in physics. [197]

shed (noun): a native English word, probably a specialized form of the word *shade*, so that a shed must originally have been named for the shadow that it casts. The Indo-European root would be *skot-* "dark, shade." A shed is a unit of area equal to 10^{-44}cm². The term appears to have been coined by analogy with the word *area* itself, which is a Latin word that meant "threshing floor." By a similar analogy, a barn had already been defined as 10^{-24}cm², so it was appropriate for a shed to be defined as an even smaller unit of area, just as the building called a shed is much smaller than the building called a barn.

sheet (noun): a native English word, from the Indo-European root *skeud-* "to shoot, chase, throw, to project" as seen in native English *shoot* and *shut-*

tle. A sheet is a piece of cloth "thrown" across a bed. When a sheet is spread out it has a large surface area, so we use the word to represent a mathematical surface, as in a hyperboloid of two sheets. Compare *nappe*. The analogical use of sheet to describe a piece of paper dates back to the 16th century, a time when paper began to be in more common use due to the spread of printing and the rise of literacy.

shell (noun): a native English word, from the Indo-European root *(s)kel-* "to cut." A shell was conceived as a piece cut off something else. Our borrowed word *scale*, as in the scale of a fish, is from the same source. From the original sense of "cut off" the meaning shifted to "outer covering," and *shell* came to apply to mollusks, turtles, eggs, and nuts. In the sense of "thin outer covering," calculus uses the method of cylindrical shells to find the volumes of certain solids of revolution. [196]

shift (verb, noun): a native English word with cognates in other Germanic languages. In Old English the word meant "to arrange." In mathematics a shift is a type of rigid transformation. Compare *translation*.

short (adjective): a native English word, from the Indo-European root *sker-* "to cut." Something short has been "cut down" from a larger size. Related words include native English *shirt*, which is cut out of cloth, and, from Old Norse, the very similar *skirt*. In a plane, the shortest distance between two points is along a straight line. On other surfaces, the shortest path may not be the one that intuition suggests. [197]

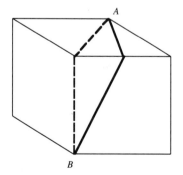

The shortest path on the surface of
the cube from *A* to *B* is along the
solid rather than the dotted lines.

shrink (verb): from Old English *scrincan* "to wither, cower, shrivel." The Indo-European root is *(s)ker-* "to turn, to bend." In mathematics, shrinking is a two-dimensional transformation that can be represented by the equations $x' = kx$ and $y' = ky$, where $0 < k < 1$. Higher-dimensional analogues are also possible. Compare *compression*. [198]

side (noun): a native English word, from a presumed common Germanic *sithaz* "extending lengthwise, long." A related cognate is *since* "extending 'lengthwise' " in time from a certain moment." The Indo-European root may be *se-* "long, late." A Moebius strip is of interest because it possesses only one side. [182]

sieve (noun): from an Indo-European root *seib-* "to pour out, drip, trickle, sieve." The root appears in native English *sift* and *soap*. The sieve of Eratosthenes is a kind of arithmetic screen that allows prime numbers to "trickle through."

sigma (noun): the eighteenth letter of the Greek alphabet, written Σ as a capital and σ in lower case. It corresponds to the Hebrew letter *samekh*, $\mathtt{O}$, meaning "post." See further explanation under *aleph*. In mathematics Σ is used to designate a sum (*q.v.*). In statistics σ designates the standard deviation. In number theory, the function $\sigma(n)$ sums all the divisors of a positive integer n.

sign (noun), **signum** (noun): Latin *signum* meant "mark, token, sign." The Indo-European root is *sekʷ-* "to follow." The linguist Calvert Watkins assumes that *signum* originally meant "a standard that one follows." Although *sign* may have referred to any mark, in mathematics the meaning is usually more restricted. It refers primarily to the plus or minus sign that indicates whether a quantity is positive or negative, and also to the signs indicating the elementary operations of arithmetic, as in a *times sign*. In algebra the signum function returns the values -1, 0, or $+1$, depending on whether the argument of the function is negative, zero, or positive; in other words, the output of the function depends only on the sign of the argument. In early 19th century American textbooks, an angle of 30° was called a *sign* because it corresponds roughly to the angle subtended by each of the twelve signs of the zodiac that encircle the celestial sphere. [186]

signature (noun): from Latin *signum* (*q.v.*) "mark, token, sign." At the end of a letter you typically write your signature, which is a set of signs, i.e., letters, identifying you. In the context of abstract mathematical systems, a signature is a set containing symbols or signs that stand for functions, relations, and symbols. A signature is often designated by the Greek letter Σ, which represents the initial sound of the word *signature*. For arithmetic, $\Sigma = \{+, \cdot, <, 0, 1\}$; the first two elements are operational symbols, the third element symbolizes a binary relationship, and the fourth and fifth elements are individual numbers. [186]

significant (adjective): from Latin *signum* (*q.v.*) "mark, token, sign," and the present participle of *facere* "to make." The Indo-European root is *dhe-* "to set," as seen in native English *do*. Something significant "makes its mark" in the world; something significant is important. When we speak of significant digits, we refer to those which are intrinsically important, as opposed, say, to zeros whose only function is to indicate where the decimal point belongs. Compare the reversed semantic role inherent in the term *mantissa*. [186, 41, 146]

silent (adjective): from Latin *silentium* "silence," from the Indo-European root *si-lo-* "silent." In mathematics a duel is a two-person zero-sum game. A duel is said to be silent if a player never knows whether or not the opponent has acted. Contrast *noisy*.

similar (adjective), **similarity** (noun), **similitude** (noun): from Latin *similis* "like, resembling, similar." The Indo-European root *sem-*, which appears also in native English *same*, meant "one," so two similar things look as if they're "one and the same" in respect to a certain property. In algebra, terms which contain the same powers of the variables involved are said to be like (*q.v.*) or similar terms; terms that aren't similar are called dissimilar. In geometry, two figures are said to be similar if they have (one and) the same shape, though not necessarily the same size. The ratio of similitude for two geometrically similar figures is the number gotten by dividing the lengths of any two corresponding parts in a given order. The symbol "$\sim$" that we use to indicate similarity is due to the German mathematician Gottfried Wilhelm Leibniz (1646–1716). [188]

simple (adjective), **simplify** (verb): from Latin *simplus*, a variant of the compound *simplex* "single, simple." (See next entry.) Latin *simplus* developed into French *simple*, which was then borrowed into English. In topology, a simple closed curve is one which doesn't intersect (= fold over) itself. A simple root of an equation is a root which occurs just one time. The suffix *-fy* in *simplify* is a much reduced version of Latin *facere* "to make," so that *simplify* means literally "to make simple." In algebra simplifying a complex fraction involves getting it into a form that has a non-fractional (= simple) numerator and denominator. Much of the work done in elementary algebra classes involves simplifying expressions. [188, 160, 41]

simplex (noun), **simplicial** (adjective): *simplex* is a Latin compound. The first element is from the Indo-European root *sem-* "one, once." The second element is from the Indo-European root *plek-* "to fold." The literal meaning of *simplex* is "folded [only] once," therefore "simple, uncomplicated." In mathematics a simplex is a generalization of the simplest closed configuration that can be made from straight line segments. For example, a triangle is a 2-simplex (because it's in two dimensions) and a tetrahedron is a 3-simplex (because it's in three dimensions). Contrast *còmplex*. In linear programming, the simplex method is a technique that the American mathematician George Danzig designed for solving simultaneous linear inequalities. It is "simple" in the sense of lending itself to computer implementation. [188, 160]

simulation (noun): from Latin *simulare* "to make like, to imitate," from the Indo-European root *sem-* "one." The concept found in Latin *simulare* is that the model you make is similar (another cognate) to or "as one with" the real thing. Applied mathematics deals with simulating real-world situations by using mathematical models. [188]

simultaneous (adjective): the Indo-European root *sem-*, which meant "one," appeared in Latin *simul* "at once, at one time." A native English cognate is *same*, which means "all of one type." Modeled after the older word *instantaneous*, *simultaneous* means "occurring at the same time." When you solve a collection of simultaneous equations, you are looking for sets of values that satisfy all of the equations at one and the same time. [188]

sine (noun): most immediately from Latin *sinus* "a curved surface," with subsidiary meanings such as "fold of a toga" and hence the "bosom" beneath the toga; "bay" or "cove." How that word came to represent a trigonometric function is quite a circuitous—and, depending on the authority you believe, contradictory—story. Howard Eves, in his *An Introduction to the History of Mathematics*, explains the origin of the word in the following way. The Hindu mathematician Aryabhata used the term *jya*, literally "chord," to represent the value of the equivalent of the sine function. When the Arabs trans-

lated Indian mathematical works, they transliterated *jya* as *jîba*, which actually meant nothing in Arabic. Now Arabic, like Hebrew, is often written with consonants only (pt n th vwls fr yrslf), so *jîba* became simply *jb*. Later readers, seeing *jb*, pronounced it as *jaib*, which was a real Arabic word meaning "cove, bay." When European mathematicians translated Arabic writings to Latin, they replaced *jaib* with the Latin word for "cove," which happened to be *sinus*. *The American Heritage Dictionary* claims that Arabic *jayb* (Eves's *jaib*) did have a meaning, namely "chord of an arc," but that Europeans confused the word with the homonym *jayb* meaning "fold of a garment," which happened to correspond to Latin *sinus*. *The Oxford Dictionary of English Etymology* claims that Arabic *jaib* meant "bosom," again translated by Latin *sinus*. For an equally intricate tale of Arabic-Latin translation, see *surd*. [192]

single (adjective): from Latin *singulus* "one apiece, separate, single." The compound consists of the Indo-European root *sem-* "one" and the Latin diminutive ending *-ulus*. The literal meaning of *single* is therefore "a little one." In mathematics, a root of an equation that occurs just one time is said to be a single root, as opposed to a double, triple, or other multiple root. [188, 238]

singleton (noun, adjective): the first element is *single* (see previous entry). The English suffix *-ton* is an unstressed version of the word *town* that normally appears in place names like *Washington* and *Hampton*. The suffix was later extended to designate people, as in *simpleton*, and things, as in *singleton*, which originally referred to a hand of cards that had just a single card from a given suit. In mathematics, a singleton set has just one member in it. [188, 238, 44]

singular (adjective), **singularity** (noun): from Latin *singulus* "separate, individual, single," from the Indo-European root *sem-* "one, as one." If there is just a single example of something, that example becomes special, so *singular* took on the meaning "out of the ordinary." In differential equations a singular point is one at which there is "trouble," as opposed to an ordinary point. The meaning "out of the ordinary, troublesome," explains why a singular matrix is a square matrix whose determinant equals 0 rather than 1, as the etymology implies. For example, in the solution of a system of dependent linear equations by Cramer's Rule, the denominators of all the fractions involved in finding the values of the variables are determinants that equal 0; the corresponding matrices of coefficients are singular. The opposite of *singular* is *nonsingular*. [188, 238]

sinistrorse or **sinistrorsum** (adjective, noun): the first component is from Latin *sinister* "left, to the left, on the left," of unknown prior origin. The second component is from Latin *versus*, past participle of *vertere* "to turn," from the Indo-European root *wer-* "to turn." Among the ancient Romans the left side was considered unlucky, probably because most people are right-handed; for that reason the borrowed adjective *sinister* means "ominous, evil." In botany, the root retains its original directional significance: a sinistrorse plant turns from right to left as it grows upward. In mathematics a sinistrorse curve is a left-handed curve, i.e., one whose torsion at a given point is positive. Contrast *dextrorse*. [251]

sink (noun): a native English word, from the Indo-European root *seng^w-* "to sink." A related borrowing from Scandinavian is *sag*. A sink is a (low) place at which water collects naturally or is collected so that it can then be removed; an example is the kitchen sink that is used to remove water from a house. By extension the word *sink* is now also used of energy as well as liquids, and the direction of removal need not be downward; a heat sink in a computer, for instance, typically sends the heat upward. By analogy, in graph theory a sink is a point in a directed graph that can be reached from all other points in the graph.

sinusoid (noun), **sinusoidal** (adjective): from Latin *sinus* (*q.v.*), used as the name of the sine function, and Greek *-oid* (*q.v.*) "having the appearance of." A sinusoid is any curve that can be represented by an equation of the type $y = a \sin(bx + c) + d$. Such a curve is like a basic sine curve but, depending on the values of a, b, c, and d, it will have been stretched, compressed, or translated. [244, 192]

six (numeral): a native English word, from the Indo-European root *s(w)eks*, of the same meaning. When compounded with two forms of the native English word *ten*, the same root gave *sixteen* and *sixty*. The Latin cognate was *sex*, a reduced version of which is seen in our borrowing *senary* "having six parts or having to do with the number six." Other borrowings from Latin include *semester*, originally six months; and *sextant*, an arc of which is one-sixth of a circle. (The English word *sex* is borrowed from Latin *sexus*, of unknown prior origin.) In prehistoric Greek the initial *s-* of the Indo-European root became *h-*, as seen in borrowed words like *hexad* (a group of six), *hexagon* and *hexahedron*. [220]

skeleton (noun): from Greek *skeletos* "dried up," from the Indo-European root *skelǝ-* "to parch, wither," because a skeleton is composed of dry bones. By analogy with a human or animal skeleton, abstract mathematics uses the term *skeleton* to refer to the set of all the vertices of a simplex. For a similar analogy see *nerve*.

skew (adjective), **skewness** (noun): from Old North French *eskuer*, a variant of Old French *eschuer* "to eschew, to avoid." French and other Romance languages had borrowed the word from a presumed Germanic verb *skiuhwan*, from *skiuhwaz*, the source of native English *shy*. In solid geometry, skew lines are lines in space that, figuratively speaking, avoid each other; in other words, they aren't parallel (because then they would stay in sight of each other) and they don't meet (because if they met they would hardly be avoiding each other). A skew (or twisted) curve is a space curve that doesn't lie in a plane. In abstract algebra, an operation * is said to be skew-commutative if $a * b = -b * a$: the term on the right "avoids" the sign of the term on the left. Compare *anti-commutative*. In statistics the skewness of a distribution is the direction (left or right) in which the distribution "avoids" being centered.

slack (adjective): a native English word from the Indo-European root *sleg-* "to be slack, loose." Related borrowings from Latin are *lax* and *relax*. If something is slack, there is extra room to tighten it up. In linear programming, slack variables are variables introduced into a system of inequalities; each slack variable stands for the amount by which a certain expression is less than a given constant. Metaphorically speaking, a slack variable tells how "loose" the expression is relative to the constant, or how much the expression needs to be "tightened up" to equal the constant. A slack variable is also known as a surplus (*q.v.*) variable.

slant (noun): a native English word of obscure origin. It may be related to other English words beginning with *sl-* that indicate slipping and sliding. In analytic geometry, a slant asymptote is an asymptote that is neither horizontal nor vertical. Solids such as a cone or a frustum of one, and a pyramid or a frustum of one, have both a height and a slant height.

slope (noun): a native English word, shortened from the earlier *aslope* "sloping, obliquely." The source may be Old English *aslupan* "to slip away, to disappear." You are likely to slip if you stand on a piece of ground that slopes downward. In mathematics the

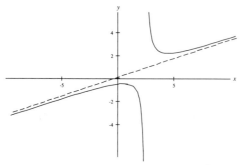

The dashed line is a slant asymptote to the curve $y = \dfrac{x^2}{3(x-2)}$.

slope of a line, defined as $\frac{\Delta y}{\Delta x}$, is a measure of its "obliqueness" or "tilt." Contrast *epols*.

slot (noun): from Old French *esclot* "hoof print of an animal, trail made by an animal's hoof prints," of unknown prior origin. In its earliest English usage, the word meant "the depression in the human breastbone," probably because of the resemblance to a line of animal hoof prints pressed into the ground. Later the word came to refer to any narrow depression or opening, and still later to the narrow opening in a machine into which we put a coin to make the machine operate. In the mathematical study of permutations and probability, the positions that objects can occupy are typically called slots. That use of the term derives from *slot* as "an opening in a sequence or schedule." By coincidence, the slot machines so commonly found in gambling casinos can also lead to the mathematical study of arrangements and probability.

small (adjective): a native English word, from the Indo-European root *(s)melo-* "small animal." In Old English *small* often meant "slender, thin, narrow," as well as "little." In calculus small changes are often symbolized by the Greek letter Δ. In spherical trigonometry, the intersection of a sphere and a plane that doesn't pass through its center is called a small circle of the sphere.

smooth (adjective): a native English word of uncertain origin; it does not appear to be related to words in any other language. In calculus a curve expressed parametrically on an interval by $x = f(t)$ and $y = g(t)$ is said to be smooth if $f'(t)$ and $g'(t)$ are continuous on the interval and aren't simultaneously 0 (except perhaps at an endpoint of the interval.) A smooth curve has no cusps on it.

snowflake (noun): a Germanic compound. Native English *snow* is from the Indo-European root

sneig^{w}h- "snow," as seen in the related Spanish word *Nevada*, literally "snowy." *Flake* is from Scandinavian, from the Indo-European root *plak-* "to be flat," an extension of *pelə-* "flat." In mathematics the snowflake curve results from an iterative process in which the middle third of each side of an equilateral triangle is replaced by two sides of a smaller equilateral triangle; the middle third of each smaller section is then replaced in the same way, and so on. The curve is named for its resemblance to a snowflake. [162]

The first three stages of a snowflake curve.

snub (verb): from Old Norse *snubba* "to turn up your nose at someone," from the Indo-European root *snu-* that appears at the beginning of many words having to do with the nose. Native English cognates include *snivel* and *snot*. Related borrowings from Dutch include *snout* and *snuff*. When you turn up your nose at someone, you metaphorically cut that person. The obsolete verb *to snub* meant "to cut or clip off." Similarly, a snub nose is one that is short and turned up (= cut off) at the tip. In solid geometry, two of the semiregular polyhedra are the snub cube and snub dodecahedron, which look like a cube and a dodecahedron whose corners have been cut off (= snubbed) in various ways.

sociable (adjective): from Latin *sociabilis* "that may be easily joined together, sociable," from *socius* "a sharer, partner, comrade, associate in an enterprise or business." The Indo-European root is *sek^w-* "to follow," since an associate follows you in your business or other activities. For an explanation of the suffix see *-able*. In mathematics a set of numbers, say $\{a, b, c, d\}$ is called a sociable group (in this case of order 4, because there are four members) if the proper divisors of a add up to b, the proper divisors of b add up to c, the proper divisors of c add up to d, and the proper divisors of d add up to a. A pair of amicable (q.v.) numbers is just a sociable group of order two. [186, 69]

solenoidal (adjective): via French, from Greek *solenoeides* "pipe-shaped." The first element is from Greek *solen* "pipe," and the suffix is from Greek-derived *-oid* (q.v.) "looking like." A solenoidal vector in a region is a vector-valued function whose integral over every reducible surface in the region is zero. [244]

solid (adjective, noun): from Latin *solidus* "firm, dense, compact, solid." *Solidus* was based on the older *sollus* "whole," from the Indo-European root *sol-* "whole." The related Latin stem *salut-* meant "health" because you're healthy when your body is whole. Related borrowings from Latin include *salutary*, *salutation*, *salvation*, and *save*. Renaissance mathematicians often used the word *solid* to refer to an algebraic expression of third degree because a solid object exists in no fewer than three dimensions; see more under *squared* and *cubed*. Solid geometry was commonly taught in American high schools through the first half of the 20th century, but has now largely disappeared from the curriculum as an independent subject. Calculus students still are asked to find volumes of various solids, especially solids of revolution. [201]

solidus (noun): a Latin word which as an adjective meant "firm, dense, compact, solid," from the Indo-European root *sol-* "whole." (See previous entry.) The *solidus* was a solid piece of money, specifically a gold coin of the Roman Empire. Because members of the military were paid with the solidus, they became known as *soldiers*. Ironically, although the etymological meaning of *solidus* is "whole," in arithmetic and algebra the solidus is the slanted bar used for writing fractions. One explanation for the seemingly unrelated meanings is that the Roman *solidus* often bore on its reverse a figure holding a spear, which was invariably slanted from lower left to upper right. Before the computer age, it was difficult for printers of books to write a fraction with the numerator directly above the denominator, so using a solidus was a practical way of writing both parts of the fraction in a single line of type. From a pedagogical point of view, however, the solidus is best avoided. Students sometimes get confused about what is on the top of the fraction and what is on the bottom. For instance, some students rewrite $\frac{1}{2}x$ as $\frac{1}{2x}$

or $\frac{1}{2}x$ which they then interpret as $\frac{1}{2x}$. Conversely, such students are likely to interpret $\frac{1}{2x}$ as $\frac{1}{2}x$. The solidus is occasionally called a virgule (q.v.). [201]

solution (noun), **solve** (verb): from Latin *solvere* "to free, to loosen." The Indo-European root *leu-* "to cut, to set free," is found in native English *lo(o)se*. In algebra, when you solve an equation or inequality for a variable x, you make one side free of everything that isn't just plain x. You have to "loosen up" the equation so all the non-x things can be moved to the other side. Many students mistakenly try to solve an expression; only an equation or an inequality can be solved. [116]

some (adjective): a disguised native English number word, from the Indo-European root *sem-* "one." If a set has some members in it, then it has at least one. In logic, *some* is the negative of *all*. The homonym *sum* (q.v.) is unrelated. [188]

sorites, singular and plural (noun): a Latin word, from Greek *soreites*, based on Greek *soros* "a heap." The Indo-European root is *teu-* "to swell," as seen in English *thigh*, the "swollen" part of the leg. In logic, a sorites is a group of "heaped up" syllogisms in which the conclusion of the first statement becomes the subject of the next, and so on until a conclusion can be drawn from the group as a whole. The word *sorites* is pronounced with three syllables, as if it were so-right-ease. [230]

source (noun): a French word meaning "fountain, spring," from Latin *surgere* "to rise up, to spring up," which is the source of our borrowed word *surge*. The Latin verb is a compound of *sub-* (q.v.) "up from under" and *regere* "to lead in a stright line, to lead in the right direction." The Indo-European root is *reg-*, of the same meaning. Related borrowings from Latin that contain the same two elements are *insurrection* and *resurrection*. Although *source* referred originally to gushing water, the word is most commonly used now with the metaphorical meaning "origin." In that sense *source* is used in graph theory to refer to a point in a directed graph from which all other points can be reached. [240, 179]

space (noun), **spatial** (adjective): via French, from Latin *spatium* "room, space, size, distance, interval, dimension." Eric Partridge believes that *spatium* is derived from Latin *patere* "to lie open," but some etymologists disagree. Latin *spatium* applied to two or three dimensions only, as does *space* in its nonmathematical sense. We speak of floor space as well as the space available in a refrigerator. In common English *space* can also refer to the dimension of time, as in "he was away from his desk for the space of two hours." As used in mathematics, *space* can refer to any number of dimensions: we have 2-space, 3-space, 4-space, etc. [202]

span (verb, noun), **spanning** (adjective): a native English word, from the Indo-European root *(s)pen-* "to draw, to stretch, to spin," as seen in English *spider* and *spindle*. From the meaning "to stretch" comes our use of the noun *span* to mean "extent, distance," as in the span of a bridge. In mathematics a spanning set of vectors in a vector space is a set such that any vector in the vector space can be written as a linear combination of the vectors in the spanning set. In other words, using the spanning vectors, you can "reach" any vector in the vector space. In older English usage, a span was a unit of length roughly equal to nine inches; it was the distance from the tip of the thumb to the tip of the little finger when the hand was spread wide. [206]

species, both singular and plural (noun): a Latin word meaning "outward appearance, shape, figure," from the Indo-European root *spek-* "to observe, to look at." Related borrowings from Latin include *spectacular*, *conspicuous*, and *spectator*. Things that look alike are of the same kind, so a species came to mean a "type" or "kind." In spherical trigonometry, two sides and/or angles of a triangle are said to be of the same species if both are acute or both are obtuse. [205]

spectrum, plural *spectra* (noun): a Latin word meaning "appearance, image, apparition," from the verb *spectare* "to watch, to look at, to observe." The Indo-European root is *spek-* "to observe." Related borrowings from French include *spectacle* and *espionage*. In physics, when a prism is used to break up white light into its visible colors, the colors can be seen one after the other, from red through violet. The range of colors being observed came to be known as a spectrum. By extension, a spectrum now refers to any range of values, not just to things that can be observed physically. In mathematics a spectrum is the set of all the eigenvalues of a matrix. This book is published in the Mathematical Association of America's Spectrum series. [205]

speed (noun): a native English word, from the Indo-European root *spe-* "to thrive, to prosper." Related borrowing from Latin include *prosperity* and its opposite, *despair*. The original English meaning of *speed* was "success," as seen in the obsolescent ex-

pressions "to wish someone "good speed" or "Godspeed." A person who prospers usually exerts himself and keeps busy, so the meaning of *speed* shifted to the hurried movement of an active person. In mathematics, if position is represented by a vector-valued function, the velocity vector is the first derivative of the position vector, whereas speed is the magnitude of the velocity vector. In nonmathematical English *velocity* is just a fancy word for *speed*.

sphere (noun), **spherical** (adjective): from Greek *sphaira* "ball, sphere," of unknown prior origin. A sphere is an ellipsoid that has all three axes equal. Technically speaking, the interior points are not part of the sphere itself. Contrast *ball*. [207]

spheroid (noun): from Greek *sphaira* "ball, sphere," of unknown prior origin, and *-oid* (*q.v.*) "having the appearance of." In mathematics, a spheroid is an ellipsoid of revolution. A spheroid has something of the appearance of a sphere, but only two of its three axes are equal. If the unequal axis is shorter than the other two, the spheroid is said to be oblate (*q.v.*). If the unequal axis is longer than the other two, the spheroid is said to be prolate (*q.v.*). [207, 244]

spinode (noun): a combination of *spine* or *spike* plus *node* (*q.v.*). *Spine* is from Latin *spina* "a thorn," while *spike* is from Latin *spica* "an ear or spike of grain." Both *spine* and *spike* are from the Indo-European root *spei-* "sharp point." English cognates include *spire*, *spoke*, and *spit* (a stick for roasting meat). In mathematics, *spinode* is a synonym of *cusp* (*q.v.*). [203, 142]

spinor (noun): a hybrid of native English *spin* and the Latin agental suffix *-or* "a person or thing that performs the indicated action." The verb *spin* is from the Indo-European root *(s)pen-* "to stretch, to spin," as seen also in English *spider* (which spins a web) and *spindle*. In mathematics a spinor is a type of two-dimensional complex vector. Its name comes from the fact that spinors are used in modern physics to describe the state of spin of electrons. [206, 153]

spiral (noun): from Latin *spira* "something twisted, coiled, or wound." The Latin word had been borrowed from Greek *speira*, with similar meanings. The Indo-European root is *sper-* "to turn, to twist." In mathematics some of the common types of spirals are Archimedean, equiangular, hyperbolic, and sinusoidal. Contrast *helix*.

spline (noun): a native English word first used in the dialect of East Anglia, a region in England. *Spline*

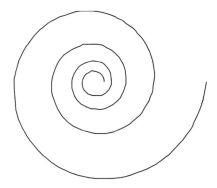

Equiangular spiral.

may be related to Old English *splin* "spindle" and to modern *splint*. A spline was originally a slat or a thin strip of wood, especially one inserted into the edges of two boards to help keep them together. A later meaning of the word was "a long, thin, flexible strip used as a guide for drawing arcs of curves." From that connection to curves comes the mathematical use of the term to mean "a function that has a specific value at a finite number of points and consists of pieces of polynomial functions joined smoothly at the known points." A spline can therefore be used to find approximate, interpolated values of a given function.

splitting (adjective): a native English word, from the Indo-European root *splei-* "to splice, split," which is an extension of the Indo-European root *spei-* "sharp point." Related borrowings from Dutch include *splice* and *splint*. In mathematics a splitting field is a minimal field such that a polynomial can be "split" (= factored) into linear factors with coefficients in that field. [203]

spur (noun): a German word pronounced as if it were spelled *shpoor* and meaning "trace, track, trail." The word *spoor* "the trail of a wild animal" is a cognate that English borrowed from Afrikaans (spoken in South Africa). The Indo-European root is *speirə-* "ankle." Native English *spur* and *spurn* are cognates. In mathematics a spur of a matrix is the same as a trace: it is the sum of the elements in the principal diagonal.

square (noun, adjective, verb), **squared** (adjective): from Old French *esquarrer*, from a presumed Vulgar Latin *exquadrare*, a compound based on Classical Latin *ex* "out" and *quadra* "a square." The Indo-European root is $k^w etwer$- "four," because a square

has four [equal] sides. The ancient Greeks and Romans conceived of the abstract quantity s^2 as the area of a square of side s. That's why something raised to the second power is said to be squared or quadratic. The words *squared* and *quadratic* are etymologically connected to the 4 sides of a square, even though what is being emphasized is the two-dimensionality of the square. In the ancient Greek system of figurate numbers, the square numbers are those which can be made by groups of dots arranged into a square. [108]

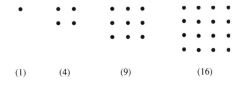

(1)   (4)   (9)   (16)

The first four square figurate numbers.

stable (adjective), **stability** (noun): a French word, from Latin *stabilis* "able to stand," from the Indo-European root *sta-* "to stand." Native English cognates include *stand* and *stud*, while borrowings from French include *store* and *stay*. For an explanation of the suffix see *-able*. In calculus a solution of a differential equation is said to be stable if the paths of all the other solutions that start at a point close to the path of the given solution stay close to that path. [208, 69]

standard (noun, adjective): from Old French *estandard*, modern French *étandard*, which in turn had been borrowed from Germanic *stand-hard* (this is one more example of a language re-importing one of its own words from a foreign borrower). The Indo-European roots are *sta-* "to stand" and *kar-* "hard." A standard is a flag used to mark a physical position because it will literally "stand hard" (= fast). A standard might also be used to indicate the presence of a person of rank. Presumably a person of rank, symbolized by the flag representing him, sets a standard of behavior or achievement for others to measure up to. The word then came to refer to any measure, not just one of behavior, as in a gold standard for money. In a different but not mutually exclusive interpretation, a standard, which stands upright and is easily visible, becomes a rallying place for people to support their causes. The régime and beliefs that they support become "standards" for everyone to live up to. In mathematics the equations of certain common curves are often given in a standard form that makes recognition easy. In statistics the standard deviation is a measure of how dispersed the data are about the mean. [208]

star (noun): a native English word, from the Indo-European root of the same meaning, *ster-*. In mathematics a star polygon is obtained by placing *n* points ($n \geq 3$) at equal intervals around a circle; starting at any of those points, every *k*th point (*k* relatively prime to *n*) is then connected. In our culture, the most commonly encountered star polygon is the pentagram (*q.v.*), obtained by connecting every second point in a set of five. In set theory, the star of a set *S* is the set of all sets that contain *S* as a subset. In number theory, a star number is a positive integer that fulfills two conditions: the integer has a certain property, and the number of digits in the integer also has the same property. For example, 17 is a star prime because both 17 and 2 (the number of digits required to write 17 in base ten) are primes. The term *star prime* was coined in 1988 by Richard L. Francis, who writes that he "considered concise descriptions which would imply 'brilliance' or a 'standing-out' feature in reference to other primes (or when generalized to various number classes). The term 'polystar' (*q.v.*) was a natural outgrowth." [212]

Star polygon formed by connecting
every third one of the seven points
spaced evenly around the circle.

state (noun): via French, from Latin *status* "the way things stand," from the Indo-European root *sta-* "to stand," which is a native English cognate. In political science, a state is a collection of people, property, institutions, i.e., all the things that "stand" in a geographic region. In mathematics a state is a vector representing the way a situation "stands" at a certain time. A state space is a list of all possible state vectors. For instance, if the weather over the next two days can be dry (*d*) or wet (*w*), the state space consists of the vectors $\langle d, d \rangle$, $\langle d, w \rangle$, $\langle w, d \rangle$, and $\langle w, w \rangle$. One characteristic of exponential functions is often succinctly abbreviated as "rate [of growth] is proportional to state [= how high the curve "stands"]. [208]

stationary (adjective): from Latin *statio*, stem *station-*, "a standing (still)," from the Indo-European root *sta-* "to stand," which is a native English cognate. Something stationary stands still. In calculus a stationary point on the curve $y = f(x)$ is a point at which the curve "stands still" in the sense of having its derivative equal to zero there. The curve neither rises nor falls at a stationary point, which may be a relative maximum, a relative minimum, or neither. Contrast *bend* and *turning*. [208]

statistic (noun), **statistics** (noun), **statistical** (adjective), **statistician** (noun): the Latin verb *stare* meant the same as its English cognate *stand*; both are descended from the Indo-European root *sta-* "to stand." From the Latin verb came the noun *status*, literally "a standing, a condition." The word *state*, which comes from *status*, is therefore a synonym of *condition*. A nation as a whole is also called a state because it is composed of the set of all conditions in a certain geographic area. As a result the term *state* came to be associated with government and politics. The German word *Statistik* was used in the 18th century to refer to political science, which has to do with states and governments. Since the study of political science involves the accumulation of data about the conditions in a country, each individual piece of data came to be known as a statistic, and the collective plural *statistics* came to refer to all the data and their collection and interpretation. Although the etymology of the word statistics is complicated, the meaning of the underlying Indo-European root is still apparent: statistics tell you how things "stand." [208]

stellated (adjective): from Latin *stella* "star," from the Indo-European root *ster-* "star." Stellated polyhedra may be formed by attaching the appropriate pyramid to each face of certain polyhedra. The result is a type of three-dimensional star. Around 1620 Johannes Kepler (1571–1630) discovered the stellated dodecahedron and the stellated icosahedron. [212]

stem (noun): a native English word, from the Indo-European root *sta-* "to stand" (which is an English cognate). The stem of a plant is the part that stands upright. In statistics a stemplot, or stem-and-leaf plot, is a type of diagram representing a relatively small, discrete distribution. It takes its name from the diagram's resemblance to the stylized stem and leaves of a plant. [208]

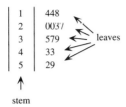

A stem-and-leaf plot
representing the numbers
14, 14, 18, 20, 20, 23, 27,
35, 37, 39, 43, 43, 52, 59.

step (noun): a native English word, from the Indo-European root *stebh-* "a post; to support, to place firmly on." The original meaning of *step* had to do with putting your foot on the ground. By transference, the word has also come to refer to an object that you step on, in particular each plank or level in a set of stairs. A related borrowing from Dutch is *a stoop*. In algebra the term *step function* has been applied to functions like $y = [x]$, the greatest integer function, because when such functions are graphed they look like a set of stairs. Those functions "step" up or down from one segment to the next. A step function is sometimes called a staircase function.

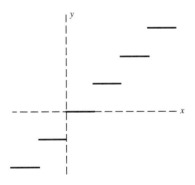

Step function: $y =$ the
greatest integer in x.

steradian (noun): a compound of *stereo* and *radian* (*q.v.*). The first component is from Greek *stereos* "solid," from the Indo-European root *ster-* "stiff," as seen also in native English *stern* and *starch*. A steradian is a type of "solid radian." In other words, the steradian is the unit that measures solid angles in three-dimensions just as the radian measures plane angles in two dimensions. The total number of stera-

dians about a point in space is 4π, which corresponds to the area of a unit sphere, just as the total number of radians about a point in a plane is 2π, which corresponds to the circumference of a unit circle. [213, 177]

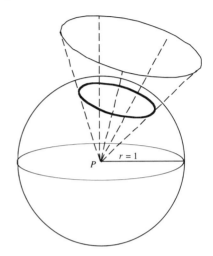

When the area inside the
dark curve on the surface
of the sphere is 1 square
unit, then the solid angle
at P is 1 steradian.

stere (noun): from Greek *stereos* "solid," from the Indo-European root *ster-* "stiff," as seen also in native English *stark*. A related borrowing from Greek is *cholesterol*, which makes your arteries "stiff." A stere is a unit of "solid" measure, i.e., volume, equal to one cubic meter. [213]

stereographic (adjective): the first component is from Greek *stereos* "solid," from the Indo-European root *ster-* "stiff," as seen in native English *stare* (a "stiff" glance). The second component is from Greek *graphein* "to write." In a stereographic projection, the points on a sphere (which is solid) are mapped (= written) onto a plane tangent to the sphere. The term seems to have been used first by the Belgian Jesuit François Aguillon (1566–1617), although the concept was already known to the ancient Greeks. In the 19th century, a stereograph was a pair of designs or photographs which, when looked at with a viewing device, merged into a three-dimensional (= solid) image in the brain of the beholder. A stereo(graphic) viewer typically had two side-by-side lenses, one for each eye. When sound systems with two separate soundtracks and two separate speakers became popular in the 1960s, the term *stereo* was borrowed from the older optical systems with their two lenses. Nowadays the only use of *stereo* that most people know is the one related to sound. [213, 68]

stereometry (noun), **stereometric** (adjective): the first component is from Greek *stereos* "solid," from the Indo-European root *ster-* "stiff," as seen also in native English *stark*. The second component is from Greek *metron* "a measure," from the Indo-European root *me-* "to measure." Solid geometry was once commonly called stereometry, though the term is little used now. [213, 126]

stochastic (adjective): from Greek *stokhos* "a stick used as a target," from the Indo-European root *stegh-* "to stick; pointed." It is possibly found in native English *sting*. The associated Greek adjective *stokhastikos* meant "capable of aiming," but in a figurative sense "capable of throwing out an idea or making a conjecture." The derived Greek noun *stokhastes* meant "diviner," a person who "aims" at the future and tries to tell what will happen. From this connection to conjecture or guessing, the adjective *stochastic* came to be applied to the study of mathematical probability because probability problems are conjectures about the likelihood of a certain event happening. In particular, *stochastic* is now used as a synonym for *random*, as in a stochastic or random variable.

stop (verb, noun): from Late Latin *stuppare* "to plug up, to stop up," from Latin *stuppa*, in turn from Greek

stuppe "coarse fiber used to caulk seams." The Indo-European root may be *steu-* "to condense, to cluster." A related borrowing that preserves the "fiber" or "material" sense is *stuff*. Only in the 15th century was the meaning of *stop* extended from the notion of plugging to that of impeding or arresting. In mathematics a stop point is one at which a curve undergoes a non-removable discontinuity. A stop point is also known as a terminus point.

straight (adjective): from Middle English *streit* or *streight*, alternate forms of the past participle of *strecche* "to stretch" (*q.v.*). When a string or other similar object is stretched, it becomes straight. Although in Old English the forms for *stretch* and *straight* were recognizably similar, the pronunciation of the words has changed sufficiently that now almost no native English speaker senses a connection between the two words. [217]

strategy (noun): as first used in English in the 15th century, a *stratagem* was an artifice used, typically by a general or a commander, to surprise an enemy. An alternate form of the word, *strategy*, developed in the 17th century. Both forms are based on Greek *strategos* "commander-in-chief." *Strategos* in turn was a compound of *stratos* "multitude, army," and *agein* "to lead." The Indo-European root underlying *stratos* is *ster-* "to spread," so that a multitude of people was conceived as a large group of people spreading out in all directions. The Indo-European root underlying *agein* is *ag-* "to drive." Many games exist for which winning mathematical strategies are known, but there are also many games for which no winning strategy has yet been found. [211, 3]

stratified (adjective): from Late Latin *stratificare* "to stratify." The first component is from Latin *stratum* "something laid down," a noun made from the past participle of *sternere* "to lay or throw down." The Indo-European root is *ster-* "to spread." A native English cognate is *to strew*. The second component is from a much reduced form of Latin *facere* "to do, to make." In statistics a stratified sample is one which is chosen from several subpopulations known as strata. [211, 41]

stretch (verb): from Old English *streccan* "to spread out, to extend." The origin of the word is uncertain. Mathematically, stretching is a two-dimensional transformation that can be represented by the equations $x' = kx$ and $y' = ky$, where $k > 1$. Higher-dimensional analogues are also possible. Compare *elongation*. [217]

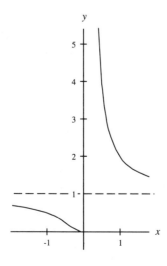

The origin is a stop point on the curve $y = 2^{1/x}$.

strict (adjective), **strictly** (adverb), **striction** (noun): via French, from Latin *strictus*, the past participle of *stringere* "to bind or tie tightly, to press together." The Indo-European root is *streig-* "to stroke, rub, press." When *strict* or *strictly* precedes a word, the meaning of that word is "bound more tightly." In mathematics, for example, a strict inequality is of the type $x > y$; the possibility that $x = y$ has been excluded. A strictly increasing function is distinguished from one which usually increases but may occasionally "level out" and have a horizontal tangent line. A line of striction is the locus of the central points of the rulings of a ruled surface. [218, 117]

strophoid (noun): from Greek *strophe* "a turning," plus *-oid* (*q.v.*) "form, shape." The Indo-European root is *streb(h)-* "to wind, to turn." A related borrowing from Greek is a *strobe*, a light that turns on and off rapidly. In mathematics, a right strophoid is a curve whose rectangular equation is

$$y^2 = x^2 \cdot \frac{a + x}{a - x}.$$

The turning indicated by the name refers to the curve's prominent loop. The name was coined by Montucci in 1846. [216, 244]

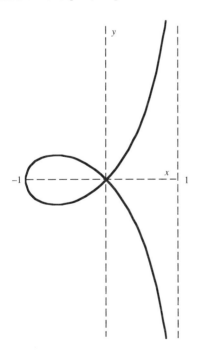

Right strophoid:

$$y^2 = x^2 \left(\frac{1+x}{1-x} \right)$$

sub- (prefix): from Latin *sub* "under." The Indo-European word *upo* meant "under," but also "up from under," and as a result the word developed opposite meanings in different Indo-European languages or even within the same language. Native English shows the upward meaning in *up* and the related verb *open* (because to open something like a box we have to lift the lid upward; notice we still often say, somewhat redundantly, *to open up* rather than just *to open*). Like Latin *sub*, Greek *hupo* retains the meaning "under," so the cognate prefixes *sub-* and *hypo-* both mean "under," as in *subscript* and *hypocycloid*. [240]

subadditive (adjective): from Latin *sub* "under" and *additive* (*qq.v.*). A function *f* is said to be subadditive if $f(x + y) \leq f(x) + f(y)$. The value on the left side of the equation is "under" (or equal to) the value obtained by adding the two terms on the right side. [240, 1, 47]

subalternate (adjective, noun): from Latin *sub* "under" and *alternate* (*qq.v.*). The Roman philosopher Anicius Manlius Severinus Boethius (480?–524?) invented the term *subaltern* to refer to a genus that is also a species of a higher genus. In logic, a subalternate is an alternate form of a given statement: the subalternate refers to a subset of the things that the original statement referred to. For example, if the original statement is "All people are smart," a subalternate is "Some people are smart." If the original statement is "No people are smart," a subalternate is "Not all people are smart." [240, 6]

subdiagonal (noun): from Latin *sub* "under, below" and *diagonal* (*qq.v.*). In a square matrix, the subdiagonal is the diagonal directly below the main one. Contrast *superdiagonal*. [240, 45, 65]

subfactorial (noun): from Latin *sub* "under," plus *factorial* (*qq.v.*). For an integer $n \geq 1$, subfactorial *n* is defined as

$$n! \left[\frac{1}{0!} - \frac{1}{1!} + \frac{1}{2!} - \frac{1}{3!} + \cdots + \frac{(-1)^{n+1}}{n!} \right].$$

For a given $n > 0$, the value of subfactorial *n* is "under" that of *n*!. If you write a letter to each of *n* friends, and if you address an envelope to each of the *n* friends, then the number of different ways that every letter can be placed in a wrong envelope is given by *n* subfactorial. [240, 41, 153]

subfield (noun): from Latin *sub* "under, below," and *field* (*qq.v.*). If a subring of a given field is itself a field, the subring is said to be a subfield. The number of elements in the subfield is "under" that of the original field. [240, 162]

submatrix (noun): from Latin *sub* "under, below," and *multiple* (*qq.v.*). A submatrix is a matrix derived from another matrix by eliminating one or more rows and/or columns. The number of rows or columns in the submatrix is "under" that of the original matrix. [240, 121, 236]

submultiple (noun): from Latin *sub* "under, below," and *multiple* (*qq.v.*). The term *multiple* obviously implies multiplication; the prefix *sub-* restricts the multiplier to a positive unit fraction, whose value is therefore less than (= under) 1. For example, the submultiples of six are 1, 2, and 3. That example might lead to the inference that a submultiple is the same as a divisor, but there is a slight difference in usage. For example, 2π is a submultiple of 6π, but 2π isn't normally called a divisor of 6π, even though it goes into 6π exactly three times. In the International System of Units, the term *submultiple* refers to prefixes like *deci-*, *micro-*, and *atto-* that represent negative powers of ten. Compare *aliquot*. [240, 132, 160]

subnormal (noun): from Latin *sub* "under" and *normal* (*qq.v.*). Given a curve $y = f(x)$, and a point P on the curve such that $y > 0$ at P, a normal line is drawn to the curve at P; the normal line is extended until it crosses the x-axis at N. Next a perpendicular is dropped from P to the x-axis, crossing the x-axis at F. The segment $\overline{NF}$ (or its length) is called the subnormal at P because it is under the part of the normal line that stretches from P to N. The subnormal is also "under" the normal in the sense of being shorter than the normal segment from P to N. [240, 75]

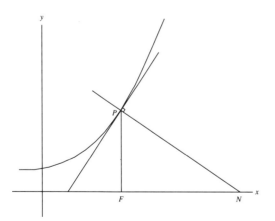

Segment $\overline{FN}$ is a subnormal.

subring (noun): from Latin *sub* "under, below," and *ring* (*qq.v.*). A subring is a subset of a given ring that has two properties: the subring is closed under addition and multiplication as defined in the original ring; if an element is in the subring, then so is the opposite of that element. A subring is a ring in its own right, but the number of elements in the subring is "under" that of the original ring. [240, 198]

subscript (noun): from Latin *sub* (*q.v.*) "under" and *scriptus*, the past participle of *scribere* "to write." The Indo-European root is *sker-* "to cut," since writing was originally carved in wood, clay tablets, etc. A subscript is a small letter or number written under (and to the right of) a letter. A mathematical subscript may serve as an index for terms in a sequence or series: in F_1, the subscript 1 indicates that we are dealing with term number 1 of the F(ibonacci) sequence. A subscript may also act as a kind of modifier of the letter that it accompanies: in $A_\bigcirc$, the circle used as a subscript indicates that we are finding the area of a circle; similarly, $A_\triangle$ designates the area of a triangle. Like *exponent*, the word *subscript* is named for its position, not for its value or its function. [240, 197]

subset (noun): from Latin *sub* "under" and *set* (*qq.v.*). A subset is a set consisting of members of a larger set. The number of members in a proper subset is "under" (= less than) the number of members in the original set. Contrast *superset*. [240, 185]

subsine (noun, adjective): from Latin *sub* "under" and *sine* (*qq.v.*). A subsine function is one which is dominated by (= is under) a function of the form $A \cos \rho x + B \sin \rho x$ [240, 192]

subspace (noun): from Latin *sub* "under, below," and *space* (*qq.v.*). A subspace of a given vector space is one whose dimension is "under" that of the original space. [240, 202]

substitute (verb), **substitution** (noun): from Latin *sub* (*q.v.*) "under" and *statuere* "to cause to stand." The Indo-European root is *sta-* "to stand," as seen in the native English cognate of the same meaning. When you substitute x for y in a mathematical expression, you are causing x to "stand in" (literally "stand under") for y. In a similar way, an *under*study may stand in for a sick actor. [240, 208]

subtangent (noun): from Latin *sub* "under" and *tangent* (*qq.v.*). Given a curve $y = f(x)$, and a point P on the curve such that $y > 0$ at P, a tangent line is drawn to the curve at P; the tangent line is extended until it crosses the x-axis at T. Next a perpendicular is dropped from P to the x-axis, crossing the x-axis at F. The segment $\overline{TF}$ (or its length) is called the subtangent at P because it is under the part of the tangent line that stretches from P to T. The subtangent is also "under" the tangent in the sense of being shorter than the tangent segment from P to T. [240, 222, 146]

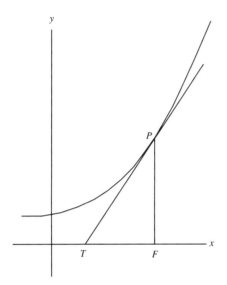

Segment $\overline{TF}$ is a subtangent.

subtend (verb): from Latin *sub* "under" and *tendere* "to stretch." The Indo-European root is *ten-* "to stretch." Related borrowings include *tender* (as in "to tender an offer") and *distend*. A native English cognate is the adjective *thin*, which describes an object that has been stretched. In geometry, when a line segment subtends an angle, the segment "stretches out underneath" the angle (assuming the angle is drawn above the line segment). [240, 226]

subtract (verb), **subtraction** (noun): from Latin *sub* (*q.v.*) "under" and *tractus*, the past participle of *trahere* "to draw, haul, drag, take." Related borrowings from Latin include *contract*, *extract*, *detract*, and *traction*. The compound *subtract* means literally "to draw away from underneath." We often use the similar English expression *take away*. When you subtract a (positive) number a from another number b, you are taking away an amount a, with the result the what's left is *under*, that is, less than, the original amount b. Other words that have been used in English for *subtract* include *extract*, *detract*, *deduct*, *diminish*, and *rebate*. In 18th century America the

variants *substract* and *substraction*, with an extra -*s*-, were in common use, and are still occasionally heard. The "−" symbol that we use to indicate subtraction is a variant of the "∼" symbol that was used for that purpose by the English mathematician William Oughtred (1574–1660). Europeans, however, sometimes use "÷" to indicate subtraction. [240, 234]

subtrahend (noun): from Latin *subtrahere* "to draw away, to subtract" (*q.v.*); the suffix -*nd*- creates a type of passive causative, so that Latin *numerus subtrahendus* meant "the number to be taken away." In the subtraction statement $5 - 3 = 2$, the 3 is the subtrahend because it is to be taken away from the 5. Discussing Latinate -*nd* terms like *minuend* and *subtrahend* in his *History of Mathematics* in 1925, D.E. Smith wrote: "It is hardly probable that such terms . . . , signifying little to the youthful intelligence, can endure much longer." But they have endured, and they still signify little to today's "youthful intelligence." [240, 234, 139]

success (noun), **successive** (adjective), **successor** (noun): from Latin *sub* (*q.v.*) "under" and *cessus*, the past participle of *cedere* "to go." The Indo-European root is *ked-* "to go, yield," as seen in many borrowings from Latin such as *recede*, *proceed*, and *intercede*. The verb *succeed* first meant "to go under or along," hence to follow, as in "the seasons succeed one another year after year." Originally a *success* was the outcome after a *success*ion of steps or measures, and that outcome wasn't necessarily favorable. Gradually the meaning of *success* was restricted to favorable outcomes. When speaking of mathematical probability, *success* has retained its older meaning of "(not necessarily favorable) outcome." For instance, in a given experiment, a success may mean contracting typhus. The adjective *successive* preserves more of the original meaning of *success*: successive terms are those that follow (= "go along after") a given term in a mathematical sequence or series. In a similar way the successor of an integer is the integer that follows. [240, 88, 153]

sufficient (adjective): from Latin *sub* "under" and the present participle of *facere* "to do, to make," from the Indo-European root *dhe-* "to put." Related borrowings from Latin include *efficient* and *deficient*. The Latin verb *sufficere* first meant "to put under, imbue with, afford, furnish." It then took on the same meaning as the Latin verb *substituere* "to substitute" (*q.v.*). If something is sufficient, it becomes a substitute for the thing originally desired: for example, "I really wanted maple syrup, but honey will suffice."

In the mathematical implication $A \rightarrow B$, A is a sufficient condition for B because A acts as a substitute for B; if A happens, B is sure to happen too. Contrast *necessary*. [240, 41, 146]

sum (noun), **summation** (noun): Latin *super* meant the same thing as its English cognate *over* (*q.v.*). A related Latin comparative adjective was *superior* (*q.v.*), applied to one who was "over" or "above" other things or people. There were two superlatives of the adjective, *supremus* and *summus*, both meaning "the highest." The noun *summa* therefore meant "the highest, the top," as seen in *summit* (*q.v.*), which English borrowed from French. When positive numbers are added, the total or sum is the highest number of all. Also, when the Greeks and Romans added numbers, they often added upward, rather than downward as we usually do; as a result, they wrote the sum literally at the highest point in their column of numbers. Although *sum* is the only current term that indicates the result of addition, *product* was occasionally used in the past. Perhaps because *sum* is such a short word, and perhaps also because it is a homophone of *some*, the redundant form *sum total* is sometimes used in nonmathematical English. The Swiss mathematician Leonhard Euler (1707–1783) introduced the Greek capital letter Σ, *sigma* (*q.v.*), equivalent to our letter S, to indicate a summation, since Latin *summa* begins with the letter *s*. [239]

summand (noun): from the Late Latin verb *summare* "to sum" (*q.v.*), with the suffix *-nd-*, which creates a type of passive causative, so that Latin *numerus summandus* means "a number that is to be added." The term *summand* is a synonym of *addend* (*q.v.*). It can also refer to part of an implicit sum, even when no overt addition is indicated. For example, with respect to a number like 1.000 000 007, the decimal portion .000 000 007 is sometimes referred to as a summand because it is being added to the integer 1. [239, 139]

summit (noun): Latin *super* meant the same thing as its English cognate *over* (*q.v.*). A related Latin comparative adjective was *superior* (*q.v.*), applied to one who was "over" or "above" other things or people. There were two superlatives of the adjective, *supremus* and *summus*, both meaning "the highest." The noun *summa* therefore meant "the highest [point], the top." The Latin word developed into Old French *som* "the highest point, end, extremity." The Old French diminutive form of that word was *somete*, from which English *summit* was borrowed. In mathematics a summit number is a number that can be

written in the form

where x is a positive integer. The x's rise upward to the right until they reach their highest point. A prime which can be represented as a sum of two summit numbers is called a summit prime. When leaders of the now-defunct Soviet Union used to meet with their counterparts from the United States, the encounters were described as superpower (*q.v.*) summits. Few journalists realized that *superpower* and *summit* are related words and that both of them have a mathematical as well as a political meaning. [239, 57]

superadditive (adjective): the first component is from Latin *super* "over, above, on top of," from the Indo-European root *uper* "over." The second component is *additive* (*q.v.*). A function f is said to be superadditive if $f(x + y) \geq f(x) + f(y)$. The value on the left side of the equation is "over" (or equal to) the value obtained by adding the two terms on the right side. [239, 1, 147]

supercircle (noun): the first component is from Latin *super* "over, above, on top of," from the Indo-European root *uper* "over." The second component is *circle* (*q.v.*). While the standard equation of a circle is $x^2 + y^2 = r^2$, the equation of a supercircle is given by $|x^n| + |y^n| = r^n$, where the value of n is "over" 2. The supercircle appears more "squared off" than a circle. The supercircle of fourth degree is also known as Lamé's special quartic. [239, 198, 238]

superdiagonal (noun): from Latin *super* "over, above," and *diagonal* (*q.v.*). In a square matrix, the superdiagonal is the one directly above the main diagonal. Contrast *subdiagonal*. [239, 45, 65]

superellipse (noun), **superellipsoid** (noun): from Latin *super* "over," plus *ellipse* or *ellipsoid* (*qq.v.*). The word *superellipse* is an unusual hybrid because a Latin prefix is attached to a Greek root; *hyperellipse* would be the expected form, but that term was already in use with a different meaning. The standard equation of an ellipse is

$$\left(\frac{x}{a}\right)^2 + \left(\frac{y}{b}\right)^2 = 1.$$

By contrast, a superellipse is a closed curve whose standard equation is

$$\left|\frac{x}{a}\right|^2 + \left|\frac{y}{b}\right|^n = 1,$$

where n is "over" 2. The prefix *super-* is explained by the value of the exponent, which is greater than that for an ellipse. A superellipse is more "rectangular" than a standard ellipse. If $a = b$, the curve becomes a supercircle. A superellipse is a special case of a Lamé curve. If the concept of a superellipse is taken into three dimensions, the equation of the resulting superellipsoid is

$$\left|\frac{x}{a}\right|^n + \left|\frac{y}{b}\right|^n + \left|\frac{z}{c}\right|^n = 1.$$

The superellipsoid is more "boxy" than a standard ellipsoid. If $a = b = c$, the surface is a supersphere. When a physical model of a superellipsoid is placed on end it will balance because the curvature at the ends is zero. [239, 52, 114, 244]

Superellipse of fourth degree.

superior (adjective): Latin *super* meant the same thing as its English cognate *over*. Both words are from the Indo-European root *uper* "over." The corresponding Latin comparative adjective was *superior*, literally "higher," but commonly applied to one who was "over" or "above" other things or people. In mathematics the term *limit superior* refers to the largest of the accumulation points of a sequence; the limit superior is "over" all other accumulation points. Compare *inferior*. [239]

superpose (verb): the first component is from Latin *super* "over, above, on top of," from the Indo-European root *uper* "over." The second component is from French *poser* "to put in a certain position," from the Latin past participle *positus* "put, placed." (See more under *component*.) In plane geometry, one figure can be superposed on another if the first can be placed on top of the second in such a way that the two coincide. Use of the verb *superpose* has been extended to three dimensions even though the positioning of one solid "on top of" another can only be carried out in the imagination. [239, 14]

superpower (noun): the first component is from Latin *super* "over, above, on top of," from the Indo-European root *uper* "over." The second component is *power* (q.v.). For positive integers x and n, x^n is

equivalent to x written as a factor n times. In the special case where $x = n$, the resulting expression x^x is called a superpower. The exponent x is written "over" the base x. A superpower is a summit (q.v.) number of second order. When leaders of the now-defunct Soviet Union used to meet with their counterparts from the United States, the encounters were described as superpower summits. Few journalists realized that *superpower* and *summit* are related words and that both of them have a mathematical as well as a political meaning. Also see *coupled*. [239, 173]

superprime (noun): the first component is from Latin *super* "over, above, on top of," from the Indo-European root *uper* "over." The second component is *prime* (q.v.). A prime which can be represented as the sum of two superpowers (see previous entry) is said to be a superprime. [239, 165]

superscript (noun): the first component is from Latin *super* "over, above, on top of," from the Indo-European root *uper* "over." The second component is from Latin *scriptum*, past participle of *scribere* "to write." The Indo-European root is *sker-* "to cut," since writing was originally carved in wood, clay tablets, etc. A subscript is a small letter, number, or symbol written above another character, usually toward the right but sometimes directly overhead or even toward the left. The most common mathematical superscript is an exponent. Other common superscripts are the single and double prime marks indicating a first and second derivative of a function. [239, 197]

superset (noun): the first component is from Latin *super* "over, above, on top of," from the Indo-European root *uper* "over." The second component is *set* (q.v.). A proper superset of a set B is a set A which contains all the elements of B and at least one other element as well. The number of elements in superset A is "over" that in set B. Contrast *subset*. [239, 185]

supersphere (noun): the first component is from Latin *super* "over, above, on top of," from the Indo-European root *uper* "over." The second component is *sphere* (q.v.). The word *supersphere* is an unusual hybrid because a Latin prefix is attached to a Greek root; *hypersphere* would be the expected form, but that term was already in use with a different meaning. While the standard equation of a sphere is $x^2 + y^2 + z^2 = r^2$, the equation of a supersphere is given by $|x^n| + |y^n| + |z^n| = r^n$, where the value of *n* is "over" 2. A supersphere appears more "boxy" than a sphere. [239, 207]

supertriangular (adjective): the first component is from Latin *super* "over, above, on top of," from the Indo-European root *uper* "over." The second component is *triangular* (*q.v.*). In Pascal's Triangle, the numbers 1, 3, 6, 10, 15, . . . , which lie on what might be designated as the 2nd diagonal, are known as triangular numbers. The numbers in the third diagonal, 1, 4, 10, 20, 35, . . . , have been called the supertriangular numbers. Whereas the triangular numbers correspond to arrays of dots in two dimensions, the supertriangular numbers correspond to arrays of dots in three dimensions, which is one dimension "over" two. They have also been called tetrahedral (*q.v.*) numbers. [239, 235, 11, 238]

supplement (noun), **supplementary** (adjective), **supplemental** (adjective): from Latin *sub* (*q.v.*) "under" or "up from under" and the Indo-European root *pelə-*, which has the same meaning as the native English cognate *full*. A supplement is a quantity that "fills up" a given amount to some predetermined level; the supplement "fills out" some lesser amount. In geometry, the supplement of an angle between 0° and 180° is another angle which, when added to the first, brings the total up to 180°. Contrast *complement*. Given a point on a circle, the two chords joining that point to the extremities of a diameter are known as supplemental chords because they "fill out" an inscribed triangle. The angle formed by supplemental chords, however, is 90°, not 180°. [240, 163]

support (noun): from Latin *sub* (*q.v.*) "up from under" and *portare* "to carry, to bear." The Indo-European root is *per-* "to lead, to pass over," as seen in the native English cognates *ford* and *to fare*. When something is supported, it is literally borne up from underneath. Given a convex planar region, a line of support is a line containing at least one point of the region in such a way that of the two open half-planes determined by the line, one of them contains no point of the region. A tangent line drawn to a convex region is a line of support, but the converse isn't necessarily true. [240, 165]

suppose (verb), **supposition** (noun): from Latin *sub* (*q.v.*) "under" and *positus*, past participle of *ponere* "to put." (See more under *component*.) Latin-derived *suppose* is a translation of the Greek word *hypothesis* (*q.v.*). In mathematics a supposition, like a hypothesis, is an idea which is "put under" scrutiny to see if it is true. The verb *suppose* has a different meaning. When a paragraph in a mathematics book begins "Suppose that . . . ," the author is saying "Let's put the following facts 'under' us as a basis on which to explore further." [240, 14]

supremum, plural *suprema* (noun): a Latin superlative of *super* "over, above," so that *supremum* means literally "the highest, the greatest." The Indo-European root is *uper* "over," as found in the native English cognate *over* and in the Greek-derived prefix *hyper-*. Related borrowings from Latin include *supremacy* and *insuperable*. Mathematically speaking, and contrary to the superlative form of the word, a supremum isn't the greatest of anything; the supremum is the *smallest* value that is still "above" (or equal to) the value of every member in the set. In other words, a *supremum* is the same as the least upper bound of a set. Contrast *infimum*. [239]

surd (noun): from Latin *surdus* "deaf." In mathematics a surd is an irrational root of an integer, or the irrational sum of such roots. Some examples of surds are $\sqrt[4]{3}$ and $\sqrt{5} + \sqrt[3]{7}$. Why should expressions like those be called "deaf"? The explanation begins with the Greek word *alogos*, a compound of *a-* "not" and *logos* "ratio, reason." (The Greek term was later literally translated into Latin, giving *irrational*.) The existence of irrational numbers once seemed illogical, since originally the ancients believed that every number was rational, even though its numerator and denominator might be quite large. Now it so happened that Greek *logos* meant not just "ratio" and "reason" but also "account, argument, discourse, saying, word." When the Arabs translated Greek *alogos*, they used the Arabic expression *jadhr asámm* "deaf root," focusing on the "discourse" sense of *logos* rather than the "rational" sense: someone who is without discourse or words is often deaf. Much later, European mathematicians translated the Arabic word for "deaf" using the equivalent Latin word *surdus*. For an equally intricate tale of Arabic-Latin

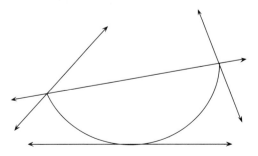

Lines of support for a convex region.

translation see *sine*. The English term *surd* came into use in the 16th century. Although quite common in textbooks through the early part of the 20th century, the word *surd* has now largely fallen out of use in mathematics. It is still used in linguistics to describe voiceless sounds. [221]

surface (noun): a French word, from Latin *superficies*, composed of *super* "over" (which is its native English cognate) and *facies* "form, shape." *Facies* is related to Latin *facere* "to make," from the Indo-European root *dhe-* "to set," as seen in native English *do*. A form or shape must be made, fashioned, or set up. A Latin *superficies* or French *surface* was thus a form put over something, like the skin that has been "put over" your body. *Surface* came to mean "thin outer covering" and finally "anything flat and thin existing in space." From Latin *facies* we also get the French word *face*, the thin covering on the front of our head. The adjective corresponding to French *surface* is Latin-derived *superficial*, but that word has taken on a negative connotation; as a result we usually speak of surface area rather than superficial area. [239, 41]

surjective (adjective): the first component is from French *sur* "onto," from Latin *super* "(from) above," from the Indo-European root *uper* "over." The prefix *sur-* is found in words like *surtax*, *surpass*, and *surrealism*, borrowed from French. The second component is from Latin *ject-*, past participial stem of *jacere* "to throw." The Indo-European root is *ye-* "to throw." Related borrowings from French include *jet* (which is thrown up into the air) and *joist*, a beam "thrown" from wall to wall. In mathematics a mapping of points from one set onto another set is called a surjective mapping. The compound *surjective* is unusual because a French prefix is attached to a learned Latin root. The expected Latin-Latin word would be *superjective*. That form isn't used, however, because Latin *super* meant "above," not "onto." Only later, as *super* evolved into French *sur*, did the meaning shift to "onto." So why not use the Latin word for "onto"? The problem is that Latin had no separate word meaning "on" or "onto." It had only the word *in-*, which also happened to mean "in" or "into." The mathematical word *injective* had already been chosen to express the idea "into," and therefore a different word had to be found to express the idea "onto." Contrast *injective*. [239, 257]

surplus (noun, adjective): via French, from Medieval Latin *superplus*. The first element is from Latin *super* "(from) above," from the Indo-European root *uper* "over." The second element is from Latin *plus* (*q.v.*) "more." Etymologically speaking, the word *surplus* is redundant: a surplus is *over* the required amount, and it is also *more* than the required amount. In linear programming, a surplus variable is the same as a slack (*q.v.*) variable. Business calculus deals with the concepts of consumers' surplus and manufacturers' surplus. [239, 163]

sursolid (noun, adjective): this now-obsolete term is an alteration of the even more obsolete *surdesolid*, which used to refer to expressions and equations of the fifth degree. The first element is from Latin *surdus* "deaf, incapable of verbal expression." The second element is from Latin *solidus* "solid," which was used by Renaissance mathematicians to refer to algebraic expressions of the third degree because solid objects are three-dimensional. (Of course a third-degree polynomial may also contain terms of lower degree, particularly squares.) The term *surdesolid* was used to describe algebraic quantities incapable of being expressed as powers of cubes or squares. The first such power is the fifth power, and so *surdesolid* came to apply especially to the fifth power. The use of *surd-* in the extended sense "incapable of being expressed" was rather strange, and people gradually lost sight of its figurative and ultimately perplexing meaning. The word was finally replaced with French *sur-* "over, above," which had developed from Latin *super* "over." People found it more sensible to regard a fifth-degree expression as being "over a solid" or third-degree expression. The term *sursolid* survived in English until the early 19th century. [221, 239, 201]

syllogism (noun): the first element is from Greek *sun-* "together with," from the Indo-European root *ksun* "with." The second element is from Greek *logismos* "reasoning," from *logos* "proportion, word." The Indo-European root is *leg-* "to collect, gather"; even in modern English "I gather" means "I think" or "I understand." In a syllogism, two statements (= "words," propositions) known as the major premise and the minor premise are "gathered together" to produce a logical conclusion. [106, 111]

symbol (noun), **symbolic** (adjective): the first element is from Greek *sun-* "together with," from the Indo-European root *ksun* "with." The second element is from Greek *bolos* "a throw," from *ballein* "to throw, to put," from the Indo-European root *gʷelə-* "throw, reach." The Greek compound meant "to put together" in the sense of "to compare." For the ancient Greeks, a symbol was a "token, sign, mark" or anything that would be comparable to the real thing whose place it took. In mathematics a symbol is a sign that stands for a quantity, an operation, or a relation. [106, 78]

symmedian (noun): a combination of *symmetric* and *median* (*qq.v.*). A symmedian of a triangle is a line that is symmetric to a median with respect to the line which bisects the angle from whose vertex the median is drawn. The three symmedians of a triangle are concurrent in a point known as the Lemoine point. [106, 127]

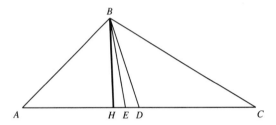

$\overline{BD}$ is a median. $\overline{BE}$ bisects angle ABC.
$\overline{BH}$, which is symmetric to median $\overline{BD}$ with respect to angle bisector $\overline{BE}$, is a symmedian.

symmetric (adjective), **symmetry** (noun): the first element is from Greek *sun-* "together with," from the Indo-European root *ksun* "with." The second element is from Greek *metron* "a measure." The Indo-European root is probably *me-* "to measure." Suppose two points are symmetric with respect to a line; if you measure the distance between one of the points and the line of symmetry, then "together with" that measurement you have simultaneously also measured the distance between the other point and the line of symmetry; the two distances are equal. [106, 126]

sympathetic (adjective): the first element is from Greek *sun-* "together with," from the Indo-European root *ksun* "with." The second element is from Greek *pathos*, originally "suffering," and by extension "feelings, emotion." The Indo-European root is *kʷent(h)-* "to suffer." Someone who is sympathetic literally "has feelings together with" someone else. Two numbers are said to be sympathetic if the proper divisors of one number add up to the other number, and vice versa. See example under *amicable*. [106]

synthetic (adjective): the first element is from Greek *sun-* "together with," from the Indo-European root *ksun* "with." The second element is from Greek *thetos*, past participle of *tithenai* "to put." The Indo-

European root is *dhe-* "to set," as seen in native English *do*. When something is synthesized, it is put together (from scratch), which is why in normal English *synthetic* has the meaning of "substitute" or "false." In mathematics synthetic geometry is the type of geometry developed by the Greeks, in which shapes were treated "together," that is, as wholes, rather than being broken up by the scale of a co-ordinate system, as in *analytic* (*q.v.*) geometry. In algebra synthetic division of polynomials is "put together" with the polynomials' coefficients only; the variables are omitted. [106, 41]

syntractrix (noun): the first element is from Greek *sun-* "together with," from the Indo-European root *ksun* "with." The second element is from Latin *tractrix* (*q.v.*). The syntractrix is the locus of a point on a tangent to the tractrix at a constant distance from each point of tangency. It is traced out "together with" the points that make up the tractrix. [106, 234, 236]

system (noun): the first element is from Greek *sun-* "together with," from the Indo-European root *ksun* "with." The second element is from Greek *histanai* "to cause to stand," from the Indo-European root *sta-*, as seen in the native English cognate of the same meaning, *stand*. A system is a collection of things all "standing together" as a group. The equations in an algebraic system of equations, for example, must be solved "together," i.e., simultaneously. [106, 208]

T

table (noun), **tabular** (adjective): *table* is a French word, from Latin *tabula* "a board, plank, writing tablet," of unknown prior origin. In addition to representing a tablet or writing table, even in Roman times the word came to stand for the document or contract being written on the tablet. In English, as early as the 14th century the word also came to refer to the arrangement of words and numbers that were occasionally included in a document. American students are most commonly taught to graph a function by making a table of values. In symbolic logic, students often make a truth table to find the truth value of a compound statement. The tabular digits of a number are all the digits in sequence from the first nonzero digit to the last nonzero digit. For example, the tabular digits of 0.03204 and 320400 are the same, namely 3204. The term *tabular* takes its name from the fact that the base-ten logarithms of 0.03204 and 320400 would both be looked up under 3204 in a table of logarithms. Now that tables of logarithms are all but extinct, the word *tabular* is rarely used in the above sense. Still in use is the tabular key on a computer keyboard, now almost always called the tab key, which moves you from column to column in a table.

tac-locus, plural *tac-loci* (noun): the first component is from Latin *tactus*, past participle of *tangere* "to touch." The Indo-European root is *tag-* "to touch." Related borrowings from Latin include *tact*, *contact*, and *tactile*. The second component is *locus* (*q.v.*). A tac-locus is a set of tac-points; in other words, a tac-locus is the set of all points where two different members of a family of curves "touch" and have a common tangent. For instance, for the family of circles tangent to each of two parallel lines, the line equidistant from the given lines is a tac-locus. [222, 215]

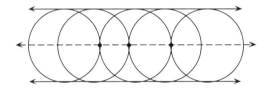

The dots are tac-points of this family of circles.
The dotted line is their tac-locus.

tacnode (noun): the first component is from Latin *tactus*, past participle of *tangere* "to touch." The Indo-European root is *tag-* "to touch." Related borrowings from Latin include *tact*, *contact*, and *tactile*. The second component is from Latin *nodus* "a knot." The Indo-European root is either *ned-* "to connect" or *gen-* "to compress, to make compact." A tacnode is a place on a curve through which two branches pass with identical tangents. The curve "touches itself" there. A tacnode is also known as a point of osculation or a double cusp. Contrast *crunode*. [222, 142]

tac-point (noun): the first component is from Latin *tactus*, past participle of *tangere* "to touch." The Indo-European root is *tag-* "to touch." Related borrowings from Latin include *tact*, *contact*, and *tactile*. The second component is *point* (*q.v.*). A tac-point is a place where two different members of a family of curves "touch" and have a common tangent, as opposed to intersecting and not having a common tangent at the point of intersection. See picture under *tac-locus*. [222, 172]

tail (noun): a native English word, from the Indo-European root *dek-*, which referred to fringes, locks of hair, and horsetails. In statistics a tail refers to the leftmost and rightmost parts of a continuous distribution, which when graphed resemble tails tapering down and trailing off in either direction. Statistical tests may be one-tailed or two-tailed.

tally (noun): from Latin *talea*, originally "a piece of a plant cut off and intended for replanting," and by extension "any pointed stake or stick." The Latin word was most likely borrowed from another language. In ancient and not-so-ancient times people have counted things by carving notches on a stick, with one notch for each item being counted. Such a stick or tally sometimes served as a record of a transaction. The Romans often used a stick which was split lengthwise so that each party had a record of the transaction: the numbers could be verified when the two halves of the stick were reunited, and any alteration would then become apparent. [223]

tangent (adjective, noun), **tangency** (noun), **tangential** (adjective): from the Latin adjective *tangens*, stem *tangent-* "touching," the present participle of *tangere* "to touch." The Indo-European root is *tag-* "to touch." Related borrowings from Latin include *tangible*, *attain*, and even *tax* (to touch [and keep!] someone else's money). In mathematics the line tangent to a circle just touches (but doesn't cross) the circle. With the invention of calculus, the notion of tangency was extended to include cases where the tangent line does indeed cross the curve that it's tangent to; an example is the tangent to the curve whose equation is $y = x^3$ at the point $(0, 0)$. In trigonometry the tangent function gets its name from the length of a line segment tangent to a unit circle. [222, 146]

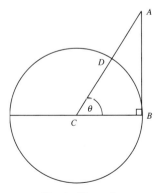

$CD = 1$, $AB = \tan\theta$.
$\overline{AB}$ is tangent to the
circle, hence its name.

tangram (noun): the first component may be from *T'ang*, the Chinese dynasty that ruled from 618 to 906. The second component is from Greek *gramma*, sometimes used as a synonym of the related word *graph* in the sense of "picture." (See more under *gram*.) A tangram is a Chinese puzzle consisting of a square divided into seven pieces; the object is to rearrange the pieces into abstract shapes or simplified figures of people, animals, etc. Some people have taken to calling each arrangement of the pieces a tangram, and each individual piece a tan. [68]

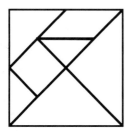

Tangram pieces.

tautochrone (noun), **tautochronous** (adjective): from Greek *tauto-* "same" and *khronos* "time." Although Greek *khronos* is of unknown prior origin, *tauto* is from the Indo-European demonstrative pronoun *to-*, as seen in native English *the*, *this*, *that*, *then*, etc. In mathematics the tautochrone is a curve that has the following property: given any two distinct points on the curve, the amount of time it takes a weighted particle to roll from either point to the lowest point on the curve is the same. In 1673 Christiaan Huygens (1629–1695) proved that the curve is a cycloid. Compare *brachistochrone* and *isochrone*. [231, 99]

tautology (noun), **tautological** (adjective): the first component is Greek *tauto-* "same," from the Indo-European demonstrative pronoun *to-* as seen in native English *the*, *there*, *thus*, *then*, etc., as well as *they*, *them*, *their*, borrowed from Old Norse. The second component is from Greek *logos* "speech, reasoning, discourse, proposition." The Indo-European root is *leg-* "to gather, collect." When you reason, you collect your thoughts and gather your statements together to reach a conclusion. In algebra a tautology is an identity, a proposition like $x = x$ that has the same value on each side of the equation. In logic a tautology is a compound statement that is true regardless of the truth values of the simplest components; the result is always the same, namely true. Etymologically speaking, a compound statement that is always false for all combinations of the truth values of the simplest components could also be called a tautology, because the truth value is always the same, namely false. That would make the meaning of *tautology* ambiguous, however, so a compound statement that is always false is called a contradiction (*q.v.*). [231, 111]

telescope (verb), **telescoping** or **telescopic** (adjective): *telescope* is a word coined in the Renaissance. The first component is from Greek *tele-* "far, at a distance," from the Indo-European root $k^w el$- "far." The second component is from Greek *skopos* "one who watches," from the Indo-European root *spek-* "to observe." A telescope is a device for making distant objects appear closer, but the use of the term *telescoping* in mathematics stems from the physical appearance of one type of portable telescope. The type of telescope in question is composed of sections, and each new section is of slightly smaller diameter than the previous section. Because of the different diameters, the narrower sections can slide inside the wider ones, and the entire telescope can be made no bigger than the biggest section. In mathematics a telescoping series is so named because a portion of each term is canceled by a portion of a later term, leaving only part of the very first and very last terms. The series "collapses" like a folding telescope. [205]

ten (numeral): a native English word, from the Indo-European root *dekm-* "ten." English *teen* is a slight variant of *ten* that is used in the numerals from thirteen through nineteen. Corresponding to the cardinal number ten is the regular ordinal *tenth*, as well as the variant *tithe* "a tenth of one's income donated to a church." Indo-European *dekm-* also had a variant *-dkm-ta-* which later lost the initial *-d-*; the resulting form developed into the *hund-* that appears in English *hundred* "ten tens." That same *hund-* also appears in English *thousand* "a "thick" hundred." The English names for 10^1, 10^2 and 10^3 are therefore all etymologically related. In Latin, Indo-European *dekm-* became *decem*; the stem *dec-* appears in many borrowings from Latin such as *decade* and *decimate*. The variant *-dkm-ta-* became Latin *centum* "hundred"; the stem *cent-* also appears in many borrowings from Latin such as *century* and *centennial*. In Greek, the basic form became *deka*, which appears in borrowings like *decagon*. See more under *hundred*. [34]

tend (verb): via French *tendre*, in turn from Latin *tendere* "to stretch," from the Indo-European root of the same meaning, *ten-*. A Native English cognate is the adjective *thin*, which describes an object that has been stretched. A related borrowing from Latin is *tension*. If you have a *tendency* to do something, you are pulled or "stretched" in a certain direction. In calculus, when a function tends to a limit, the values of the function are "stretched" toward that limit. [226]

tensor (noun): a modern Latin word composed of *tensus*, past participle of *tendere* "to stretch," plus the agental suffix *-or* "the one that does." In anatomy a tensor is a muscle that tightens or tenses something. In mathematics a tensor is an abstract object with a specified system of coordinates. Tensors were first created by the mathematician William Rowan Hamilton (1805–1865), the inventor of quaternions. The term is explained in the 1886 edition of *Solid Geometry*, by W.S. Aldis: "Since the operation denoted by a quaternion consists of two parts, one of rotating *OA* into the position *OB* and the other of *extending OA* into the length *OB*, a quaternion . . . may be represented as the product of two factors, . . . the versor . . . and . . . the tensor of the quaternion." [226, 153]

tera-, abbreviated *T* (numerical prefix): from Greek *teras*, stem *terat-*, "a marvel, a prodigy," and ultimately "a monster." The Indo-European root is $k^w er$- "to make," a monster being something that is "made to an extreme size." A related borrowing from Sanskrit is *karma*, the future that you make for yourself based on your current actions. A related borrowing from Greek is *teratology* "the biological study of monstrosities or abnormal formations." Because a monster is often very large, the metric system has given the prefix *tera-* a value of 10^{12}, or one trillion. If an English word were used, we could speak of a monster-meter rather than a terameter. The prefix *tera-* became a part of the International System of Units in 1960.

term (noun), **terminal** (adjective), **terminate** (verb): from Latin *terminus* "boundary line, limit." The Indo-European root is *ter-* "peg, post, boundary." The Latin word eventually acquired the extended meaning "something bounded," rather than the boundary itself; for example, we use a phrase like "in the short term." Later *term* came to refer to each member in a collection of things, because each member has a distinct boundary separating it from the other members in the collection. With reference to a mathematical series, the terms are the expressions that are isolated or "bounded" by plus or minus signs. Also, since things are often grouped together by type, in algebra we speak of like terms. In geometry the terminal side of an angle "sets a

limit" for the angle. A terminating decimal is one that "has a boundary"; in other words, the sequence of digits eventually ends. [227]

terminus, plural *termini* (noun): a Latin word meaning "boundary line, limit." Extended meanings are "farthest part" and "end." The Indo-European root is *ter-* "peg, post, boundary." As used in number theory, a terminus is the final digit of an integer expressed in a given base. For example, the one-digit terminus of the number 327 is 7. In base ten, 00 and 44 are the only two-digit termini with repeated digits that can correspond to a perfect square. In analysis a terminus point is the same as a stop (*q.v.*) point. [227]

ternary (adjective): from Late Latin *ternarius*, from *terni* "three each, three by three," from *ter* "three times." The Indo-European root is *trei-* "three." In mathematics a ternary operation on a set E is a function from E^3 into E. A ternary mathematical operation or function requires three arguments. A ternary system of numerical notation uses base three. Compare *binary*, *denary*, *octonary*, *quinary*, *senary*, and *vicenary*. [235]

tessellate (verb), **tessellation** (noun): from Latin *tessera* "a square tablet" or "a die used for gambling." Latin *tessera* may have been borrowed from Greek *tessares*, meaning "four," since a square tile has four sides. The diminutive of *tessera* was *tessella*, a small, square piece of stone or a cubical tile used in mosaics. Since a mosaic extends over a given area without leaving any region uncovered, the geometric meaning of the word tessellate is "to cover the plane with a pattern in such a way as to leave no region uncovered." By extension, space or hyperspace may also be tessellated. [108, 238]

tesseract (noun): the first component is from Greek *tessares* "four." The second component is from Greek *aktis* "ray of light." A tesseract is a four-dimensional hypercube (*q.v.*). In other words, a tesseract is to four-dimensions what a cube is to three and a square is to two. In the same way that a cube can be projected onto a plane, a tesseract can be projected (as if with rays of light) onto space; we can visualize the three-dimensional projection of the four-dimensional tesseract, even if we can't see the tesseract itself. [108]

test (noun, verb): Classical Latin *testa* meant "a piece of burned clay, a brick, a tile." In later Latin and Old French the word took on the meaning "pot." English borrowed the word with the meaning "cupel," a small, porous, cup-like vessel used to treat

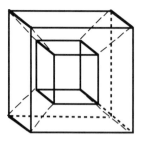

A tesseract projected into
three dimensions (and
rendered on paper in two).

gold and silver alloys. When you assay gold and silver, you often want to find out the purity of the metal, and so *test* came to mean "check for validity." In calculus some common tests for convergence of infinite series are the ratio test and the comparison test (and of course the test which tests students to see if they have learned how to use those convergence tests).

tetracuspid (noun): the first component is from Greek *tetra-*, from the Indo-European root $k^w etwer$- "four." The second component is from Latin *cuspis*, stem *cuspid-*, originally "the point of a lance," and later "a (whole) lance, a spike, a spear," and ultimately "any pointy thing." The word is of unknown prior origin but is assumed to have been borrowed from another language, as were many Roman military terms. In mathematics a tetracuspid is a hypocycloid of four cusps; it is now more commonly called an astroid (*q.v.*). [108, 29]

tetrad (noun): from Greek *tetra-*, from the Indo-European root $k^w etwer$- "four." A tetrad is any group of four things. In mathematics a harmonic tetrad is the set of points $\{A, B, C, D\}$ such that the cross ratio $(ABCD)$ equals -1. [108]

tetradecagon (noun): the first component is from Greek *tetra-*, from the Indo-European root $k^w etwer$- "four." The second component is from Greek *deka-*, from the Indo-European root *dekm-* "ten." The ending is from Greek *gonia* "angle," from the Indo-European root *genu-* "angle, knee." A tetradecagon is a fourteen-angled (and therefore also fourteen-sided) polygon. [108, 34, 65]

tetragon (noun), **tetragonal** (adjective): the first component is from Greek *tetra-*, from the Indo-European root $k^w etwer$- "four." The second component is from Greek *gonia* "angle," from the Indo-European root *genu-* "angle, knee." A tetragon

is a four-angled, hence also four-sided, polygon. Greek-derived words for two-dimensional figures like *hexagon* and *decagon* are in common use in American mathematics books, but although the term *tetragon* is in some current dictionaries, it has been almost entirely replaced by the Latin-based *quadrilateral*. The adjective *tetragonal* is still used in crystallography. [108, 65]

tetrahedron, plural *tetrahedra* (noun), **tetrahedral** (adjective): the first component is from Greek *tetra-*, from the Indo-European root $k^w etwer$- "four." The second component is from Greek *hedra*, from prehistoric Greek *sedra* "base." The Indo-European root is *sed-* "to sit," as seen in native English *sit* and *seat*. A tetrahedron is a four-based, i.e., four-faced polyhedron. A regular tetrahedron is one of the five regular polyhedra. The tetrahedral numbers are $1, 4, 10, 20, 35, \ldots$. They correspond to the number of dots or spheres that can be arranged into a tetrahedron, as for instance in a pile of cannonballs. Compare *supertriangular*. [108, 184]

tetrakis- (numerical prefix): a Greek form meaning "four times," based on *tetra-*, from the Indo-European root $k^w etwer$- "four." The prefix is used as part of the name of a stellated regular solid. It indicates that each original face became four faces. A tetrakishexahedron, for example is a polyhedron with four times six, or 24, faces. [108]

tetriamond (noun): see *polyiamond*.

tetromino (noun): the first component is from Greek *tetra-*, from the Indo-European root $k^w etwer$- "four." For the second component and further explanation see *polyomino*. [108, 37]

theorem (noun), **theory** (noun), **theoretical** (adjective): from Greek *theorema*, from *theorein* "to look at," of unknown prior origin. A related borrowing is *theater*, since you go to the theater to look at a play. A theorem was originally a sight or the act of seeing. Something that is looked at for any amount of time becomes an object of study. In mathematics, after studying a situation or class of objects, a person hopes to make speculations and then prove them, so *theorem* came to mean the proof of a speculation that has been arrived at by looking at something. The related word *theory*, on the other hand, stops one step short of a theorem, since a theory hasn't yet been proved.

therefore (noun): a native English compound of *there* and *for*, meaning "for that [reason]." The first element is from Indo-European *per-* "forward, through." The second element is from the Indo-European demonstrative pronoun *to-*, as also found in *then*, *that*, *the*, *this*, and *thus*. In mathematics, when a statement begins with *therefore*, you are really saying "for all that I've said there [in my previous statements]." In mathematics, the easily written symbol $\therefore$ is used in place of the longer *therefore*. [231, 165]

theta (noun): the eighth letter of the Greek alphabet, written Θ as a capital and θ in lower case. The Greeks borrowed the letter from the Phoenicians, who took their alphabet from the Canaanites. Greek theta corresponds to the Hebrew letter *teth*, $\mathfrak{D}$, meaning "snake." In trigonometry the letter θ is often used to denote an angle. See further explanation under *aleph*.

thick (adjective), **thickness** (noun): native English words with cognates in the Germanic languages but of unknown prior origin. In mathematics the thickness of an oval is defined as the minimum value of the breadth (*q.v.*). Contrast *length*.

third (adjective): the ordinal number that corresponds to the native English cardinal number *three*, from the Indo-European root *trei-* "three." The current English form developed in a part of England known as Northumberland by metathesis from what used to be more common forms such as *thrid* and *thredd*. (In a similar way, modern *bird* developed from older *brid*.) Up until the 19th century, a third of an inch was known as a *barleycorn*, i.e., the seed of a grain of barley; three barleycorns placed side by side must have measured approximately an inch. In calculus, given a function representing the position of a particle, the first derivative may be interpreted as velocity and the second derivative as acceleration; there is no common or simple interpretation for the third derivative other than the rate of change of the second derivative. [235]

thousand (numeral): actually an English compound, *thus-hund*. The first component is related to English *thumb* and *thigh*, and means "swollen, large." The Indo-European root is *teu-* "to swell." Related borrowings from Latin are *tumor* and *tumulus*. The second component is the root found in *hundred* (*q.v.*), which is based on the Indo-European root *dekm-* "ten." The literal meaning of *thousand* is "a swollen or big hundred" because it is ten times a hundred. The English words for 10^1, 10^2 and 10^3 are therefore all based mathematically and etymologically

on the notion of "ten-ness." Compare *million* for the same type of augmentative construction seen in *thousand*. In a similar way, using additive rather than multiplicative augmentation, author Mary Blocksma refers to a meter as a "fat yard" and a liter as a "fat quart." [230, 34]

three (numeral): a native English word, from the Indo-European root *trei-*, of the same meaning. When compounded with two forms of the native English word *ten*, the same root gave *thirteen* and *thirty*. The corresponding ordinal number is *third* (*q.v.*). In Latin the original Indo-European root developed into the stem *tri-*; related borrowings include *trinity* and *trio*. In Greek the original Indo-European root developed into the stem *tri-*; related borrowings include *triptych* and *tripod*. In older treatments of elementary algebra the Rule of Three made use of the fact that in a proportion the product of the means equals the product of the extremes; in other words, given any three of the members of the proportion, the fourth member can always be found. [235]

tilde (noun): a Spanish word derived from Old Catalan *title*, which developed from Latin *titulus* "a superscription, inscription, label, title." The Indo-European root may be *tel-* "floor, board." In Old Spanish the practice arose of rewriting the sequence of letters *nn* as *ñ*, where the $\sim$ symbol was a simplified or stylized *n*. Because the symbol was written above another letter, it was given the Latin name *titulus*. A tilde appears as part of the current geometrical symbol indicating congruence, $\cong$. In geometry the lowered tilde, $\sim$, conveys the meaning "similar to," and in logic it conveys the meaning "not."

tile (noun, verb), **tiling** (noun), **tiler** (noun): from Old English *tigele*, from Latin *tegula* "a tile." The Indo-European root is *(s)teg-* "to cover," so the Latin diminutive *tegula* meant literally "a little covering." An English cognate is *thatch*. Related borrowings from Latin are *toga* and *protect*. In mathematics a tiling is the same as a two-dimensional tessellation, a covering of the plane with variously shaped pieces (= tiles) in such a way that no region remains uncovered. A tile that can be used to tile the plane is known as a tiler, and a tile that can't by itself tile the plane is called a nontiler. [209, 238]

time (noun), **times** (preposition): surprisingly, the Indo-European root *da-* that is the source of our native English word *time* meant "to divide"! The infinite flow of events and occurrences that we call existence was conceived as being divided up into periods, each of which was a short division or "time." When we say "five times three," we are saying that a larger amount, in this case fifteen, can be divided up into five groups of three; in other words, groups of three occur five times. Because multiplication and division are inverse operations, every division implies a corresponding multiplication. In English, the meaning of *time* shifted from the divisional to the multiplicative aspect of the operation. As a verb we properly say *to multiply*, although some children say *to times* instead, for obvious reasons. From the same root as *time* comes native English *tide*, which used to mean what *time* now means, as can be seen in the expression "time and tide wait for no man" and in the obsolescent term *eventide* "evening time." [30]

ton (noun): since the 15th and 16th centuries a ton has been a unit of weight, now equivalent in the United States to 2000 lbs. However, as late as the 14th century, a ton was the space occupied in a ship's hold by a tun of wine, which was the equivalent of 4 hogsheads. In nautical terminology, a ton still means 100 cubic feet. A tun was originally a tub or vat, but later came to refer to a cask. The English word is from Medieval Latin *tunna*, which is probably of Celtic origin. A metric ton is 1000 kilograms.

top (noun): a native English word with cognates in other Germanic languages. English cognates include *tip* and *tap*. The original meaning seems to have been "a plug, a small projection," and by extension "the part of an object that sticks out," i.e., the top. Also related is French *toupée*, whose stem was originally borrowed from Germanic; a *toupée* is placed on top of the head. The kind of spinning toy known as a top is from the same Germanic source. The curve represented by the equation $y^4 + a^2x^2 = 2ay^3$ is known as a top because of its resemblance to the shape of

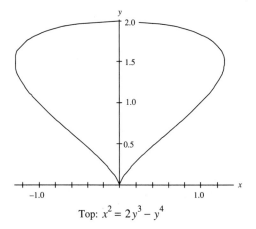

Top: $x^2 = 2y^3 - y^4$

the toy. Although the upper part of a fraction is "officially" called the numerator (*q.v.*), many English speakers call it the top, which is a much simpler word. In a complex fraction, combinations of the words *top* and *bottom* behave like positive and negative numbers under multiplication. For example, the bottom of the top of a complex fraction ends up on the bottom when the complex fraction is simplified. Compare the picture under *isomorphism*.

topology (noun), **topological** (adjective): the first component is from Greek *topos* "place, region," of unknown prior origin. The topic of a report is its subject (= semantic region). In medicine a topical solution is applied to a "region" of the human body, as opposed to being taken internally. The second component of *topology* is from Greek *logos*, a word that meant many things, among them "speech, reason, account." The Indo-European root is *leg-* "to gather." Topology is a branch of mathematics that gives an account of regions and shapes. [232, 111].

toroid (noun): a compound of *torus* and Greek-derived *-oid* "looking like" (*qq.v.*). A torus can be generated by rotating a circle about a line that doesn't intersect the circle. If any other closed curve is used instead of a circle, the resulting surface is known as a toroid because, being a ring, it looks like a torus. [233, 244]

torus (noun): a Latin word of unknown prior origin that had many meanings: "knot, bulge, protuberance, raised ornament, boss (as in *emboss*)." In architecture, a torus is a type of molding on a column. In Latin one of the things *torus* meant was the bulging of a muscle, and modern biology uses the word in a similar way. Because of its typically bulging, rounded shape, a torus in mathematics came to refer to a solid or surface that is shaped like a doughnut. (Note that *doughnut* was originally a *dough knot*.) A torus can be generated by rotating a circle about a line that doesn't intersect the circle. If the rotated circle has a radius a, and if the distance from the center of the rotated circle to the axis of rotation is b, then the volume of the torus is $4\pi^2ab$. In English a torus has also been called an anchor ring. [233]

total (adjective, noun, verb): from Latin *totus* "the whole, entire." When you find the total of a group of numbers, you combine values of the whole group. Latin *totus* may be from the Indo-European root *teuta-* "tribe," in which case *total* can be interpreted as "a tribe's worth" or "relating to the entire tribe." The Indo-European root is the source of *Dutch*, the name of the people (= tribe) living in Holland; it is also the source of the word *Teutonic*. A look at anthropology reveals that many ethnic groups around the world have called themselves simply "the tribe" or "the people."

totient (noun), **totitive** (noun): from Latin *tot* "that many, so many." It correlates with the Latin *quot* "how many," as seen in *quotient*. (The correlation between *qu*- words and *t*- words evidenced in Latin is also found in English, both natively and through borrowing from Old Norse. Most of our question words begin with *wh*- (or *h . . . w*), and can often be answered with a *th*- word: Where? There. What? That. When? Then. Who? They. How? Thus.) The Swiss mathematician Leonhard Euler (1707–1783) created his totient function to answer the following question: for a given positive integer n, how many smaller positive integers are relatively prime to it? Each time the question "how many?" is asked, the function, often designated $\phi(n)$, answers "that many." For example, since 1 and 5 are the only two smaller positive integers that have no factor except 1 in common with 6, $\phi(6) = 2$. In this example, the numbers 1 and 5 are called totitives of 6. [231, 146]

tournament (noun): from Old French *tourneiement* (modern *tournoiement*), literally "a turning around." The main part of the word is from Latin *tornus* "a lathe," from *tornare*, originally "to turn on a lathe," and then "to turn" in general. The Latin word had been borrowed from Greek *tornos* "a tool for drawing a circle, a lathe." The Indo-European root is *tera*- "to rub, to turn." In graph theory, a tournament is a directed graph in which every pair of points is joined by exactly one arc. That configuration corresponds to the type of real-life tournament in which each contestant competes against each other contestant just once. The "turning" or "circular" sense is even carried over into English when we speak of the various *rounds* of a tournament. [228]

trace (verb, noun): from French *tracer* "to trace," from a presumed early Romance *tractiare*, from Latin *tractus*, past participle of *trahere* "to drag, to draw." The Indo-European root is *tragh*- "to draw, to drag." In fact Native English *draw* and *drag* are themselves cognates, via an Indo-European root *dhragh*- which was a variant of *tragh*-. When you drag something, especially over open ground, it leaves behind a trace on the ground; therefore *trace* came to mean a mark left by the passage of something. When we use tracing paper we mark the passage of an original image underneath the tracing paper. As a result,

draw (in the artistic sense of dragging a pencil over a piece of paper) and *trace* are related words. In mathematics the traces of a surface are the "marks" left on the coordinate axes by the "passage" of the surface. [234]

tractrix or **tractory** (noun): *tractrix* is a Latin feminine noun meaning "the one that pulls," a meaning evident in the corresponding Latin masculine noun *tractor*. Both nouns are based on *tractus*, past participle of Latin *trahere* "to drag, to draw." The Indo-European root is *tragh*- "to draw, to drag." In mathematics the tractrix is a specific curve: it is the locus of a point which starts out on the *y*-axis and that is dragged by one end of a rod whose other end moves along the *x*-axis. Since the word for *line* or *curve* in Latin is feminine, we usually use *tractrix* as the name of the curve, although the more masculine *tractory* has also occasionally been used. The tractrix was studied by the Dutch mathematician Christiaan Huygens (1629–1695) in 1692. [234, 236, 153]

Tractrix

trajectory (noun): the first component is from Latin *trans* "across, beyond," from the Indo-European root *tera*- "to cross over, pass through." A native English cognate is *thorough*: The second component is from Latin *iactus*, past participle of *iacere* "to throw." The Indo-European root of the same meaning is *ye*-. A trajectory is the path an object follows when it is thrown through (= across) the air. [229, 257]

trammel (noun): from the Late Latin compound *tramaculum*, presumably for *tremaculum*, from *tres* "three" and *macula* "mesh." *Tres* is from the Indo-European root *trei* "three," but *macula* is of unknown prior origin. A trammel is a vertical fishing net made up of a finely woven mesh in between two more loosely woven meshes. Perhaps because the inner layer is trapped between the two outer layers, *trammel* also came to refer to a device used for hobbling horses to keep them from straying or kicking. The verb *to trammel* developed the figurative meaning "to hamper, to impede, to confine." In mathematics a trammel is a device that draws curves. The so-called trammel of Archimedes, for example, consists of a bar whose ends are confined (= trammeled) to move in two perpendicular grooves; as the bar moves, any fixed point between the two ends traces out an ellipse. [235]

tranjugate (adjective, noun): a hybrid made from the first part of *transpose* and the last part of *conjugate* (*qq.v.*). Relative to a given matrix whose elements are complex numbers, the tranjugate is obtained by obtaining the transposed matrix and then replacing each element with its complex conjugate. [229, 259]

transcendental (adjective): the first component is from Latin *trans* "across, beyond," from the Indo-European root *terə-* "to cross over, pass through." A native English cognate is *thorough*. The second component is from Latin *scandere* "to climb." The Indo-European root is *skand-* "to climb, to leap," as seen in *ascend* and *descend*, borrowed from French. A borrowing from Greek is *scandal*, originally "a trap, a snare (= something to leap over)." Within the field of complex numbers, an algebraic number is one which is the root of a polynomial equation having rational coefficients. Transcendental numbers, however "climb beyond" algebraic numbers, because transcendental numbers cannot be expressed using only roots of such equations. In 1844, Joseph Liouville (1809–1882) first proved the existence of transcendental numbers. [229, 194]

transfinite (adjective): the first component is from Latin *trans* "across, beyond," from the Indo-European root *terə-* "to cross over, pass through." A native English cognate is *through*. The second component is *finite* (*q.v.*). The transfinite numbers express the cardinality of sets "beyond" finite sets, i.e., of infinite sets. [229, 60]

transform (verb), **transformation** (noun): the first component is from Latin *trans* "across, beyond," from the Indo-European root *terə-* "to cross over, pass through." A native English cognate is *nostril*: (a hole passing through the nose). The second component is from Latin *forma* "shape, form" (*q.v.*). In geometry, when a figure undergoes a transformation, it goes "beyond" its original form or orientation to a new one. [229, 137]

transitive (adjective), **transitivity** (noun): the first component is from Latin *trans* "across, beyond," from the Indo-European root *terə-* "to cross over, pass through." A native English cognate is *through*. The second component is from Latin *itus*, past participle of *ire* "to go," from the Indo-European root *ei-* "to go." Related borrowings from Latin are *exit* and *transition*. In mathematics the transitive property of equality says that if $a = b$ and $b = c$, then $a = c$; in other words, the equality goes from a beyond b and all the way to c. [229, 50]

translate (verb), **translation** (noun): the first component is Latin *trans* "across, beyond," from the Indo-European root *terə-* "to cross over, pass through." A native English cognate is *thorough*. The second component is from Latin *latus* a past participle meaning "borne, carried." The word represents a metathesis of the Indo-European root *telə-* "to lift, to support." When you translate a text, you carry the meaning over from the original language to a new language. In mathematics, when you *translate* a curve you "carry it beyond" where it used to be to a new place (without, however, stretching or rotating it). A translation is also known as a shift (*q.v.*). [229, 224]

transmultiplication (noun): the first component is from Latin *trans* "across, beyond," from the Indo-European root *terə-* "to cross over, pass through." A native English cognate is *through*. The second component is *multiplication* (*q.v.*). If the last digit of a number is moved "over" to the first position, the new number is not usually an integral multiple of the original number. For example, 421 is not an integral multiple of 214. If the new number is a multiple of the original one, the phenomenon is known as transmultiplication. For instance, 714285 is five times as great as 142857. The term transmultiplication includes cases where several digits are moved as a block, whether from the end to the beginning or from the beginning to the end of a number. [229, 132, 160]

transpose (verb, noun), **transposition** (noun): the first component is from Latin *trans* "across, beyond," from the Indo-European root *terə-* "to cross over, pass through." A native English cognate is *thorough*. The second component is from Latin *positus*, past participle of *ponere* "to put." (See more under *component*.) In algebra, when a term of an equation is transposed, it is "put over" onto the other side of the equation (with its sign changed, of course). When a matrix is transposed, the rows are "put over" where the columns used to be, and vice versa. [229, 14]

transversal (noun) **transverse** (adjective): the first component is Latin *trans* "across, beyond," from the Indo-European root *terə-* "to cross over, pass through." A native English cognate is *thrill*, which originally meant "to pierce." The second component is from Latin *versus*, past participle of *vertere* "to turn" or simply "to go." The Indo-European root is *wer-* "to turn, to bend." A native English cognate is *wrinkle*. In geometry, a transversal is a line that crosses over a pair of parallel lines. The transverse axis of a hyperbola is the one that "goes across" from vertex to vertex. [229, 251]

trapezohedron, plural *trapezohedra* (noun): from *trapezium* (q.v.), in its most literal sense "a four-sided polygon," and *-hedron* (q.v.). A trapezohedron is a polyhedron each of whose faces is a quadrilateral and whose equator is a staggered polygon. [108, 158, 184]

trapezoid (noun); **trapezoidal** (adjective); **trapezium**, plural *trapezia* (noun): the Greek word *trapeza* "table," was composed of *tetra* "four" and the Indo-European root *ped-* "foot." A Greek table must have had four feet (= legs). The suffix *-oid* (q.v.) means "looking like," so that a trapezoid is a figure that looks like a table (at least in somebody's imagination). Some Americans define a trapezoid as a quadrilateral with at least one pair of parallel sides. Under that definition, a parallelogram is a special kind of trapezoid. For other Americans, however, a trapezoid is a quadrilateral with one and only one pair of parallel sides, in which case a parallelogram is not a trapezoid. The situation is further confused by the fact that in Europe a trapezoid is defined as a quadrilateral with no sides equal. Even more confusing is the existence of the similar word *trapezium*, which in American usage means "a quadrilateral with no sides equal," but which in European usage is a synonym of what Americans call a trapezoid. Apparently to cut down on the confusion, *trapezium* is not used in American textbooks. The trapeze used in a circus is also related, since a trapeze has or must once have had four "sides": two ropes, the bar at the bottom, and a support bar at the top. [108, 158, 244]

traversable (adjective): the first component evolved from Latin *trans* "across, beyond," from the Indo-European root *tera-* "to cross over, pass through." The second component is from Latin *versus*, past participle of *vertere* "to turn" or simply "to go," from the Indo-European root *wer-* "to turn, to bend." To traverse something is to go across it. For an explanation of the suffix see *-able*. In topology, a network is said to be traversable if you can start at a point on it and go over every arc of the network exactly once. [229, 251, 69]

tredecillion (numeral): patterned after *billion* by replacing the *bi-* with the root of Latin *tredecim* "thirteen," a compound of *tres* "three" and *decem* "ten." Since in most countries a billion is the second power of a million, a tredecillion was defined as the thirteenth power of a million, or 10^{78}. In the United States, however, a billion is 10^9, and a tredecillion adds twelve groups of three zeroes, making a tredecillion equal to 10^{42}. [235, 34, 72, 151]

tredo- (numerical prefix): according to the 1990 *Guinness Book of Records*, this prefix is the smallest one in the metric system. It supposedly multiplies the unit to which it is attached by 10^{-30}. *Tredo-* is from Danish *tredive* "thirty," from *tre* "three" and *ti* "ten." The final *-o* of *tredo-* has been added for uniformity: all of the numerical prefixes in the International System of Units have two syllables and end in a vowel, which in the case of submultiples created in recent times is always an *-o*. The only problem is that the prefix *tredo-* doesn't seem to exist. In a letter dated 1 December 1992, an editor at *Guinness* wrote that "[w]hile the prefixes dea and tredo did appear in earlier editions of the *Guinness Book of Records*, their origin appears to be a mystery. Subsequently, they were replaced by the officially recognised extremes, namely yotta and yocto, which were officially adopted by the International Committee on Weights and Measures in 1991." [235, 34]

tree (noun): a native English word, from the Indo-European root *deru-* or *dreu-* "to be firm, solid, steadfast." Related English cognates are *tray* and *trough*, both of which were originally made of wood. Another cognate is *true*, because a true friend is, metaphorically speaking, as firm and steadfast as a tree. In mathematics a tree diagram shows the sequence of decisions that a person can make under certain circumstances. The diagram gets its name from the branching decision lines that resemble the branches on a tree. Similarly, in graph theory a tree is a connected graph that contains no cycles, just as a branch of a real tree can't circle around and grow back into itself. [39]

trefoil (noun): the first component is from Latin *tres* "three," from the Indo-European root *trei-*, of the same meaning. For the second component and further explanation see *multifoil*. In the study of knots, the trefoil is the familiar knot also known as the cloverleaf. Contrast the etymologically but not mathematically identical *trifolium*. [235, 21]

trend (noun): a native English word related to Old English *trinda* "round lump, ball." When pushed, a ball rolls in a certain direction, so the meaning of *trend* shifted from the ball itself to the direction in which the ball rolls. The word later came to mean the direction in which any phenomenon progresses. A related expression is *trundle bed*, which has rollers under its legs so it can be easily moved around. These words are of Germanic origin but are not found in common Indo-European. In statistics a trend line is a line that represents the "general drift" of the data.

triacontahedron, plural *triacontahedra* (noun): the first element is from the Greek compound *triakonta*, from the Indo-European roots *trei-* "three" and a not-easily-recognized form of *dekm-* "ten," so that *triakonta* means "thirty." The second element is from Greek-derived *-hedron* "seat, base," used to indicate the face of a polyhedron. A triacontahedron is a polyhedron with thirty faces. [235, 34, 184]

triad (noun): from Greek *tri-* "three," from the Indo-European root *trei-* "three." A triad is any group of three things. In a three-dimensional vector space, the basic triad is the set of vectors $\langle 1, 0, 0 \rangle$, $\langle 0, 1, 0 \rangle$, and $\langle 0, 0, 1 \rangle$, also known respectively as **i**, **j**, and **k**. [235]

triakis- (numerical prefix): a Greek form meaning "three times," based on *tri-* "three." The prefix is used as part of the name of a stellated regular solid. It indicates that each original face became three faces. A triakisoctahedron, for example is a polyhedron with three times eight, or 24, faces. [235]

trial (noun): a Norman French word, from French *trier* "to sort out, to separate." The French word may be from Vulgar Latin *tritare*, from the past participle of Classical Latin *terere* "to grind, to rub to pieces." The Indo-European root would be *terə-* "to rub, turn, pierce," as seen in native English *thresh*. When a person is on trial, the evidence is being "sifted" and "sorted out" so that the truth can be determined. The verb *to try* originally meant "to make an attempt to separate the false from the true," but it has now come to mean "make an attempt" at anything. In the study of probability, a trial is an experiment. When you extract a square root using the once-standard manual algorithm, at each new stage of the process you resort to a trial divisor. American schools commonly teach the trial-and-error method of factoring non-monic trinomials. The method is aptly named: it requires a great deal of trial and tribulation and results in lots of errors. [228]

triamond (noun): see *polyiamond*.

triangle (noun), **triangular** (adjective), **triangulation** (noun): the first component is from Latin *tri-* "three," from the Indo-European root *trei-* "three." The second component is *angle* (*q.v.*). In geometry, a triangle is a polygon that has three angles, and therefore also three sides. The Latin word was a translation of Greek *trigon* (*q.v.*), but *trigon* is no

longer commonly used as an English noun, except as part of the word *trigonometry*. Since a triangle has three sides, it could equally well be called a *trilateral* (compare *quadrilateral*), but *trilateral* (*q.v.*) is used as a noun in mathematics in a somewhat different way. Although a triangle is most commonly defined as a polygon, a triangle may also be a plane figure whose "sides" are curves. In spherical trigonometry, not only aren't the "sides" of a triangle straight, but the triangle is no longer even a plane figure. Among the figurate numbers, the triangular numbers are the set $\{1, 3, 6, 10, 15 \ldots\}$, because those numbers can be represented by a triangular pattern of dots in which each new row contains one dot more than the previous row. See picture under *figurate*. Triangulation is the partitioning of a polygon into non-overlapping triangles all of whose vertices are vertices of the polygon. [235, 11, 238]

triaxial (adjective): the first component is from Latin *tri-* "three," from the Indo-European root *trei-* "three." The second component is *axial* (*q.v.*). In solid analytic geometry, a coordinate system containing three axes is said to be triaxial. [235, 5]

trichotomy (noun), **trichotomous** (adjective): the first component is from Greek *trikha* "in three (parts)," from the Indo-European root *trei-* "three." The second component is from Greek *temnein* "to cut," from the Indo-European root *tem-* "to cut." Related borrowings from Greek include *tome*, a volume "cut" out of a larger set of books, and *entomology*, the study of insects, whose bodies are "cut" into distinct sections. The word *trichotomy* was modeled after the similar *dichotomy* (*q.v.*) "a division into two [usually] mutually exclusive parts." In mathematics the trichotomy principle says that, given two real numbers a and b, exactly one of the following three relationships is true: $a > b$, $a = b$, or $a < b$. [235, 225]

tricuspid (noun): the first component is from Latin *tri-*, from the Indo-European root *trei-* "three." The second component is from Latin *cuspis*, stem *cuspid-*, "a point, a spike, a spear," of unknown prior origin. The tricuspid is an epicycloid of three cusps; it is also known as a deltoid (*q.v.*). [235, 29]

tridecagon (noun): the first component is from Greek *tri-*, from the Indo-European root *trei-* "three." The second component is from Greek *deka-*, from the Indo-European root *dekm-* "ten." The ending is from Greek *gonia* "angle," from the Indo-European root *genu-* "angle, knee." A tridecagon is a thirteen-angled (and therefore also thirteen-sided) polygon. [235, 34, 65]

trident (noun): the first component is from Latin *tri-*, from the Indo-European root *trei-* "three." The second component is from Latin *dens*, stem *dent-*, "tooth." The Indo-European root is *dent-* "tooth," from the more basic *ed-* "to bite," hence "to eat." English cognates include *tooth* and *eat*. A trident is a spear-like weapon with three prongs (= teeth) at its tip; the most famous trident was the one wielded by Poseidon, the Greek god of the sea. In mathematics the trident is a curve whose equation is

$$y = \frac{ax^3 + bx^2 + cx + d}{x}.$$

The upper portion of the graph is made up of divergent branches that somewhat resemble the three prongs on a trident. [235]

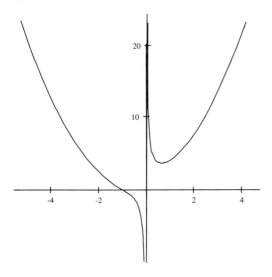

Trident: $y = \dfrac{x^3 + x^2 + x + 1}{x}$

tridiagonal (adjective): the first component is from Greek *tri-*, from the Indo-European root *trei-* "three." The second component is *diagonal* (*q.v.*). Contrary to what might be assumed from the etymology, a tridiagonal matrix is not a matrix that has three diagonals. A tridiagonal matrix is a square matrix all of whose nonzero elements occur in the subdiagonal, the main diagonal, and the superdiagonal. Outside of those three diagonals, every element is a 0. [235, 45, 65]

trifolium (noun): the first component is from Latin *tri-*, from the Indo-European root *trei-* "three." The second component is from Latin *folium* "leaf"; the Indo-European root is *bhel-* "to thrive, to bloom." The same root is found in Greek-derived *chlorophyll*. In mathematics the trifolium is a curve defined by the polar equation $r = \sin\theta \left(4a\sin^2\theta - b\right)$, where $0 < b < 4a$. Its name refers to the curve's three loops or "leaves." Unlike the three-petaled rose, whose petals are all congruent, only two of the trifolium's "leaves" are congruent. Compare the etymologically but not mathematically identical *trefoil*. Also compare *folium*, *bifolium*, and *quadrifolium*. [235, 21]

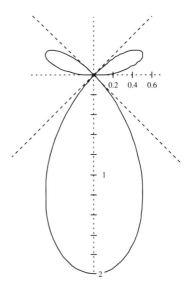

Trifolium: $r = \sin\theta\,(4\cos^2\theta - 2)$

trigon (noun), **trigonal** (adjective): the first component is from Greek *tri-*, from the Indo-European root *trei-* "three." The second component is from Greek *gonia* "angle," from the Indo-European root *genu-* "angle, knee" (an English cognate). Greek *trigon* means the same as its Latin translation, *triangle*. The noun *trigon* is almost never used as a separate word in English any more, but it appears as a main component in *trigonometry*. The word *trigon* may also still be used to refer to an ancient Greek harp that was named for its triangular shape. The adjective *trigonal* is very much alive in crystallography. [235, 65]

trigonometry (noun), **trigonometric** (adjective): the first part of the word is Greek *trigon* "triangle" (see previous entry). The second part of *trigonometry* is from Greek *metron* "a measure." The Indo-European root is probably *me-* "to measure." Trigonometry is literally the measuring (of angles and sides) of triangles. Historically speaking, the triangular approach to trigonometry is ancient, whereas the circular approach now taught in our schools is relatively recent. Compare *polygonometry*. [235, 65, 126]

trihedron, plural *trihedra* (noun), **trihedral** (adjective, noun): the first component is from the Greek *tri-*, from the Indo-European root *trei-* "three." The second component is from Greek *hedra*, (prehistoric Greek *sedra*), meaning "base." The Indo-European root is *sed-* "to sit," as seen in native English *sit* and *seat*. A trihedral angle is an angle formed by three intersecting "bases," i.e., planes. The (moving) trihedral or trihedron of a space curve or surface at a given point is the configuration made up of three mutually perpendicular planes: the tangent plane, the principal normal plane, and the binormal plane. There is no danger that the word *trihedron* will be taken to mean a polyhedron with three faces, because the smallest number of faces a polyhedron can have is four. [235, 184]

trilateral (adjective, noun): the first component is from Latin *tri-*, from the Indo-European root *trei-* "three." The second component is from Latin *latus*, stem *later-* "side," of unknown prior origin. A trilateral is a configuration of three lines (i.e., "sides") that intersect in pairs only. [235, 110]

trillion (numeral): patterned after *billion* by replacing the *bi-* with *tri-* "three." Since in most countries a billion is the second power of a million, a trillion was defined as the third power of a million, or 10^{18}. In the United States, however, a billion is 10^9, and a trillion adds one group of three zeroes, making a trillion equal to 10^{12}. [235, 72, 151]

trimetric (adjective), **trimetry** (noun): the first component is from Greek *tri-*, from the Indo-European root *trei-* "three." The second component is from

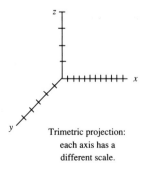

Trimetric projection:
each axis has a
different scale.

Greek *metron* "a measure." The Indo-European root is probably *me-* "to measure." When a three-dimensional graph is projected onto a flat piece of paper, the positive *x*-, *y*-, and *z*-axes are typically drawn radiating from a single point, and then a scale is placed on each axis. If the two-dimensional projections of all three axes have different scales, the projection is said to be a trimetry. Trimetry is one type of axonometry (*q.v.*). Contrast *dimetry* and *isometry*. [235, 126]

trimorphic (adjective): the first component is from Greek *tri-*, from the Indo-European root *trei-* "three." The second component is the latter part of *automorphic* (*q.v.*). In recreational number theory, an automorphic number is one whose second power ends with the same sequence of digits as the number itself; both endings have the "same shape." A trimorphic number is one whose third power ends with the same sequence of digits as the number itself. An example is 49, whose cube is 117649, which also ends in 49. [235, 137]

trinode (noun), **trinodal** (adjective): the first component is from Latin *tri-*, from the Indo-European root *trei-* "three." The second component is *node* (*q.v.*). A trinode is a point on a curve at which three nodes coincide. [235, 142]

trinomial (noun): the first part of the compound is from Greek *tri-*, from the Indo-European root *trei-* "three." One explanation for the second part of the compound involves Greek *nomos*, which meant many things: usage, custom, law, division, portion, part. In that case, a trinomial is a mathematical expression consisting of three parts. A second and more likely correct explanation involves Latin *nomen*, cognate to the English *name*, so that a trinomial is an expression involving three names, i.e., terms. In biology a trinomial is a three-part name consisting of a genus, a species, and a subspecies. [235, 143, 145]

triomino or **tromino** (noun): from Greek *tri-* "three," plus Latin *domino*. Whereas a standard domino contains only two square sections with numbers on them, a triomino contains three square sections with no numbers on them. With higher-order -ominoes such as pentominoes, a player is challenged to combine the pieces to produce simple pictures of certain shapes or recognizable objects. However, because there are only two distinct triominoes, triominoes are rather unchallenging. See explanation under *pentomino*. [235, 37]

triple (verb, adjective, noun), **triply** (adverb): from Latin *triplus* "threefold." The first component is from Latin *tri-*, from the Indo-European root *trei-* "three." The second component is from the Indo-European root *pel-* "to fold." To triple something is to make it three-fold, i.e., to multiply it by 3. When the number of children born at one time is three times as great as usual, the babies are called triplets. A triple of numbers (a, b, c) is a set of three numbers in a given order. A triple root of an equation is a root which occurs three times. A triple point on a curve is a point through which three branches pass. Compare *double, quadruple, quintuple, sextuple, septuple*, and *octuple*. A triply orthogonal system of surfaces is one involving three families of surfaces such that each surface in one family is orthogonal to every member of the other two families. [235, 160, 117]

trirectangular (adjective): the first component is from Latin *tri-*, from the Indo-European root *trei-* "three." The second component is *rectangular (q.v.)* "involving a right angle." In spherical trigonometry, a trirectangular triangle contains three right angles. [235, 179, 11, 238]

trisect (verb), **trisection** (noun), **trisector** (noun): the first component is from Latin *tri-*, from the Indo-European root *trei-* "three." The second component is from Latin *sectus*, past participle of *secui* "to cut." The Indo-European root is *sek-* "to cut," as seen in native English *saw*. When you trisect a line segment you cut it into three equal parts. The trisection of an arbitrary angle using only compass and unmarked straightedge was one of the three great unsolved problems of Greek geometry; in the 19th century that trisection was proven to be impossible. When the Latin agental suffix *-or* is added to *trisect*, the result is *trisector* "something (usually a line) which trisects." Contrast *trisectrix*. [235, 185, 153]

trisectrix (noun): the first component is from Latin *tri-*, from the Indo-European root *trei-* "three." The second component is from Latin *sectus*, past participle of *secui* "to cut." To trisect something is to cut it into three equal parts. The Latin agental suffix *-trix*, when added to a root, meant "the female person or thing that does [whatever the root indicates]." The feminine suffix was chosen for this word because a trisectrix is a curve, and Latin *curva* is a feminine noun. The trisectrix is a curve that can be used to trisect any angle. (The trisection of an arbitrary angle using only a compass and unmarked straight-

edge is impossible.) The rectangular equation of the trisectrix is $x^3 + xy^2 + ay^2 - 3ax^2 = 0$. Contrast *trisector*. [235, 185, 236]

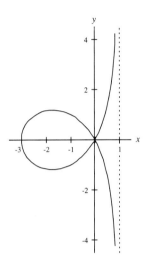

Trisectrix:

$$x = \frac{t^2 - 3}{t^2 + 1}, \quad y = \frac{t^2 - 3t}{t^2 + 1}.$$

triskelion (noun): the first component is from Greek *tri-*, from the Indo-European root *trei-* "three." The second component is from Greek *skelos* "leg." The Indo-European root is *(s)kel-* "crooked." In ancient Greece, a triskelion was a design consisting of three bent legs radiating from a common center; the angle at which any two of the legs met was 120°. In modern mathematics, the triskelion has been simplified to six line segments. It is used as an example of a configuration possessing three-way rotational symmetry but no other common type of symmetry such as bilateral or point symmetry. [235, 195]

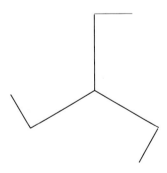

Triskelion.

trivial (adjective): the first component is from Latin *tri-*, from the Indo-European root *trei-* "three." The second component is from Latin *via* "way, road," from the Indo-European root *wegh-* "to go, to transport." Since it was common in Roman times for three roads to meet, something commonplace or insignificant came to be called trivial. In mathematics, as in common English, the "threeness" of the word has been lost. A trivial (= insignificant) solution of the equation $x^2 + y^2 = z^2$ is $(0, 0, 0)$, as opposed to $(5, 12, 13)$, which is a nontrivial solution. [235, 243]

trochoid (noun): from Greek *trokhos* "a wheel" and the suffix *-oid* (*q.v.*) "looking like." The Indo-European root is *dhregh-* "to run." The related Greek *trokhileia* "a system of wheels and pulleys," is the source of English *truckle* and ultimately of *truck*. In mathematics a trochoid is the locus of a point on the (possibly extended) radius of a circle (= wheel) as that circle rolls along a straight line. Depending on whether the fixed point is within, on, or outside the circle, the resulting locus will be a curtate cycloid, a cycloid, or a prolate cycloid, respectively. See pictures under *epitrochoid* and *hypotrochoid*. [43, 244]

tromino: see *triomino*.

true (adjective), **truth** (noun): *true* is a native English word, originally meaning "loyal, trustworthy," as in "a true friend." The Indo-European root is *deru-* or *dreu-* "to be firm, solid, steadfast." From the meaning "steadfast" come derivatives referring to trees. Native English cognates are *tree* itself, *tray* (= wooden board), and *trim* (= strong, healthy). Related borrowings from Latin include *durum* wheat (which is hard) and *duress* (= hardship). Also related is *Druid*, an ancient worshiper of trees. In logic, a proposition is a statement which may be true or false. The truth value of a compound statement may be determined with a truth table. In algebra, when we solve an equation we try to find all values of the variable that make the equation true. [39]

truncate (verb), **truncation** (noun): from Latin *truncus* "stem, trunk (of a tree)." The verb *truncare*, with past participle *truncatus*, meant literally "to reduce to a trunk" by cutting off the other parts of the tree. In geometry, a truncated cone or pyramid is what remains after a part containing the vertex is cut off by a plane that is not parallel to the base. (Contrast *frustum*.) In the realm of calculators and computers, truncation is the cutting off and discard-ing of digits beyond a certain point. Although the Latin word originally referred to the trunk of a tree, it later came to mean the trunk of a human body, and then any object resembling or having to do with the trunk of a body, as in the kind of trunk you pack clothes in or the swimming trunks a man wears to cover the lower trunk of his body. The Indo-European root is *terə-* "to cross over, pass through," though it's hard to see how those meanings relate to the trunk of a tree. Perhaps because the trunk of a tree is straight, it can be made to pass through an opening. Native English cognates include *through*, *thorough* (= going all the way through) and nos*tril* (the hole going through the nose). [229]

turn (verb), **turning** (adjective): via French *tourner* "to turn" from Latin *tornus* "a lathe," from *tornare*, originally "to turn on a lathe," and then "to turn" in general. The Latin word was borrowed from Greek *tornos* "a tool for drawing a circle, a lathe." The Indo-European root is *terə-* "to rub, to turn." Native English cognates include *thresh* (which was often done by walking in circles) and *throw* (which originally meant "to twist," as happens when you throw your back out). A related borrowing from Greek is *trauma* (a wound, because skin is rubbed away). In calculus, a turning point is a point on a curve at which the derivative turns from positive to negative, or vice versa: a turning point is a relative maximum or minimum. Compare *bend*; contrast *stationary*. [228]

twelve (numeral): a native English word with a surprising meaning that will delight the hearts of elementary teachers involved with regrouping. The word *twelve* comes from a presumed Germanic compound *twa-lif* "two-leave," since 2 units are left over after you take away the base 10. The Indo-European roots are *dwo-* "two" and *leik*^{w}- "to leave." Compare the same sort of regrouping inherent in *eleven*. Also compare the two-ness in English words like *twin* and *twain* (as in Mark Twain and "never the twain shall meet"). [48, 114]

twin (noun, adjective): a native English word, from the Indo-European root *dwo-* "two." In number theory, twin primes are primes that are just two units apart. For example, 17 and 19 are twin primes. So are 1,000,000,000,061 and 1,000,000,000,063. Just as biological twins look similar, twin primes "resemble" each other because of their numerical closeness. [48]

two (noun): a native English word descended from the Indo-European *dwo-*, of the same meaning. Although the *w* is no longer pronounced in the basic English word, it continues to be pronounced in quite a few other related words: *twain*, *twice*, *twenty*, *twelve* (*q.v.*), *twilight* (= the interval between two times of day), *twine* (double thread), *twill* (also double thread), *twig* (a little branch that splits in two), *betwixt* and *between* (*q.v.*). In Latin the original Indo-European root developed into *duo* as well as *bi-* (originally *dui-*); each form appears in many words that English has borrowed from Latin. Also from Latin is *dubious* (trying to decide between two choices) and the related noun *doubt*. In Greek, the Indo-European root developed into *duo* and related forms. From Greek English has borrowed words like *dyad* and *dodecagon*. The ancient Greeks believed 2 to be the first number because they considered 1 a generator of numbers but not a number itself. The integer 2 is the first and only even prime. [48]

type (noun): from Greek *tupos* "a mold, a die," from the Indo-European root *(s)teu-* "to push, stick, knock, beat." A mold is an object which is "knocked or beaten out." In traditional printing, each piece of movable type is made from a mold. When we say metaphorically that a certain object "fits the mold," we are saying that it is of a certain type. Many mathematical objects are grouped by type. For example, the conic sections are of three types, depending on eccentricity. In tilings using polygonal tiles, there are 21 possible types of vertex. [214]

U

ultrafilter (noun): the first component is from Latin *ultra* "beyond," from the Indo-European root *al-*, of the same meaning. The second component is *filter* (*q.v.*). An ultrafilter is a filter that isn't a proper subset of any filter. [6, 161]

umbilic (noun, adjective), **umbilical** (adjective): from Latin *umbilicus* "navel." The Indo-European root underlying both *umbilicus* and native English *navel* is *nobh-*, alternate form *ombh-*, "navel." An umbilic point of a surface is a point at which the surface bends in like a navel.

unary (adjective): from Latin *unus* "one," from the Indo-European root *oi-no-* "one." In mathematics a unary operation or function acts on one number at a time. Taking the absolute value is a unary operation;

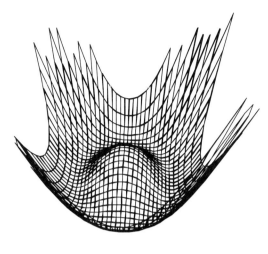

Umbilic.

in contrast, a binary operation like multiplication requires two numbers. [148]

undecillion (numeral): patterned after *billion* by replacing the *bi-* with the root of Latin *undecim* "eleven," a compound of *unus* "one" and *decem* "ten." Since in most countries a billion is the second power of a million, an undecillion was defined as the eleventh power of a million, or 10^{66}. In the United States, however, a billion is 10^9, and an undecillion adds nine groups of three zeroes, making an undecillion equal to 10^{36}. [148, 34, 72, 151]

under (adverb, preposition): a native English word, from the Indo-European root *ndher-* "under." Related borrowings from Latin include *inferior* and *infra-*. In common English, *under* often means "less than." For example, 1 and 2 are the only positive integers under 3. [140]

unicursal (adjective): the first component is from Latin *uni-*, from the Indo-European root *oi-no-* "one." The second component is from Latin *cursus*, past participle of *currere* "to run"; the Indo-European root is *kers-* "to run." A unicursal graph can be traced in its entirety and with no duplications by running a pencil over it one time (i.e., without lifting the pencil and resuming the tracing elsewhere). [148, 95]

uniform (adjective): the first element is from Latin *unus*, from the Indo-European root *oi-no-* "one." The second element is from Latin *forma* "form, shape," which may be borrowed from Greek *morphe* "shape"; another hypothesis links it to Latin *ferire* "to strike," since a construction form may be struck from a mold. When people wear uniforms, they are all dressed in one and the same form or fashion. In mathematics, the word *uniform* is sometimes used in the sense of "equal"; for example, uniform acceleration is acceleration in which for equal changes in time there are equal changes in velocity. [148, 137]

unilateral (adjective): the first element is from Latin *unus*, from the Indo-European root *oi-no-* "one." The second element is from Latin *latus*, stem *later-*, "side," of unknown prior origin. A surface is unilateral if it has only one side. An example of a unilateral surface is a Möbius band. [148, 110]

Unilateral surface: a Möbius band.

unimodal (adjective), **unimodular** (adjective): the first element is from Latin *unus*, from the Indo-European root *oi-no-* "one." The second element is from Latin *modulus* (*q.v.*), a diminutive of *modus* "measure, standard." In a histogram, the mode is the value that has the highest bar. By analogy, a function is said to be unimodal if it has just one "highest point," i.e., just one relative maximum. A unimodular matrix is a square matrix whose "measure," is 1, i.e., whose determinant has a value of 1. [148, 128, 238]

union (noun): from Latin *unio*, stem *union-* "a oneness," from *unus* "one." The Indo-European root is *oi-no-* "one." (The similar-sounding English *onion* is, via French, from the same source; the Latin word came to mean "a single large pearl" (= "a big one"), and then "an onion," because of an onion's resemblance to a pearl.) In mathematics the union of several sets is a set in which each element is listed just one time, no matter how many times that element actually appears in the original group of sets. The union of two sets is represented by the symbol $\cup$, which coincidentally resembles the first letter of the word *union*. The symbol is sometimes referred to as a cup (*q.v.*). [148]

unipotent (adjective): the first component is from Latin *unus*, from the Indo-European root *oi-no-* "one." The second component is from Latin *potens*, stem *potent*, "able, powerful," from the Indo-European root *poti-* "powerful; lord." In mathematics a square matrix is said to be unipotent if, when raised to some power, the result is the corresponding identity matrix. [148, 173, 146]

$$\begin{bmatrix} 0 & 0 & 6 \\ \frac{1}{2} & 0 & 0 \\ 0 & \frac{1}{3} & 0 \end{bmatrix}^3 = \begin{bmatrix} 1 & 0 & 0 \\ 0 & 1 & 0 \\ 0 & 0 & 1 \end{bmatrix}$$

The matrix on the
left is unipotent
of order 3.

unique (noun): a French word, from Latin *unicus* "one and only one, one of a kind." The Indo-European root is *oi-no-* "one." Many mathematical proofs involve showing that a certain quantity is unique. In the realm of real numbers, for example, 0 is the unique identity element for addition, and 1 is the unique identity element for multiplication. [148]

unit (noun), **unitary** (adjective) **unity** (noun): from Latin *unus*, the number "one," from the Indo-European root *oi-no-* "one." The units place in a decimal number is reserved for the number of ones the number contains. A unit fraction is a fraction whose numerator is the number 1. The number 1 has often been called unity, as in the statement "there are three complex third roots of unity." A unitary matrix is a square, complex matrix which when multiplied by its tranjugate yields a square matrix having a 1 everywhere on the main diagonal (and a 0 everywhere else). [148]

universal (adjective), **universe** (noun): the first component is from Latin *unus*, from the Indo-European root *oi-no-* "one." The second component is from Latin *versus*, past participle of *vertere* "to turn." The Indo-European root is *wer-* "to turn, to bend." The universe is everything that exists, all "turned into one"; putting it differently, the universe is the one and only thing that consists of everything else. In mathematics the universal set is the set consisting of all the elements that are allowed to be considered in that one situation. [148, 251]

unknown (adjective, noun): a native English compound, the first element being from the Indo-European root *ne* "not." The verb *know* is from the Indo-European root *gno-* "to know." Other native English cognates are *canny* (= knowing how to), *ken*, and *cunning*. European mathematicians who wrote in Latin often referred to an unknown quantity with the Latin word *res* "a thing." Compare *republic*, literally the "public thing" that we allow to govern us. Italian mathematicians of the Renaissance used their word for "thing," which was *cosa*, as in the Mafia-style organization known as the *cosa nostra* "our thing." English mathematicians even borrowed the Italian term as the noun *coss*, which was in use until about 1800; the Rule of Coss was another name for algebra. As an Englishman named Digges wrote in 1671: "The ingeniouse Student, having any meane taste of cossicall numbers, shall finde them playne and easie." Aside from the quaint language, that sounds strikingly similar to current ads for high school algebra textbooks. [141, 75]

unless (conjunction): a native English word. The first component is from *on*, from the Indo-European root *an-* "on." The initial vowel was altered by the influence of the negative prefix *un-*, since the word *unless* is semantically a negative. The second component is *less* (*q.v.*). In logic the statement "*p* unless *q*" is defined as $\sim q \rightarrow p$. In other words, the negation of *q* implies *p*. [10, 115]

untouchable (noun): the prefix *un-* is native English, from the Indo-European root *ne* "not." *Touch* is from French *toucher* "to touch," from an assumed Vulgar Latin *toccare* "to strike a bell," and then more generally "to touch." The Vulgar Latin verb is presumed to have been imitative, in much the same way that we use the conventionalized term "tick-tock" in English to mimic the sound of a clock or watch. The mathematician Paul Erdős has used the term *untouchable* to refer to a number that can never be the sum of the proper divisors of another number. (For an explanation of the suffix see *-able*.) The first five untouchable numbers are 2, 5, 52, 88, and 96. Compare *deficient, abundant, perfect, amicable*, and *sociable*. [141, 222, 69]

up (adverb), **upper** (adjective): native English words, from the Indo-European *upo* "under," but also "up from under." The Indo-European root developed opposite meanings in different languages. Whereas English shows the upward meaning, not only in *up* but also in the related verb *to open (up)*, the Greek cognate *hupo* and the Latin cognate *sub* retain the meaning "under." In a standard two-dimensional rectangular coordinate system, the *y*-coordinate indicates how far up or down from the *x*-axis a point is located. An upper bound of a sequence is a number whose value is always "up

above" (or equal to) the value of every term in the sequence. [240]

urn (noun): via French, from Latin *urna* "a water pot, water jar, urn." The word may be related to Latin *urere* "to burn," because urns used to be made of burnt earth. In ancient Greek elections men voted by putting black or white beans into an urn. In modern probability problems, an urn is often used as the container from which balls, chips, or other objects are to be randomly chosen. The modern obsession with urns seems to stem from their use in antiquity, but a jar or coffee can or hat or other container might be more appropriate for students nowadays, few of whom know what an urn is or have had any contact with one.

V

vacuously (adverb): from Latin *vacuus* "empty, clear, devoid of." The Indo-European root is *eu-* "lacking, empty." Related borrowings from Latin include *vain* and *vacant*, and native English *want* (as in the phrase "to be wanting") is a cognate. In mathematics a proposition is said to be vacuously true when there is nothing to contradict it. For example, in combinatorics, if 0 objects were chosen from a group of objects, the choosing may be considered to have taken place vacuously with replacement. [58, 117]

valence (noun): via French, from Latin *valentia* "capacity," from the present participle of *valere* "to be strong, vigorous, healthy, worthy." The Indo-European root is *wal-* "to be strong." Related borrowings from French include *prevail* and *avail*. In mathematics the valence of a node in a network is the number of paths that meet at that node. [241, 146]

valid (adjective), **validity** (noun): from Latin *valere* "to be strong, vigorous, healthy, worthy." The Indo-European root is *wal-* "to be strong." A native English cognate is *wield* (= to use power). Related borrowings from Latin are *valor* and *valiant*. In mathematics an argument is said to be valid as long as the rules of logic are followed, even if the statements involved are ludicrous. [241]

value (noun): from Latin *valere* "to be strong, vigorous, healthy, worthy." The Indo-European root is *wal-* "to be strong." The Latin verb evolved into

French *valoir* "to be worth," with feminine past participle *value*. The value of something is how much it is worth. A native English cognate is *wield* (= to use power). Related borrowings from French are *prevail* and *countervail*. In mathematics the value of a variable expression depends on the numbers substituted for the variable(s). [241]

vanish (verb): from Old French *esvanir* "to disappear, grow weak," from Classical Latin *evanescere* "to vanish, disappear, pass away." The Indo-European root is *eu-* "lacking, empty." A native English cognate is *to wane*, and a borrowing from Old Norse is *to want*, as in "to find someone wanting." Related borrowings from Latin include *vain*, *vacant*, and *devastate*. In mathematics, the verb *vanish* is a synonym of "to equal zero." When an expression vanishes, it becomes zero. The mathematical use of *vanish* was quite common in textbooks up to the middle of the 20th century; although a lot less common now, it has not entirely vanished. [58]

vary (verb), **variable** (adjective, noun), **variance** (noun), **variation** (noun): the Indo-European root *wer-* referred to a raised spot on the skin, as seen in the native English *wart*. According to the linguist Calvert Watkins, since a wart appears mottled or variegated, and is a change from the normal appearance of skin, the root also took on the meaning of "change." Both senses can be seen in Latin *variare* "to fleck, diversify, look spotty, alter, change," as seen in our borrowed word *to vary*. From Latin *variabilis* "changeable," comes *variable*. For an explanation of the suffix see *-able*. In mathematics a variable is a quantity, usually represented by a single letter, that can change its values. The German mathematician Gottfried Wilhelm Leibniz (1646–1716) was apparently the first to use the term *variable* in that way. There are many other uses of *variable* in mathematics; to avoid confusion we should probably assign a different word to each meaning, but changing tradition is difficult. The word *va-ri-a-ble* has four syllables; pronouncing—or worse, spelling—*variable* as if it had three syllables is substandard. In statistics the variance is a measure of how the values being measured are dispersed about (= vary from) the mean. [250, 69]

vector (noun): a Latin word meaning "carrier," from the verb *vehere* "to carry." The Indo-European root is *wegh-* "to go, to transport," as found in native English *way* and *weight*. Related borrowings from Latin include *vehicle* and *convection*. In mathematics a vector is a segment that is "carried off" for a certain distance in a certain direction. In other words, a vector is a carrier of two pieces of information: a length and a direction. [243, 153]

vel (conjunction): a Latin word meaning "or," used originally when there was no inherent opposition between the things being mentioned. The word is from Latin *velle* "to want." Latin *vel* must have developed from questions like "Do you want an apple [or] do you want a peach?" The Indo-European root is *wel-* "to wish, to will"; *will* is an English cognate. Related borrowings from Latin include *voluntary* and *volition*. Mathematics sometimes uses the Latin word *vel* because English *or* is ambiguous. It can be inclusive: this thing or that thing or both things. It can be exclusive: this thing or that thing but not both things. In logic, to avoid ambiguity, *vel* indicates the inclusive *or*. It corresponds to the possibility that a hungry person might answer the fruit question by saying "I'll take one of each." The symbol that usually stands for *vel* is ∨, which happens to be the first letter of the word. Contrast *aut*. [249]

velocity (noun): from Latin *velox*, stem *veloc-*, "swift, active." The same root appears in the word *vegetable*, because a vegetable is something that grows and shows activity. Strangely enough, in modern English a vegetable has come to symbolize inactivity because, although a vegetable grows, it stays in the same place for its whole life. In the mathematics of vector-valued functions, the velocity vector is the first derivative of the position vector. Contrast velocity, which has both magnitude and direction, with speed, which has only magnitude. In nonmathematical English *velocity* is just a fancy word for *speed* (q.v.).

versed (adjective): from Latin *versus*, past participle of *vertere* "to turn." The Indo-European root is *wer-* "to bend, to turn." In trigonometry, the versed sine of θ, abbreviated vers θ, is defined as $1 - \cos \theta$. The name is explained by the figure below. $DB =$ versin θ. It is the segment reached by "turning" seg-

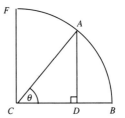

$CA = CB = CF = 1.$ $AD = \sin \theta.$
$CD = \cos \theta.$ $DB = $ versin $\theta.$

ment $\overline{DC}$ (whose length is cos θ) about segment $\overline{AD}$ (whose length represents sin θ). Two related borrowings from Latin, *convert* and *version*, further help explain the term *versed sine*. If the expression $1 - \cos\theta$ is multiplied by its conjugate, it is converted into $\sin^2\theta$, which is a "version" of sin θ. Compare *coversine*, *exsecant*, and *haversine*. [251]

versiera (noun): a name coined by the Italian mathematician Guido Grandi (1671–1742) for the curve whose equation is $x^2 y = 4a^2(2a - y)$. He derived the name from the Latin phrase *sinus versus* "a curve that is 'turned.' " Latin *versus* is the past participle of *vertere* "to turn," from the Indo-European root of the same meaning, *wer-*. The turning in question locates points on the curve in the following way: as point A moves (= turns) on the fixed circle, ZA intersects line m at B. A line dropped vertically from B meets a horizontal line containing A at the point C, which is on the versiera. See more under *witch*. [251]

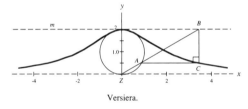

Versiera.

vertex, plural *vertices* (noun): a Latin word meaning originally "whirl, eddy, vortex" (in fact *vortex* is just a mistakenly modified spelling of *vertex*). The noun is from the Latin verb *vertere* "to turn." The Indo-European root is *wer-* "to turn, to bend." Native English cognates include *wrist* and *wrestle*. A related borrowing from Latin is *vertigo*, "a turning, dizzy feeling." Even in Roman times Latin *vertex* came to mean "the pivot point of the heavens," "the top of a person's head," and "that which is in the highest place." In mathematics the vertex is the highest (or lowest) point on a curve or polygon because that's the place where the curve or polygon turns to go back down (or up) after having risen (or fallen) to reach that point. In terms of calculus, the derivative "turns" from positive to negative when a point moving on a curve crosses through a vertex of that curve. [251]

vertical (adjective): from Latin *vertex*, stem *vertic-*, "the high point of an object." Since the vertex (*q.v.*) of an object is overhead, a person who wants to see it has to look straight up, i.e., vertically. The word *vertical* now describes anything that goes up and down, particularly a straight line. In geometry vertical angles are, so to speak "straight up and down"

from one another. Another way of looking at it is that you've done 180° of turning to get from one of the angles to the other one; you've turned through a straight angle, which is the equivalent of a straight line. [251]

vicenary (adjective), **vigesimal** (adjective): *vicenary* is from Latin *viceni* "twenty each," while *vigesimal* is from Latin *vigesimus* "twentieth." Both words are related to Latin *viginti* "twenty," and all are from the Indo-European root *wikmti-*, a compound of *wi-* "in half, in two parts," and a shortened form of *dekm* "ten." The original concept was that 20 consists of two parts of 10 each. A vicenary or vigesimal number system is one that uses a base of twenty. The best known historical example is the Mayan numeration system. For *vicenary* compare *binary*, *denary*, *octonary*, *quinary*, *senary*, and *ternary*. For *vigesimal* compare *decimal*. [253, 34]

vigintillion (numeral): from Latin *viginti-* "twenty," from the Indo-European root *wikmti-*, a compound of *wi-* "in half, in two parts," and a shortened form of *dekm-* "ten." The original concept was that 20 consists of two parts of 10 each. The word vigintillion is patterned after *billion* by replacing the *bi-* with *viginti-*. Since in most countries a billion is the second power of a million, a vigintillion is defined as the twentieth power of a million, or 10^{120}. In the United States, however, a billion is 10^9, and a vigintillion adds eighteen groups of three zeroes, making a vigintillion equal to 10^{63}. [253, 34, 72, 151]

vinculum (noun): from Latin *vincire* "to bind, to tie," of unknown prior origin. The diminutive *vinculum* referred to things that were used to bind people, such as ropes, bonds, or fetters. In mathematics textbooks that were used through the early years of the 20th century, writers often put a bar over terms that were intended to be grouped; nowadays we would use parentheses, brackets, braces, or other grouping symbols. The bar that was written over two or more terms came to be known as a *vinculum* because it bound the terms together. For example, what we would now write as $(x + y)$ could be written $\overline{x + y}$. Historically speaking, when the vinculum was first used by Nicolas Chuquet in 1484, he put it under rather than over the terms being grouped. Some recent authors have extended the definition of *vinculum* to include the bar between the numerator and the denominator of a fraction, given that the fraction bar often acts as a grouping symbol. [238]

virgule (noun): a French word, from the Latin diminutive *virgula* "a small rod or wand." The more basic Latin *virga* "twig, sprout, rod," is of unknown prior origin. In mathematics and science the virgule is the slanted bar used to indicate division, primarily division of units, as in km/hr. When used in numerical fractions, the slanted bar is usually called a *solidus* (*q.v.*). [238]

volume (noun): from Latin *volumen* "something that is rolled up," from the verb *volvere* "to roll." The Indo-European root is *wel-* "to turn, to roll." A Latin *volumen* was a roll of writing, i.e., a scroll. When books replaced scrolls as the most common way of disseminating written works, the term *volume* was carried over to books. Beginning in the 16th century, *volume* came to refer to the bulkiness of a book or the amount of space it occupies. In the 17th century *volume* was first used to mean the size or mass of an object in general, no longer necessarily a book. Only as recently as the late 19th century did the word develop the extended sense "amount." We can now say that the volume of traffic is up this year, or that a student takes voluminous notes. We can also ask someone to turn down the volume (= amount of sound) on a radio. In mathematics, volume is a three-dimensional measure of space or substance. [248]

vulgar (adjective): from Latin *vularis* "of or belonging to the multitude; usual, general, common," from the noun *vulgus* "the masses, the public, the common people." Originally *vulgar* referred to masses of people in a neutral way; later the word acquired negative connotations associated with the worst traits of multitudes and common people. In arithmetic, a vulgar fraction is the same as a common or simple fraction, i.e., one in which both the numerator and denominator are whole numbers. A vulgar fraction is contrasted with a complex (*q.v.*) fraction. In mathematics a vulgar fraction is now almost always called a common fraction.

W

washer (noun): a native English word of uncertain origin. Many people assume it is connected to the verb *wash* because washers are placed in hoses and faucets to prevent liquids from "washing" or spilling out. Nevertheless, no clear historical evidence exists to link the words *washer* and *wash*. In calculus volumes of revolution of certain solids may be calculated using the washer method, which relies on summing up the volumes of thin elements that look like washers.

wave (noun): a native English word, from the Indo-European root *wegh-* "to go, to transport." A wave is literally water "going somewhere." The development of *wave*, the noun, was influenced by *wave*, the verb "to move a hand up and down"; the two words are actually from different Indo-European sources. In mathematics waves are typically represented by sinusoidal functions. [243]

weak (adjective): from the Old Norse adjective *veikr* "pliant, flexible," which supplanted the cognate native English adjective *wac*. The Indo-European root is *weik-* "to bend, to wind." In mathematics various abstract concepts may be modified by the adjective *weak*; some examples are weak compactness, weak completeness, weak convergence, and weak topology. [246]

wedge (noun): a native English word with cognates in other Germanic languages; probably from the Indo-European root *wog^w-ni-* "plowshare, wedge." In plane geometry an infinite wedge is any of the four regions created by two intersecting lines. In solid geometry an infinite wedge is any of the four regions created by two intersecting planes. A finite wedge is a type of polyhedron.

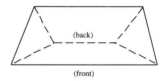

(back)

(front)

A wedge.

week (noun): a native English word, from the Indo-European root *weik-* "to bend, to wind." The original meaning of *week* seems to have been "a 'turning over' or succession of days." There is no reason to believe that the Germanic peoples used the word to represent a specific period of time until they came into contact with the Romans; only after that encounter did *week* come to have its modern meaning of "seven days." [246]

weigh (verb), **weight** (noun), **weighted** (adjective): native English words, from the Indo-European root *wegh-* "to go, to transport." Something of the original meaning can still be seen in the nautical expressions "to weigh anchor" and "anchors aweigh." When an object is weighed, it is moved onto a scale. The

British *wee* "small," is from the same source; it was originally used in the phrase *a little wee* "a little weight." In mathematics a weighted average is one in which some values "weigh more heavily in the result" than others; for example, a final exam in a course may "weigh" twice as much as any chapter test. [243]

weird (noun): a native English word with a weird development in meaning. The Indo-European root is *wer-* "to turn, to bend." The word then developed the sense "turn into, become, befall." What befalls you or what becomes of you is your destiny, so *wyrd*, as the word used to be spelled, developed the meaning "fate." By the 14th century, the word meant "controlling the destinies of men." Because Shakespeare used *weird* in *Macbeth* to describe one of the play's witches (who controlled men's destinies), the word took on its modern meaning of "strange, mysterious." In number theory, an abundant (*q.v.*) number is said to be weird if there is no subset of its proper divisors that adds up to the number itself. The term is used because almost every abundant number does have a subset of proper divisors that adds up to the number itself. The smallest weird number is 70. No odd weird number has ever been found. [251]

well (adverb): a native English word, from the Indo-European root *wel-* "to wish, to will." When something goes well, it goes as you would wish it to. In mathematics a well-ordered set is a linearly ordered set for which every subset has a first element. We might say that that's how we would wish a linearly ordered set to behave in the best of all possible worlds: every subset has a definite beginning. [249]

whole (adjective): a native English word, from the Indo-European root *kailo-* "whole, uninjured." Other native English cognates include *hale*, *holy*, and *heal(th)*. One of Euclid's axioms is that the whole equals the sum of its parts. The set of whole numbers includes 0 and all the positive integers.

wide (adjective), **width** (noun): the native English adjective is cognate to similar words in other Germanic languages. A plausible source for the word is Indo-European *wi-* "apart, in half," since something wide has its opposite edges far apart. Also related is the common preposition *with*, the original meaning of which was *against* (as when you fight with your enemy, the person you are far apart from in purpose or ideas). In pairs of opposite adjectives like *wide* and *narrow*, the "positive" member of the pair also serves as a generic for the entire range of values:

the question "how wide?" can be answered with a quantity that is relatively wide, such as a mile, or very narrow, such as a millimicron. By contrast, the question "how narrow?" already presumes that the answer will be narrow. For the *-th* suffix in *width*, see *length*. The width of a rectangle is generally taken to be the shorter of the rectangle's two dimensions. [253]

winding (adjective): from the native English verb *to wind*. The Indo-European root is *wendh-* "to turn, to wind, to weave," as seen in the English cognates *wander* and *wend* (whose original past tense, *went*, has taken over for the now obsolete past tense of *go*). In mathematics a winding number tells how many times a closed plane curve passes counterclockwise around a designated point in that plane.

witch (noun): the modern descendant of Old English *wicce*, of the same meaning. A related Old English word was *wigle* "divination, sorcery," akin to modern *guile*. In mathematics the witch is a curve whose equation in rectangular coordinates is $x^2 y = 4a^2(2a - y)$. The curve was studied and named the *versiera* "the turned or twisted curve," by the Italian mathematician Guido Grandi (1671–1742). Another Italian mathematician, Maria Gaetana Agnesi (1718–1799), later discussed the curve at some length. When her work was translated into English, a problem arose with the word *versiera*, which, in addition to the original "turning" meaning had developed the meaning "a witch," presumably because a witch twists or distorts the truth, and also because a witch was held to be an *adversoria*, or adversary, of righteous people. The English translator of Agnesi's work said that in Italian the curve was vulgarly called "the witch," and the name took hold in English. See picture under *versiera*.

work (noun): a native English word, from the Indo-European root *werg-* "to work." Other English cognates are *wrought*, as in *wrought* (= worked) iron, and *wright*, as in *wheelwright*, a man who makes wheels. From the cognate Greek *ergon* "work," come *energy*, *erg*, and the suffix *-urgy*, as in *metallurgy*. From the related Greek *organon* "tool," come both *organ* and *orgy*. In calculus, work is calculated by integrating expressions for force times distance.

wrong (adjective): a native English word that meant originally "crooked, twisted." Related English cognates are *wreath*, *wry*, and *writhe*. The Indo-European root is *wer-* "to turn, bend." When something is twisted out of its proper position or condition, that thing is incorrect. The current meaning dates back to the 14th century. [251]

Y

yard (noun): a native English word that no longer resembles its presumed ancestor, the Indo-European root *ghasto-* "a rod, a staff." In Old English the forerunner of *yard* meant "a straight shoot, a twig, a rod, a staff." Only after about the year 1000 did the term come to designate a specific length of 3 feet, although a yard was also sometimes defined as 16.5 feet, a length that later came to be called a *rod*. The word *yard* referring to a length is only coincidentally spelled and pronounced the same as the historically different word *yard* referring to an enclosure, as in the front yard of a house.

year (noun): a native English word, from the Indo-European root *yer-*, of the same meaning. The Indo-European word also developed the meaning "season," as seen in Old Slavonic (the predecessor of modern Russian) *yara* "spring" and in Greek *hora* "season." The designated amount of time grew even shorter when the Romans borrowed the Greek word and used it to mean *hour*. [258]

yocto-, abbreviated *y* (numerical prefix): in the International System of Units the prefix *yocto-* multiplies the unit to which it is attached by 10^{-24}, which can be rewritten as $(10^{-3})^8$, the eighth power of one one-thousandth. *Yocto-* is from Greek *okto* or Latin *octo* "eight," the final *-o* of which already complies with the current system: all of the numerical prefixes in the International System have two syllables and end in a vowel, which in the case of submultiples created in recent times is always an *-o*. The initial *y-* of *yocto-* was added to avoid confusion with the prefix *octo-* that occurs in words like *octillion*. The *y-* was also added to avoid having the first letter *o* used as an abbreviation, because it could easily be confused with the numeral 0. The prefix *yocto-* was proposed in 1990 and was officially adopted as part of the International System of Units in 1991. Compare *zepto-*. [149]

yotta-, abbreviated *Y* (numerical prefix): in the International System of Units the prefix *yotta-* multiplies the unit to which it is attached by 10^{24}, which can be rewritten as $(10^3)^8$, the eighth power of a thousand. *Yotta-* is ultimately from Latin *octo* "eight," though the *-tt-* is reminiscent of Italian *otto* "eight." The fi-

nal *-o* has been replaced by *-a* for uniformity: all of the numerical prefixes in the International System have two syllables and end in a vowel, which in the case of magnifying prefixes created in recent times is always an *-a*. The initial *y-* of *yotta-* was added to avoid confusion with the prefix *octo-* that occurs in words like *octillion*. The *y-* was also added to avoid having the first letter *O* used as an abbreviation, because it could easily be confused with the numeral 0. The prefix *yotta-* was proposed in 1990 and was officially adopted as part of the International System of Units in 1991. Compare *zetta-*. [149]

Z

zepto-, abbreviated *z* (numerical prefix): in the International System of Units the prefix *zepto-* multiplies the unit to which it is attached by 10^{-21}, which can be rewritten as $(10^{-3})^7$, the seventh power of one one-thousandth. *Zepto-* is from the Latin stem *sept-* "seven," with the final *-o* added for uniformity: all of the numerical prefixes in the International System of Units have two syllables and end in a vowel, which in the case of submultiples created in recent times is always an *-o*. The initial *s-* of the Latin word was changed to *z-* to avoid confusion with the prefix *sept-* that occurs in words like *septillion*; furthermore, the abbreviation *s* was already in use for *seconds*, but *z* was still available. The prefix *zepto-* was proposed in 1990 and was officially adopted as part of the International System of Units in 1991. Compare *yocto-*. [191]

zero, plural *zeros* or *zeroes* (cardinal number): a variant of the Arabic word *çifr* "empty," translated from the corresponding Hindu word *sunya*, since the Hindus apparently were the first to develop the concept of zero. When you use a zero as part of a number written in place-value notation, you are keeping a particular place "empty." The Arabic word took on varying forms in European languages, including *zephirum*, *zefiro*, and *cipher* (*q.v.*). The modern form *zero* seems to have appeared in print for the first time in 1491 in *De Arithmetica Opusculum*, by Philippi Calandri. Although the number zero is, logically speaking, neither singular or plural, English has to pick one of the two when a countable noun follows. In that case it treats zero as a plural: "I allow zero pets in my house." Since zero is a late addition to the English numbers, the corresponding ordinal is *zeroth*, which borrows the suffix from the "higher" ordinals *fifth*, *sixth*, *seventh*, *eighth*, *ninth*, *tenth*, etc., rather than from the arithmetically closer but linguistically more irregular *first*, *second*, or *third*. In algebra we can make the seemingly contradictory statement that 4 is a zero of a function, meaning that when 4 is plugged into the function, the value taken on by the function is 0. [26]

zetta-, abbreviated *Z* (numerical prefix): in the International System of Units the prefix *zetta-* multiplies the unit to which it is attached by 10^{21}, which can be rewritten as $(10^3)^7$, the seventh power of a thousand. *Zetta-* is ultimately from Latin *septem* "seven," though the *-tt-* is reminiscent of Italian *sette* "seven." The final *-a* of *zetta-* was added for uniformity: all of the numerical prefixes in the International System have two syllables and end in a vowel, which in the case of magnifying prefixes created in recent times is always an *-a*. The initial *s-* of *sept-* was changed to *z-* to avoid confusion with the prefix *sept-* that occurs in words like *septillion*; furthermore, the abbreviation *s* was already in use for seconds, but *Z* was still available. The prefix *zetta-* was proposed in 1990 and was officially adopted as part of the International System of Units in 1991. Compare *yotta-*. [191]

zone (noun): Latin *zona* was borrowed from Greek *zone* "girdle," from an Indo-European root *yos-* "to gird." In solid geometry a zone of a sphere is that part of the surface contained between two parallel planes that intersect the sphere; a zone is like a girdle around the sphere. If one of the parallel planes is tangent to the sphere, the resulting zone is said to be of one base. (The solid bounded by the zone and the two parallel planes is called a layer of the sphere.) The concept of a zone may be extended to any surface of revolution cut by two planes parallel to the axis of revolution. In nonmathematical English, *zone* has come to mean a region in general, not necessarily one that encircles an object, as in the Twilight Zone. [260]

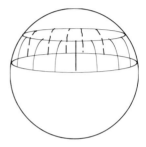

Zone of a sphere.

zonohedron, plural *zonohedra* (noun): the first component is from Greek *zone* "girdle" (see previous entry). The second component is from Greek-derived *-hedron* (*q.v.*) "seat, base," used to indicate the face of a polyhedron. A zonohedron is a convex polyhedron bounded by parallelograms. The parallelograms are so situated as to form "zones" or "ribbons" that completely encircle the polyhedron. The cube is the simplest of all zonohedra. [260, 184]

Appendix

Mathematical entries that are etymologically related, grouped by roots.

IE = Indo-European; other sources are spelled out.

A question mark (?) indicates that the term may or may not be from the given root. Within a given entry, question-marked items may be clearly related to each other, but there can be doubt about whether that entire group of related words is from the given root. For example, *cone*, *conic*, *conical*, *conicoid*, and *conoid* are all related to each other, but may not be from the Indo-European root *ko-* [102] under which they are listed.

[1] *ad-* (IE) "to, at, near": abridged, accelerate, acceleration, accrue, accumulate, accumulation, accuracy, accurate, ad-, add, addend, addition, additive, adherent, adjacent, adjoin, adjoint, adjugate, admissible, affine, aggregation, alligation, amortization, amortize, amount, annihilate, annihilator, applied, appreciate, appreciation, approach, approximate, approximation, arbitrary (?), arrange, array, ascending, associative, associativity, assume, assumption, attractor, reductio ad absurdum, subadditive, superadditive.

[2] *æqu-* (Latin) "even, level": equal, equation, equant, equator, equiangular, equiareal, equiconjugate, equidistant, equilateral, equilibrium, equimultiple, equipollent, equipotent, equivalence, equivalent, inequality.

[3] *ag-* (IE) "drive": actuarial, actuary, ambiguous, axiom, axiomatic, exact, strategy.

[4] *ak-* (IE) "sharp": acnode, acute, edge.

[5] *aks-* (IE) "axis": axial, axis, axonometric, axonometry, coaxial, triaxial.

[6] *al-* (IE) "beyond": aliquot, allometric, alternate, alternating, alternation, anallagmatic, antiparallel, parallel, parallelepiped, parallelepipedal, parallelogram, parallelotope, subalternate, ultrafilter.

[7] *aleph* (Hebrew) "ox; first letter of the alphabet": aleph, alpha, alphametic, pentalpha.

[8] *all* (English) "all": all, almost.

[9] *ambhi* (IE) "around": ambiguous, amphicheiral, because, belong, between.

[10] *an* (IE) "on": analogy, analysis, analytic, analyticity, analyze, anharmonic, ANOVA, on, onto, unless.

[11] *ang-* (IE) "bend": angle, angular, birectangular, equiangular, pentangle, quadrangle, quadrangular, rectangle, rectangular, supertriangular, triangle, triangular, trirectangular, triangulation.

241

[12] *ant-* (IE) "front, forehead": answer, antecedent, anti-, anticommutative, antiderivative, antigonal, antilogarithm, antimorph, antinomy, antiparallel, antipodal, antipodes, antiprism, enantiomorph, enantiomorphic, enantiomorphous, endpoint.

[13] *ap-* (IE) "take, reach": apex, couple, decouple.

[14] *apo-* (IE) "away": a posteriori, a priori, abscissa, absolute, absorb, absorption, abstract, abundant, apothem, component (?), composite (?), composition (?), compound (?), contrapositive (?), exponent (?), exponential (?), exponentiation (?), modus ponens (?), opposite (?), position (?), positive (?), postmultiplication (?), postmultiply (?), proposition (?), propositional (?), puzzle (?), reductio ad absurdum, superpose (?), suppose (?), transpose (?), transposition (?).

[15] *ar-* (IE) "fit together": anharmonic, antilogarithm, arithmetic, arm, cologarithm, coordinate, cryptarithm, harmonic, hundred, logarithm, logarithmic, order, ordinal, ordinary, ordinate, poset, rate, ratio, rational, rationalize, reason, semilogarithmic.

[16] *are-* (Latin) "threshing floor": are, area, areal, equiareal.

[17] *at-* (IE) "go": annuity, millennium.

[18] *aug-* (IE) "increase": augment, auxiliary.

[19] *aut-* (Greek) "self": automaton, automorphic, automorphism.

[20] *awi-* (IE) "egg": oval, ovoid.

[21] *bhel-* "blow, swell; thrive, bloom," and the extension *bhleu-* "swell up, overflow" (IE): ball, bifolium, cinquefoil, fluxion, folium, hexafoil, multifoil, quadrifolium, quatrefoil, trefoil, trifolium.

[22] *bhendh-* (IE) "bind": band, bend, bundle.

[23] *bher-* (IE) "carry": circumference, diffeomorphism, differ, difference, differential, differentiate, infer, inference, reference.

[24] *bhreg-* (IE) "break": broken, fractal, fractile, fraction, fractional.

[25] *caten-* (Latin) "chain": catenary, catenoid, chain, chaînette, concatenate, concatenation.

[26] *çifr* (Arabic) "empty, zero": cipher, zero.

[27] *clocc-* (Medieval Latin) "bell": clockwise, counterclockwise.

[28] *cruc-* (Latin) "cross": cross, cruciform, crunode.

[29] *cuspid-* (Latin) "point, spike, spear": cusp, cuspidal, cuspitate, tetracuspid, tricuspid.

[30] *da-* (IE) "divide": geodesic, time, times.

[31] *daleth-* (Hebrew) "door; fourth letter of the Hebrew alphabet": atled, del, delta, deltahedron, deltoid.

[32] *de-* (IE), root of demonstrative pronouns: antiderivative (?), avoirdupois (?), de- (?), decidable (?), decouple (?), decrease (?), decrement (?), deduce (?), deduction (?), deductive (?), defective (?), deficient (?), define (?), definite (?), definition (?), deform (?), deformation (?), degenerate (?), degree (?), denial (?), denominate (?), denominator (?), denumerable (?), dependent (?), depreciate (?), depressed (?), derivative (?), derive (?), descending (?), describe (?), detach (?), detachment (?), determinant (?), determine (?), developable (?), deviate (?), deviation (?), diminish (?), indegree (?), independent (?), into, onto, outdegree (?), Q.E.D. (?).

[33] *deik-* (IE) "show": biconditional, bit, condition, conditional, contradict, contradiction, contradictory, digit, digital, disc, index, indicator, indicatrix, monodigit, pandigital, predicate.

[34] *dekm-* (IE) "ten": atto-, cent, centi-, centesimal, centillion, century, dea-, deca-, decade, decagon, decahedral, decahedron, decakis , December, deci-, decile, decillion, decimal, deka-, denary, dime, dodecagon, dodecahedron, dozen, duodecagon, duodecahedron, duodecillion, duodecimal, enneacontahedron, femto-, hectare, hecto-, hendecagon, hendecahedron, heptadecagon, hexacontahedron, hexadecagon, hexadecimal, hundred, icosagon, icosidodecahedron, icosahedron, icositetrahedron, interdecile, novemdecillion, octadecagon, octodecillion, pentadecagon, percent, percentage, percentile, quattuordecillion, quindecillion, quintal, rhombicosidodecahedron, septendecillion, sexagesimal, sexdecillion, ten, tetradecagon, thousand, tredecillion, tredo-, triacontahedron, tridecagon, undecillion, vicenary, vigesimal, vigintillion.

[35] *del-* (IE) "long": belong, elongation, furlong, length, long, oblong, prolong.

[36] *demə-* (IE) "break, tame": diamond, polyiamond.

[37] *demə-* (IE) "home": codomain, domain, dominant, dominate, heptomino, hexomino, monomino, octomino, pentomino, polyomino, tetromino, triomino, tromino.

[38] *der-* (IE) "run, walk, step": loxodrome, loxodromic, monodromy, palindrome, palindromic, paradromic.

[39] *deru-* (IE) "to be firm": tree, true, truth.

[40] *deuk-* (IE) "lead": deduce, deduction, deductive, induction, inductive, produce, product, reduce, reductio ad absurdum, reduction.

[41] *dhe-* (IE) "put": apothem, bias (?), coefficient, cofactor, defective, deficient, face, factor, factorable, factorial, factorize, feasible, freedom, homothesis, homothetic, hypothesis, idemfactor, multiperfect, parenthesis, parenthesize, perfect, profit, pseudoperfect, Q.E.F., quantify, quantifier, quasi-perfect, rectifiable, rectification, rectify, satisfy, semiperfect, significant, simplify, stratified, subfactorial, sufficient, surface, synthetic.

[42] *dheigh-* (IE) "form, build": figurate, figure.

[43] *dhregh-* (IE) "run": epitrochoid, hypotrochoid, trochoid.

[44] *dhuno-* (IE) "fortified place": down, singleton.

[45] *dia-* (Greek) "through": diabolic, diacaustic, diagonal, diagonalize, diagram, dialytic, diameter, diametral, pandiagonal, pseudodiagonal, subdiagonal, superdiagonal, tridiagonal.

[46] *dis-* (Latin) "apart, away": diffeomorphism, difference, differential, differentiate, digraph, dilation, dimension, direct, directed, direction, director, directrix, disconnected, discontinuity, discontinuous, discrete, discriminant, disjoint, disjunction, dispersion, displacement, dissect, dissection, dissimilar, dissipate, dissipative, distance, distribute, distribution, distributive, distributivity, diverge, divergent, divide, dividend, divisibility, divisible, division, divisor, equidistant.

[47] *do-* (IE) "give": add, addend, addition, additive, data, die, subadditive, superadditive.

[48] *dwo-* (IE) "two": between, bi-, bias (?), biconditional, bicontinuous, bicorn, bifolium, bifurcate, bifurcation, bilinear, billion, bimodal, binary, binomial, binormal, bipolar, biquadrate, biquadratic, birectangular, bisect, bisection, bisector, bisymmetric, bit, biunique, byte, combination, combinatorics, combine, diakis-, dichotomous, dichotomy, digon, digonal, dihedral, dihedron, dimetric,

dimetry, dipyramid, ditonic, dodecagon, dodecahedron, double, dozen, dual, duel (?), duodecagon, duodecahedron, duodecillion, duodecimal, duplation, duplicate, dyad, dyadic, icosidodecahedron, rhombicosidodecahedron, twelve, twin, two.

[49] *eghs* (IE) "out": coefficient, counterexample, eccentric, eccentricity, elevation, eliminand, eliminant, eliminate, elongation, enumerable, enumerate, enumeration, enunciation, escribe, evaluate, event, evolute, evolution, ex-, exact, example, exceed, excenter, except, exception, excess, exchange, excircle, exclude, exhaustion, exist, existence, exogenous, exotic, expand, expected, explicit, exponent, exponential, exponentiation, expression, exsecant, extend, extension, extent, exterior, extract, extraneous, extrapolate, extrapolation, extreme, extremum, sample.

[50] *ei-* (IE) "go": circuit, initial, transitive, transitivity.

[51] *em-* (IE) "take, distribute": assume, assumption, counterexample, example, sample.

[52] *en-* (IE) "in": and (?), coincide, coincident, ellipse, ellipsis, ellipsoid, ellipsoidal, empirical, enantiomorph, enantiomorphic, enantiomorphous, endogenous, endomorphism, entail, entailment, envelope, hyperellipse, hyperellipsoid, im grossen, im kleinen, implication, implicit, imply, in- (prepositional prefix), incenter, incidence, incircle, inclination, include, increase, increment, indegree, index, indicator, indicatrix, induction, inductive, infer, inference, inflection, information, initial, injective, inner, inscribe, inside, instantaneous, intension, intercept, interdecile, interest, interior, intermediate, interpolate, interpolation, interquartile, intersect, intersection, interval, into, intrinsic, intuitionism, inverse, invert, invertible, involute, involution, nand (?), osculinflection, parenthesis, parenthesize, reentrant, semiellipse, superellipse, superellipsoid.

[53] *epi* (IE) "near, at, against": cover, covering, epicycloid, epimorphism, epitrochoid, oblate, oblique, oblong, obtuse, ogive (?), opposite, parallelepiped, parallelepipedal.

[54] *er-* (IE) "move, set in motion": orient, orientable, orientation, origin.

[55] *erə-* (IE) "to separate": hermit, rare.

[56] *es-* (IE) "be": entity, essential, interest, present, Q.E.D., Q.E.F.

[57] *-et(te)* (French) "diminutive suffix": bracket, chaînette, facet, glissette, rosette, roulette, summit.

[58] *eu-* (IE) "lacking, empty": vacuously, vanish.

[59] *fals-* (Latin) "false": fallacy, false, regula falsi.

[60] *fin-* (Latin) "boundary, border": affine, cofinite, define, definite, definition, finite, infinitary, infinite, infinitesimal, infinity, transfinite.

[61] *flec-* (Latin) "bend": circumflex, flecnode, flexagon, flexion, inflection, osculinflection, reflect, reflection, reflex.

[62] *foc-* (Latin) "fireplace, hearth": confocal, focal, focus.

[63] *ge-* (Greek) "earth": gematria, gematry, geodesic, geometer, geometric, geometry, hypergeometric, pangeometry.

[64] *gen-* (IE) "give birth": degenerate, endogenous, exogenous, general, generalize, generalization, generator, generatrix, genus, homogeneous, monogenic, natural.

[65] *genu-* (IE) "angle, knee": antigonal, apeirogon, decagon, diagonal, diagonalize, digon, dodecagon, duodecagon, enneagon, flexagon, -gon, hendecagon, heptadecagon, heptagon, heptagonal, hexadecagon, hexagon, hexagonal, icosagon, isogon, isogonal, monogonal, nonagon,

octadecagon, octagon, octagonal, orthogonal, orthogonality, pandiagonal, pentadecagon, pentagon, pentagonal, perigon, polygon, polygonal, polygonometry, pseudodiagonal, subdiagonal, superdiagonal, tetradecagon, tetragon, tetragonal, tridecagon, tridiagonal, trigon, trigonal, trigonometry.

[66] *ger-* (IE) "curving, crooked": group, grouping, groupoid, gruppoid, semigroup.

[67] *ger-* (IE) "gather": aggregation, categorical.

[68] *gerbh-* (IE) "scratch": diagram, digraph, gram, graph, hexagram, histogram, hodograph, lexicographic, lexicographically, multigraph, nomograph, pantograph, parallelogram, pentagram, pictogram, program, programming, relagraph, scattergram, stereographic, tangram.

[69] *ghabh-* (IE) "give; receive," in Latin *habilis* "capable of, able to be": -able, admissible, amicable, avoirdupois, commensurable, compatible, conformable, constructible, countable, decidable, denumerable, developable, divisibility, divisible, enumerable, factorable, feasible, -ible, invertible, metrizable, multivariable, orientable, permissible, probabilistic, probability, reachability, reachable, rectifiable, reducible, reliability, reliable, removable, separable, sociable, stability, stable, traversable, untouchable, variable.

[70] *ghais-* (IE) "to adhere": adherent, coherently.

[71] *gher-* (IE) "scratch, scrape, and its extension *ghreu-* "rub, grind": characteristic, chromatic, congruence (?), congruent (?), great, greater, greatest, im grossen.

[72] *gheslo-* (IE) "thousand": billion, centillion, decillion, duodecillion, kilo-, mile, mill, millennium, milli-, milliard, million, nonillion, novemdecillion, octillion, octodecillion, quadrillion, quattuordecillion, quindecillion, quintillion, septemdecillion, septillion, sexdecillion, sextillion, tredecillion, trillion, undecillion, vigintillion.

[73] *ghredh-* (IE) "walk, go": cogredient, contragredient, degree, grad, gradient, indegree, multigrade, outdegree, progression, regression.

[74] *glid-* (Germanic) "glide": glide, glissette.

[75] *gno-* (IE) "know": binormal (?), conormal (?), gnomon, lognormal (?), norm (?), normal (?), normalize (?), notation, orthonormal (?), subnormal (?), unknown.

[76] *gre-no-* (IE) "grain": grain, kernel.

[77] *$g^w a$-* or *$g^w em$-* (IE) "come": arbitrary (?), base, basis, eigenbasis, event, revenue, outcome.

[78] *$g^w el\partial$-* (IE) "throw, reach": diabolic, hyperbola, hyperbolic, hyperboloid, parabola, parabolic, paraboloid, problem, symbol, symbolic.

[79] *horos-* (Greek) "boundary, limit": horizontal, horopter.

[80] *i-* (IE), stem found in pronouns: idemfactor, idempotent, identity, iterate, iteration, pseudoidentity.

[81] *imagin-* (Latin) "image, imitation, copy": cis, image, imaginary, preimage.

[82] *iso-* (Greek) "equal": isochronal, isochrone, isochronous, isocline, isogon, isogonal, isohedral, isometric, isometry, isoperimetric, isoperimetry, isoptic, isosceles, isotopic, isotopy, isotoxal, isotropic.

[83] *kad-* (IE) "fall": case, chance, coincide, coincident, incidence.

[84] *$ka\partial$-id-* (IE) "to strike": abscissa (?), decidable, precision.

[85] *kan-* (Semitic) "rod, reed": canonical, channel.

[86] *kap-* (IE) "grasp": capacity, except, exception, intercept, principal, principle.

[87] *kata-* (IE) "something thrown down": catacaustic, catastrophe, categorical.

[88] *ked-* (IE) "go, yield": antecedent, exceed, excess, necessary, precedent, success, successive, successor.

[89] *kel-* (IE) "cover": cell, hole, hollow, hull, pigeonhole.

[90] *kelə-*(IE) "shout": calendar, class (?), clear.

[91] *kent-* (IE) "prick, jab": barycentric, center, central, centroid, circumcenter, concentric, eccentric, eccentricity, excenter, incenter, orthocenter.

[92] *ker-* (IE) "grow": accrue, concrete, crescent, decrease, decrement, increase, increment.

[93] *ker-* (IE) "horn, head": bicorn, corner, cornicular, keratoid.

[94] *kerd-* (IE) "heart": cardioid, concordantly.

[95] *kers-* (IE) "run": carrier, carry, concurrent, recurring, recursion, recursive, unicursal.

[96] *keu-* (IE) "bend": cube, cubed, cubic, cubical, cuboctahedron, cuboid, cup, height, high, hypercube, rhombicuboctahedron, semicubical.

[97] *keu-* (IE) "burn": catacaustic, caustic, diacaustic.

[98] *keuə-* (IE) "swell": accumulate, accumulation, cavity, concave, concavity, cumulant, cumulative.

[99] *khron-* (Greek) "time": brachistochrone, isochronal, isochrone, isochronous, tautochrone, tautochronous.

[100] *klei-* (IE) "lean": clinometer, inclination, isocline.

[101] *kleu-* (IE) "hook, peg": close, closed, closure, conclude, exclude, exclusive, include, inclusive.

[102] *ko-* (IE) "sharpen": cone (?), conic (?), conical (?), conicoid (?), conoid (?).

[103] *kom-* (IE) "with," or its derivative *contra* (Latin) "against": anticommutative, biconditional, bicontinuous, cis, co-, coaxial, cobordism, codomain, coefficient, cofactor, cofinite, cogredient, coherently, coincide, collect, collinear, collinearity, collineation, collocation, cologarithm, combination, combinatorics, combine, commensurable, common, commutative, commutativity, commute, compact, compactum, compare, compass, compatible, complement, complete, complex, component, composite, composition, compression, compute, computer, concatenate, concatenation, concave, concavity, concentric, conclude, concordantly, concrete, concurrent, concyclic, condition, conditional, confidence, confocal, conformable, conformal, congruence, congruent, conjecture, conjugate, conjunction, connected, connectivity, conormal, consecutive, consequent, consistent, constant, constellation, constraint, construct, constructivism, contact, contain, conterminous, contingency, continuity, continuous, contour, contradict, contradiction, contradictory, contragredient, contrapositive, contrary, contravariant, control, converge, convergent, converse, convex, convolution, coordinate, coplanar, copolar, coprime, co-punctal, correct, correlation, corresponding, cosecant, coset, cosine, cost, cotangent, coterminal, count, countable, counting, counterclockwise, counterexample, couple, covariant, cover, covering, coversine, decouple, disconnected, discontinuous, either, equiconjugate, hypercomplex, inconsistent, metacompact, neighbor, neither, nonconjunction, paracompact, paracompactum.

[104] *konk(h)o-* (IE) "mussel, shellfish": cochleoid, cochloid, conchoid.

[105] *krei-* (IE) "sieve, discriminate": criterion, critical, discrete, discriminant.

[106] *ksun* (IE) "with": asymmetric, asymmetry, asymptote, asymptotic, asymptotically, bisymmetric, cyclosymmetric, syllogism, symbol, symbolic, symmedian, symmetric, symmetry, sympathetic, synthetic, syntractrix, system.

[107] *kʷel-* (IE) "move around": acyclic, bipolar, concyclic, copolar, cycle, cyclic, cyclides, cycloid, cycloidal, cyclomatic, cyclosymmetric, cyclotomic, epicycloid, hemicycle, hypocycloid, orthopole, palindrome, palindromic, polar, pole.

[108] *kʷetwer-* (IE) "four": biquadrate, biquadratic, four, heterosquare, icositetrahedron, interquartile, quadrangle, quadrangular, quadrant, quadrantal, quadratic, quadratrix, quadrature, quadric, quadrifolium, quadrilateral, quadrillion, quadrivium, quadruple, quart, quarter, quartic, quartile, quaternary, quaternion, quatrefoil, quattuordecillion, square, squared, tessellate (?), tessellation (?), tesseract, tetracuspid, tetrad, tetradecagon, tetragon, tetragonal, tetrahedral, tetrahedron, tetrakis-, tetriamond, tetromino, trapezium, trapezohedron, trapezoid, trapezoidal.

[109] *kʷo-* (IE), base of relative and interrogative pronouns: aliquot, either, neither, Q.E.D., Q.E.F., quantic, quantifier, quantify, quantity, quasi-perfect, quotient, quotition.

[110] *later-* (Latin) "side": equilateral, lateral, latus rectum, quadrilateral, trilateral, unilateral.

[111] *leg-* (IE) "collect": analogy, antilogarithm, collect, cologarithm, ecological, heterological, homologous, homology, lexicographic, lexicographically, log, logarithm, logarithmic, logic, logical, logician, logicism, logistic, lognormal, semilogarithmic, syllogism, tautological, tautology, topological, topology.

[112] *legh-* (IE) "lie": law, layer, low, outlier.

[113] *leig-* (IE) "bind": alligation, reliability, reliable.

[114] *leikʷ-* (IE) "leave": eleven, ellipse, ellipsis, ellipsoid, ellipsoidal, hyperellipse, hyperellipsoid, semiellipse, superellipse, superellipsoid, twelve.

[115] *leis-* (IE) "less": least, less, unless.

[116] *leu-* (IE) "loosen, divide, cut apart": absolute, analysis, analytic, analyticity, analyze, ANOVA, dialytic, resolvent, solution, solve.

[117] *lik-* (Germanic) "body, form": asymptotically, coherently, concordantly, lexicographically, like, likelihood, likely, linearly, mutually, relatively, respectively, strictly, triply, vacuously.

[118] *lim-* (Latin) "threshold, borderline": eliminand, eliminant, eliminate, limit.

[119] *lino-* (IE) "flax": bilinear, collinear, collinearity, collineation, curvilinear, line, lineal, linear, linearity, linearly, midline, rectilinear.

[120] *lithr-* (Mediterranean) "scale": equilibrium, lb., level, liter.

[121] *ma-* (IE) "mother": material, matrix, matroid, submatrix.

[122] *mag-* (IE) "knead, fashion, fit": mass, match.

[123] *magh-* (IE) "be able, have power": magic, main, panmagic.

[124] *mappa* (Latin) "napkin, towel, cloth": map, mapping, nappe.

[125] *me-* (IE) "big": almost, more, most.

[126] *me-* (IE) "measure": allometric, asymmetric, asymmetry, axonometric, axonometry, bisymmetric, clinometer, commensurable, cyclosymmetric, diameter, diametral, dimension, dimetric, dimetry, gematria, gematry, geometer, geometric, geometry, hypergeometric, isometric, isometry, isoperimetric, isoperimetry, measure, mensuration, meter, metric, metrical, metrizable, month, pangeometry, parameter, parametric, perimeter, polygonometry, semiperimeter, stereometric, stereometry, symmetric, symmetry, trigonometric, trigonometry, trimetric, trimetry.

[127] *me-* and its extension *medhyo-* (IE) "middle": intermediate, mean, medial, median, mediant, mediation, mediator, meridian, mesokurtic, metacompact, metamathematical, metamathematics, method, midline, midpoint, semimean, symmedian.

[128] *med-* (IE) "take appropriate measures": bimodal, empty, meter (and related words under [126] (?)), mod, modal, mode, model, modular, modulo, modulus, modus ponens, modus tollens, molding, multimodal, unimodal, unimodular.

[129] *meg-* (IE) "great": magnitude, major, maximal, maximize, maximum, mega-, minimax, omega, semimajor.

[130] *mei-* (IE) "(ex)change, go, move": anticommutative, common, commutative, commutativity, commute, mutually, permute, permutation.

[131] *mei-* (IE) "small": diminish, minimax, minimize, minimum, minor, minuend, minus, minute, semiminor.

[132] *mel-* (IE) "strong, great" in Latin *multi* "many": equimultiple, multifoil, multigrade, multigraph, multimodal, multinomial, multiperfect, multiple, multiplicand, multiplicative, multiplicity, multiplier, multiply, multivariable, multivariate, postmultiplication, postmultiply, premultiplication, premultiply, submultiple, transmultiplication.

[133] *men-* (IE) "small, isolated": mon(o)-, monic, monodigit, monodromy, monogenic, monogonal, monohedral, monoid, monomial, monomino, monomorph, monomorphic, monotonic.

[134] *men-* (IE) "think": cyclomatic, mnemonic, money, Q.E.D.

[135] *mendh-* (IE) "learn": ethnomathematics, math, mathematical, mathematician, mathematicize, mathematics, mathematization, mathematize, mathlete, mathophile, mathophilia, metamathematical, metamathematics.

[136] *meuə-* (IE) "push away": moment, removable.

[137] *morph-* (Greek) "shape, form": antimorph, automorphic, automorphism, conformable, conformal, cruciform, crunode, deform, deformation, diffeomorphism, enantiomorph, enantiomorphic, enantiomorphous, endomorphism, epimorphism, form, formal, formalism, formula, holomorphic, homeomorphic, homeomorphism, homomorphic, homomorphism, information, isomorphic, isomorphism, meromorphic, monomorph, monomorphic, morphism, piriform, polymorph, pyriform, transform, trimorphic, uniform.

[138] *mregh-u-* (IE) "short": abridged, brace, brachistochrone, branch (?).

[139] *-nd-* (Latin) "which is to be [verb]ed": addend, dividend, eliminand, integrand, minuend, multiplicand, operand, Q.E.D., Q.E.F., radicand, repetend, retinend, subtrahend, summand.

[140] *ndher-* (IE) "under, below": inferior, infimum, under.

[141] *ne* (IE) "not": acyclic, anallagmatic, anarboricity, annihilate, annihilator, anomalous, anomaly, apeirogon, aperiodic, asymmetric, asymmetry, asymptote, asymptotic, asymptotically, atomic,

atomism, denial, diamond, entire, heptiamond, hexiamond, imprimitive, improper, in- (negative prefix), incompatible, inconsistent, indeterminate, inequality, infinitary, infinite, infinitesimal, infinity, integer, integral, integrand, integrate, integration, nand, naught, necessary, negate, negation, negative, neither, nil, nilpotent, nonconjunction, none, nonresidue, nonsingular, nontiler, nor, not, nought, null, nullity, pentiamond, polyiamond, tetriamond, triamond, unknown, untouchable.

[142] *ned-* (IE) "bind, tie": acnode, connected, connectivity, crunode, disconnected, flecnode, net, network, nodal, node, spinode, tacnode, trinodal, trinode.

[143] *nem-* (IE) "assign, allot, take": antinomy, binomial (?), denumerable (?) enumerate (?), monomial (?), multinomial (?), nim, nomograph, number (?), numeracy (?), numeral (?), numerate (?), numeration (?), numerator (?), numerical (?), polynomial (?), trinomial (?).

[144] *newn-* (IE) "nine": enneacontahedron, enneagon, enneahedral, enneahedron, nine, nonagon, nonillion, noon, November, novemdecillion.

[145] *no-men-* (IE) "name": binomial (?), denominate, denominator, monomial (?), multinomial (?), nominal, polynomial (?), trinomial (?).

[146] *-nt-* (Latin) "present participle": abundant, adherent, adjacent, antecedent, circulant, coefficient, cogredient, coherently, coincident, complement, complementary, component, concordantly, concurrent, congruence, congruent, consequent, consistent, constant, contingency, contragredient, contravariant, convergent, cosecant, cotangent, covariant, crescent, cumulant, deficient, dependent, determinant, differential, differentiate, discriminant, divergent, dominant, eliminant, equant, equidistant, equipollent, equipotent, equitangential, equivalent, essential, exponent, exponential, exponentiation, exsecant, gradient, idempotent, incident, independent, instantaneous, latent, mediant, modus ponens, modus tollens, nilpotent, octant, orient, orientable, orientation, orthant, persistence, potency, precedent, present, pseudotangent, quadrant, quadrantal, quotient, redundant, reentrant, reference, resolvent, resultant, salient, secant, sequence, sequential, significant, subtangent, sufficient, tangency, tangent, tangential, totient, unipotent, valence.

[147] *odd-* (Old Norse) "uneven": odd, odds, oddorial.

[148] *oi-no-* (IE) "one": any, biunique, eleven, inch, nonconjunction, none, nonresidue, nontiler, null, nullity, one, ounce, repunit, unary, undecillion, unicursal, uniform, unilateral, unimodal, unimodular, union, unipotent, unique, unit, unitary, unity, universal, universe.

[149] *okto(u)-* (IE) "eight": atto-, cuboctahedron, eight, octadecagon, octagon, octagonal, octahedral, octahedron, octakis-, octal, octant, octic, octillion, octimal, October, octodecillion, octomino, octonary, octonion, octuple, rhombicuboctahedron, yocto-, yotta-.

[150] *ok^w-* (IE) "see": horopter, isoptic, orthoptic.

[151] *-on-* (Latin), augmentative suffix: billion, centillion, decillion, duodecillion, echelon, festoon, gallon, limaçon, million, nonillion, novemdecillion, octillion, octodecillion, pattern, quadrillion, quattuordecillion, quindecillion, quintillion, septemdecillion, septillion, sexdecillion, sextillion, tredecillion, trillion, undecillion, vigintillion.

[152] *op-* (IE) "work": operand, operate, operation, operator, optimal, optimization, optimize, optimum.

[153] *-or* (Latin) "a male person or thing that does the indicated action": annihilator, attractor, bisector, cofactor, denominator, director, divisor, eigenvector, equator, factor, factorable, factorial, factorize, functor, generator, idemfactor, indicator, mediator, numerator, oddorial, operator, primorial, protractor, sector, spinor, subfactorial, successor, tensor, tractory, trisector, vector.

[154] *os-* (IE) "mouth": oscillate, oscillation, osculating, osculation, osculinflection.

[155] *othth-* (Old English) "or": nor, or.

[156] *pag-* (IE) "fasten": compact, compactum, metacompact, paracompact, paracompactum.

[157] *pant-* (IE) "all": pandiagonal, pandigital, pangeometry, panmagic, pantograph.

[158] *ped-* (IE) "foot": antipodal, antipodes, foot, hippopede, parallelepiped, parallelepipedal, pedal, trapezium, trapezohedron, trapezoid, trapezoidal.

[159] *peig-* (IE) "cut, incise": pictogram, pint (?).

[160] *pel-* (IE) "fold" , and its extension *plek-* "plait: applied, complex, double, duplation, duplicate, duplication, equimultiple, explicit, googolplex, hypercomplex, implication, implicit, imply, manifold, multiple, multiplicand, multiplication, multiplicative, multiplicity, multiplier, multiply, octuple, postmultiplication, postmultiply, premultiplication, premultiply, quadruple, quintuple, septuple, sextuple, simple, simplex, simplicial, simplify, submultiple, transmultiplication, triple, triply.

[161] *pel-* (IE) "thrust, strike, drive": extrapolate, extrapolation, filter, interpolate, interpolation, ultrafilter.

[162] *pelə-*(IE) "flat; to spread," and its extension *plat-* "flat": coplanar, displacement, field, flat, hyperplane, place, planar, plane, planted, platykurtic, replace, replacement, snowflake, subfield.

[163] *pelə-* (IE) "fill": cis, complement, complementary, complete, cosecant, cosine, cotangent, plus, polygon, polygonal, polygonometry, polyhedral, polyhedron, polyhex, polyiamond, polymorph, polymorphic, polynomial, polyomino, polystar, polytope, supplement, supplemental, supplementary, surplus.

[164] *penkʷe* (IE) "five": cinquefoil, femto-, five, pentacle, pentadecagon, pentagon, pentagonal, pentagram, pentahedral, pentahedron, pentakis-, pentalpha, pentangle, pentiamond, pentomino, peta-, quinary, quindecillion, quintic, quintile, quintillion, quintuple.

[165] *per-* (IE) "forward, before, in front of, through," with extensions meaning "around; to lead, pass over; to traffic in, to distribute; to try, to risk": a priori, antiparallel, apeirogon, aperiodic, appreciate, appreciation, approach, approximate, coprime, depreciate, depreciation, empirical, first, imprimitive, improper, isoperimetric, isoperimetry, multiperfect, parabola, parabolic, paraboloid, paracompact, paracompactum, paradox, paradoxical, paradromic, parallel, parallelepiped, parallelepipedal, parallelogram, parallelotope, parameter, parametric, parenthesis, parenthesize, percent, percentage, percentile, perfect, perigon, perimeter, period, periodic, permissible, permute, permutation, perpendicular, perpendicularity, persistence, perspective, piercing, porism, precedent, precision, predicate, preimage, premise, premultiplication, premultiply, present, previous, price, primal, primality, prime, primitive, primorial, principal, pro-, probabilistic, probability, problem, produce, product, profit, program, programming, progression, project, projection, projective, prolate, prolong, proof, proper, property, proportion, proportional, proportionality, proposition, propositional, protasis, prototile, protractor, prove, pseudoperfect, pseudoprime, quasi-perfect, reciprocal, reciprocity, semiperfect, semiperimeter, semiprime, separable, separate, separatrix, superprime, support, therefore.

[166] *per-* (IE) "strike": compression, depressed, depression, expression.

[167] *perə-*(IE) "grant, allot": compare, comparison, pair (?), pairwise (?), parity (?), part, partial, particular, partition, poset, proportion (?), proportional (?), proportionality (?), separate, separatrix.

[168] *pet-* (IE) "rush, fly": asymptote, asymptotic, asymptotically, repeat, repdigit, repetend, reptile, repunit.

[169] *petə-* (IE) "spread out": compass, expand, expansion, fathom, petal.

[170] *petti-* (Gaulish) "piece, share": patch, piecewise.

[171] *peu-* (IE) "cut, strike": compute, computer, count, countable, counting.

[172] *peuk-* (IE) "prick": co-punctal, cut, cutpoint, endpoint, midpoint, pivot (?), point, puncture, tac-point.

[173] *poti-* (IE) "powerful, lord": equipotent, idempotent, nilpotent, potency, power, superpower, unipotent.

[174] *pri-* (Greek) "saw": antiprism, prism, prismatic, prismatoid, prismoid, prismoidal.

[175] *pseud-* (Greek) "false": pseudodiagonal, pseudoidentity, pseudoperfect, pseudoprime, pseudosphere, pseudospherical, pseudotangent, pseudovertex.

[176] *puramid-* (Greek) "pyramid": dipyramid, pyramid, pyramidal.

[177] *radi-* (Latin) "rod, staff": radian, radial, radiate, radius, ray, steradian.

[178] *re-* (Latin) "backward": corresponding, nonresidue, reciprocal, reciprocity, recurring, recursion, recursive, reduce, reductio ad absurdum, reduction, redundant, reentrant, reference, reflect, reflection, reflex, regression, relagraph, relation, relative, relatively, reliable, remainder, removable, repdigit, repeat, repetend, replace, replacement, reptile, repunit, residual, residue, resolvent, respectively, result, resultant, retinend, retract, revenue, reverse, reversion, revolve.

[179] *reg-* (IE) "move in a straight line": birectangular, correct, direct, directed, direction, director, directrix, latus rectum, rectangle, rectangular, rectifiable, rectification, rectify, rectilinear, region, regula falsi, regular, regulus, right, rule, ruler, ruling, semiregular, source, trirectangular.

[180] *rei-* (IE) "flow, run": antiderivative, derive, derivative, random (?), run.

[181] *ret-* (IE) "run, roll": control, rotate, rotation, rotund, roulette, round.

[182] *se-* (IE) "long, late": inside (?), outside (?), side (?).

[183] *sed-* (IE) "go": aperiodic, ergodic, hodograph, method, period, periodic.

[184] *sed-* (IE) "sit": cuboctahedron, decahedral, decahedron, deltahedron, diakisdodecahedron, dihedral, dihedron, dodecahedron, duodecahedron, enneacontahedron, enneahedral, enneahedron, hendecahedron, -hedron, heptahedral, heptahedron, hexacontahedron, hexahedral, hexahedron, hosohedron, icosahedral, icosahedron, icosidodecahedron, icositetrahedron, isohedral, monohedral, nested, nonresidue, octahedral, octahedron, pentahedral, pentahedron, polyhedral, polyhedron, residual, residue, rhombicosidodecahedron, rhombicuboctahedron, rhombohedron, saddle, scalenohedron, tetrahedral, tetrahedron, trapezohedron, triacontahedron, trihedral, trihedron, zonohedron.

[185] *sek-* and its extensions *skei-* and *sked-* (IE) "cut": abscissa (?), bisect, bisection, bisector, cosecant, coset, dissect, dissection, exsecant, heteroscedastic (?), heteroscedasticity (?), homoscedastic (?), homoscedasticity (?), intersect, intersection, poset, scatter, scattergram, secant, section, sector, segment, set, subset, superset, trisect, trisection, trisector, trisectrix, exsecant.

[186] *sekʷ-* (IE) "follow": associative, associativity, consecutive, consequent, intrinsic, second, sequence, sequential, sign, signature, significant, signum, sociable.

[187] *sel-* (IE) "jump": result, resultant, salient, saltus.

[188] *sem-* (IE) "one, as one": anomalous, anomaly, hecto-, hendecagon, hendecahedron, heterological, heteroscedastic, heteroscedasticity, heterosquare, homeomorphic, homeomorphism,

homogeneity, homogeneous, homologous, homology, homomorphic, homomorphism, homoscedastic, homoscedasticity, homothesis, homothetic, homotopic, homotopy, nonsingular, same, similar, similarity, similitude, simple, simplex, simplicial, simplify, simulation, simultaneous, single, singleton, singular, singularity, some.

[189] *semi-* (IE) "half": hemicycle, hemicylinder, hemicylindrical, hemisphere, hemispherical, semi-, semicircle, semicircular, semicubical, semiellipse, semigroup, semilogarithmic, semimajor, semimean, semiminor, semiperfect, semiperimeter, semiprime, semiregular.

[190] *sent-* (IE) "head for": sense (?), sentence (?), sentential (?).

[191] *septm-* (IE) "seven": heptadecagon, heptagon, heptagonal, heptahedral, heptahedron, heptiamond, heptomino, September, septendecillion, septillion, septuple, seven, zepto-, zetta-.

[192] *sin-* (Latin) "hollow, bend, curve": cis, cosine, coversine, haversine, sine, sinusoid, sinusoidal, subsine.

[193] *(s)kamb-* (IE) "curve, bend": change, exchange.

[194] *skand-* (IE) "climb": ascending, descending, echelon, scalar, scale, transcendental.

[195] *(s)kel-* (IE) "crooked": cylinder (?), cylindrical (?), cylindroid (?), hemicylinder (?), hemicylindrical (?), heteroscedastic (?), heteroscedasticity (?), homoscedastic (?), homoscedasticity (?), isosceles, triskelion.

[196] *(s)kel-* (IE) "cut": half, halve, haversine, scalene, scalenohedron, shell.

[197] *sker-* (IE) "cut": circumscribe, curtate, describe, escribe, inscribe, score, shear, short, subscript, superscript.

[198] *(s)ker-* (IE) "bend, turn": arrange, arrangement, circle, circuit, circulant, circular, circulating, circumcenter, circumference, circumflex, circumscribe, corollary, crisp, curvature, curve, curvilinear, excircle, incircle, kurtosis, leptokurtic, mesokurtic, platykurtic, range, rank, ring, semicircle, semicircular, shrink, subring, supercircle.

[199] *sme-* (IE) "smear": micro-, micron.

[200] *(s)meit(ə)-* (IE) "throw": admissible (?), permissible (?), premise (?).

[201] *sol-* (IE) "whole": holomorphic, solid, solidus, sursolid.

[202] *spat-* (Latin) "room, space": hyperspace, space, spatial, subspace.

[203] *spei-* (IE) "sharp point": épi, spinode, splitting.

[204] *(s)peik-* (IE) "woodpecker, magpie": pico- (?), pie (?).

[205] *spek-* (IE) "observe": expected, perspective, respectively, scope, species, spectrum, telescopic, telescoping.

[206] *(s)pen-* (IE) "draw, stretch, spin": avoirdupois, dependent, independent, perpendicular, perpendicularity, pound, span, spanning, spinor.

[207] *sphair-* (Greek) "ball": hemisphere, hemispherical, hypersphere, pseudosphere, pseudospherical, sphere, spherical, spheroid, supersphere.

[208] *sta-* (IE) "stand": consistent, constant, cost, distance, equidistant, exist, existence, histogram (?), inconsistent, instantaneous, persistence, stability, stable, standard, state, stationary, statistic, statistical, statistician, statistics, stem, substitute, substitution, system.

[209] *(s)teg-* (IE) "cover": nontiler, tile, tiler, tiling, prototile, reptile.

[210] *stelə-* (IE) "extend": dilation, lamina (?).

[211] *ster-* (IE) "spread": construct, construction, constructivism, strategy, stratified.

[212] *ster-* (IE) "star": asterisk, astroid, constellation, polystar, star, stellated.

[213] *ster-* (IE) "stiff": steradian, stere, stereographic, stereometric, stereometry.

[214] *(s)teu-* (IE) "push, beat": obtuse, piercing, type.

[215] *stloc-* (Latin) "place": collocation, local, locate, location, locus, tac-locus.

[216] *streb(h)-* (IE) "turn, wind": catastrophe, strophoid.

[217] *strecc-* (Old English) "spread out, extend": straight, stretch.

[218] *streig-* (IE) "stroke, rub, press": constraint, strict, strictly, striction.

[219] *s(w)e-* (IE) "self; apart": ethnomathematics, quasi-perfect, separable, separate, separatrix.

[220] *s(w)eks-* (IE) "six": exa-, hex, hexacontahedron, hexadecagon, hexadecimal, hexafoil, hexagon, hexagonal, hexagram, hexahedral, hexahedron, hexakis-, hexiamond, hexomino, polyhex, senary, sexagesimal, sexdecillion, sextic, sextillion, sextuple, six.

[221] *swer-* (IE) "buzz, whisper": reductio ad absurdum, surd, sursolid.

[222] *tag-* (IE) "touch": contact, contingency, cotangent, entire, equitangential, integer, integral, integrand, integrate, integration, pseudotangent, subtangent, tac-locus, tacnode, tac-point, tangency, tangent, tangential, untouchable.

[223] *tale-* (Latin) "cutting, twig": entail, entailment, tally.

[224] *telə-* (IE) "lift, support": correlation, modus tollens, oblate, prolate, relagraph, relation, relative, relatively, translate, translation.

[225] *tem-* (IE) "cut": atomic, atomism, cyclotomic, dichotomy, dichotomous, trichotomous, trichotomy.

[226] *ten-* (IE) "stretch": bicontinuous, contain, continuity, continuous, continuum, discontinuous, discontinuity, ditonic, extend, extension, extent, hypotenuse, intension, monotonic, protasis, retinend, subtend, tend, tensor.

[227] *ter-* (IE) "peg, post, boundary": conterminous, coterminal, determinant, determine, indeterminate, term, terminal, terminate, terminus.

[228] *terə-* (IE) "rub, turn": contour, tournament, trial (?), turn, turning.

[229] *terə-* (IE) "cross over, pass through": trajectory, tranjugate, transcendental, transfinite, transform, transformation, transitive, transitivity, translate, translation, transmultiplication, transpose, transposition, transversable, transversal, transverse, truncate, truncation.

[230] *teu-* (IE) "swell": sorites, thousand.

[231] *to-* (IE), stem of demonstrative pronouns: im grossen, im kleinen, tautochrone, tautochronous, tautological, tautology, therefore, totient, totitive.

[232] *topos-* (Greek) "place, region": homotopic, homotopy, isotopic, isotopy, orthotope, parallelotope, polytope, topological, topology.

[233] *tor-* (Latin) "bulge": toroid, torus.

[234] *tragh-* (IE) "draw, drag": abstract, attractor, extract, protractor, retract, subtract, subtraction, subtrahend, syntractrix, trace, tractory, tractrix.

[235] *trei-* (IE) "three": supertriangular, ternary, third, three, trammel, tredecillion, tredo-, trefoil, triacontahedron, triad, triakis-, triamond, triangle, triangular, triangulation, triaxial, trichotomous, trichotomy, tricuspid, tridecagon, trident, tridiagonal, trifolium, trigon, trigonal, trigonometric, trigonometry, trihedral, trihedron, trilateral, trillion, trimetric, trimetry, trimorphic, trinodal, trinode, trinomial, triomino, triple, trirectangular, trisect, trisection, trisector, trisectrix, triskelion, trivial.

[236] *-trix* (Latin) "the female person or thing that does": directrix, generatrix, indicatrix, matrix, quadratrix, separatrix, submatrix, syntractrix, tractrix, trisectrix.

[237] *ud-* (IE) "up, out": outcome, outdegree, outer, outlier, outside.

[238] *-ul-* (Latin), a diminutive particle: angle, angular, annulus, birectangular, calculate, calculus, circle, circulant, circular, circulating, cornicular, corollary, couple, decouple, equiangular, excircle, formula, incircle, level, model, modular, modulo, modulus, molding, nonsingular, nontiler, null, nullity, oscillate, oscillation, osculating, osculation, osculinflection, particular, pearl, pentacle, pentangle, perpendicular, perpendicularity, quadrangle, quadrangular, rectangle, rectangular, regula falsi, regular, regulus, roulette, rule, ruler, semiregular, single, singleton, singular, singularity, supercircle, supertriangular, tessellate, tessellation, tile, tiler, tiling, triangle, triangular, trirectangular, triangulation, unimodal, unimodular, vinculum, virgule.

[239] *uper* (IE) "over": hyper-, hyperbola, hyperbolic, hyperboloid, hypercomplex, hypercube, hyperellipse, hyperellipsoid, hypergeometric, hyperplane, hyperspace, hypersphere, over, overlap, sum, summand, summation, summit, superadditive, superdiagonal, supercircle, superellipse, superellipsoid, superior, superpose, superpower, superprime, superscript, superset, supersphere, supertriangular, supremum, surface, surjective, surplus, sursolid.

[240] *upo* (IE) "under, up from under": assume, assumption, hypo-, hypocycloid, hypotenuse, hypothesis, hypotrochoid, open, source, sub-, subadditive, subalternate, subdiagonal, subfactorial, subfield, submatrix, submultiple, subnormal, subring, subscript, subset, subsine, subspace, substitute, substitution, subtangent, subtend, subtract, subtraction, subtrahend, success, successive, successor, sufficient, supplement, supplemental, supplementary, support, suppose, supposition, up, upper.

[241] *wal-* (IE) "to be strong": eigenvalue, equivalence, equivalent, evaluate, valence, valid, validity, value.

[242] *wed-* (IE) "water, wet": abundant, redundant.

[243] *wegh-* (IE) "go, transport": convex, deviate, deviation, eigenvector, ogive (?), previous, quadrivium, trivial, vector, wave, weigh, weight, weighted.

[244] *weid-* (IE) "see": alysoid, arcwise, astroid, cardioid, catenoid, centroid, cissoid, clockwise, clothoid, cochleoid, cochloid, conchoid, conicoid, conoid, counterclockwise, cuboid, cycloid, cycloidal, cylindroid, deltoid, ellipsoid, ellipsoidal, epicycloid, epitrochoid, groupoid, gruppoid, helicoid, histogram (?), hyperboloid, hyperellipsoid, hypocycloid, hypotrochoid, ideal, keratoid, matroid, monoid, nephroid, -oid, ovoid, pairwise, paraboloid, piecewise, prismatoid, prismoid, prismoidal, ramphoid, rhomboid, sinusoid, sinusoidal, solenoidal, spheroid, strophoid, superellipsoid, toroid, trapezoid, trapezoidal, trochoid.

[245] *weidh-* (IE) "divide, separate": divide, dividend, divisibility, divisible, division, divisor.

[246] *weik-* (IE) "bend, wind": weak, week.

[247] *weik-* (IE) "clan": ecological, sandwich.

[248] *wel-* (IE) "turn, roll": convolution, developable, envelope (?), evolute, evolution, helical, helicoid, helicoidal, helix, involute, involution, revolve, revolution, volume.

[249] *wel-* (IE) "wish, will": vel, well.

[250] *wer-* (IE) "raised spot": ANOVA, contravariant, covariant, multivariable, multivariate, variable, variance, variation, vary.

[251] *wer-* (IE) "turn": converge, convergent, converse, coversine, dextrorse, dextrorsum, diverge, divergent, haversine, inverse, invert, invertible, pseudovertex, reverse, reversion, rhomb, rhombic, rhombicosidodecahedron, rhombicuboctahedron, rhombohedron, rhomboid, rhombus, sinistrorse, sinistrorsum, transversal, traversable, transverse, universal, universe, versed, versiera, vertex, vertical, weird, wrong.

[252] *werə-* (IE) "find": eureka, heuristic.

[253] *wi-* (IE) "apart, in half": icosagon, icosahedron, icosidodecahedron, icositetrahedron, rhombicosidodecahedron, vicenary, vigesimal, vigintillion, wide (?), width (?).

[254] *wrad-* (IE) "branch, root": radical, radicand, radix, root.

[255] *wrod-* (Unknown) "rose": rhodonea, rose, rosette.

[256] *wrodh-* (IE) "grow straight": orthant, orthic, orthocenter, orthogonal, orthogonality, orthonormal, orthopole, orthoptic, orthotope.

[257] *ye-* (IE) "throw": adjacent, bijective, conjecture, injective, project, projection, projective, surjective, trajectory.

[258] *yer-* (IE) "year": hour, year.

[259] *yeug-* (IE) "join": adjoin, adjoint, adjugate, conjugate, conjunction, disjoint, disjunction, equiconjugate, joint, nonconjunction, tranjugate.

[260] *yos-* (IE) "gird": zone, zonohedron.

References

Abbott, David, editor. 1986. *Mathematicians*. New York: Peter Bedrick Books.

Adams, Daniel. 1814. *The Scholar's Arithmetic*. Keene (New Hampshire): John Prentiss.

Anderson, Sabra S. 1970. *Graph Theory and Finite Combinatorics*. Chicago: Markham.

Bailey, Herb R., and Roger G. Lautzenheiser. 1993. "A Curious Sequence," in *Mathematics Magazine*, volume 66, number 1. Washington: Mathematical Association of America.

Baker, C.C.C.T. 1966. *Dictionary of Mathematics*. New York: Hart Publishing Company.

Ballou, Donald H., and Frederick H. Steen. 1943. *Plane and Spherical Trigonometry*. New York: Ginn and Company.

Barnhart, Robert K. 1988. *Barnhart Dictionary of Etymology*. Bronx, N.Y.: H.H. Wilson.

Beiler, Albert H. 1964. *Recreations in the Theory of Numbers*. New York: Dover Publications.

Berge, Claude. 1985. *Graphs*. Amsterdam: North-Holland.

Blocksma, Mary. 1989. *Reading the Numbers*. New York: Penguin Books.

Bôcher, Maxime. 1907. *Introduction to Higher Algebra*. New York: Macmillan.

Bridgwater, William, and Seymour Kurtz, editors. 1969. *The Columbia Encyclopedia*. New York: Columbia University Press.

Brink, Raymond W. 1937. *A First Year of College Mathematics*. New York: D. Appleton-Century.

Cajori, Florian. 1928. *A History of Mathematical Notations*. La Salle (Illinois): Open Court Publishing Company.

Campbell, Jeremy. 1982. *Grammatical Man*. New York: Simon and Schuster.

Carroll, Lewis. [1897] 1958. *Symbolic Logic*. New York: Dover.

Carson, R.A.G. 1978. *Principal Coins of the Romans*. London: British Museum Publications.

Chantraine, Pierre. 1968. *Dictionnaire étymologique de la langue grecque*. Paris: Klincksieck.

Ciardi, John. 1983. *A Second Browsers Dictionary*. New York: Harper and Row.

COMAP. 1988. *For All Practical Purposes*. New York. W.H. Freeman.

Corominas, Joan. 1983. *Breve diccionario etimológico de la lengua castellana*. Madrid: Gredos.

Coxeter, H.S.M. 1973. *Regular Polytopes*, 3rd edition. New York: Dover.

Daintith, John, et al., editors. 1987. *The Macmillan Book of Quotations*. New York: Macmillan.

Dauzat, Albert; Jean Dubois; and Henri Mitterand. 1971. *Nouveau dictionnaire étymologique et historique*. Paris: Larousse.

Davies, Charles. 1858. *New University Arithmetic*. New York: A.S. Barnes.

Davis, Philip J. 1961. *The Lore of Large Numbers*. New York: Random House.

Davis, Philip J., and Reuben Hersh. 1981. *The Mathematical Experience*. Boston: Birkhäuser.

de Michele, Vincenzo. 1973. *Crystals*. New York: Crescent Books.

Dickson, Leonard Eugene. 1939. *New First Course in the Theory of Equations*. New York: John Wiley & Sons.

Eiss, Harry Edwin. 1988. *Dictionary of Mathematical Games, Puzzles, and Amusements*. New York: Greenwood Press.

Ekeland, Ivar. 1988. *Mathematics and the Unexpected*. Chicago: University of Chicago.

Encyclopedia Britannica, 15th edition. 1986. Chicago.

Ernout, A., and A. Meillet. 1951. *Dictionnaire étymologique de la langue française*. Paris: Klincksieck.

Evans, Jacqueline P. 1970. *Mathematics: Creation and Study of Form*. Reading (Massachusetts): Addison-Wesley.

Eves, Howard. 1976. *An Introduction to the History of Mathematics*. New York: Holt, Rinehart and Winston.

———. 1981. *Great Moments in Mathematics after 1650*. Mathematical Association of America.

Flood, W.E. 1977. *Scientific Words*. Westport (Connecticut): Greenwood Press.

Francis, Richard L. 1990. "Superpowers," in *The AMATYC Review*, 12, no. 1 (Fall 1990).

———. 1993. "Star Numbers and Constellations," in *Mathematics Teacher*, 86, no. 1 (January 1993).

Freedman, David, et al. 1991. *Statistics (2nd edition)*. New York: W.W. Norton.

Gardner, Martin. 1988. *Hexaflexagons and Other Mathematical Diversions*. Chicago: University of Chicago.

———. 1975. *Mathematical Carnival*. New York: Alfred A. Knopf.

Gellert, W., et al., editors. 1977. *The VNR Concise Encyclopedia of Mathematics*. New York: Van Nostrand Reinhold Company.

Gibson, Carol, ed. 1981. *The Facts on File Dictionary of Mathematics*. New York: Facts on File.

Glenn, John, and Graham Littler, editors. 1984. *A Dictionary of Mathematics*. Cambridge (Massachusetts): Harper and Row.

Gove, Philip Babcock, editor. 1976. *Webster's Third New International Dictionary*. Springfield (Massachusetts): G. & C. Merriam Company.

Greenstein, Carol Horn. 1978. *Dictionary of Logical Terms and Symbols*. New York: Van Nostrand Reinhold Company.

Greimas, A.-J. 1980. *Dictionnaire de l'ancien français*. Paris: Larousse.

Grünbaum, Branko, and G.C. Shephard. 1987. *Tilings and Patterns*. New York: W.H. Freeman.

Guralnik, David B., and Joseph H. Friend, editors. 1966. *Webster's New World Dictionary of the American Language*. Cleveland: World Publishing Company.

Hall, Leon M. 1992. "Trochoids, Roses, and Thorns—Beyond the Spirograph," in *College Mathematics Journal*, volume 23, number 1 (January 1992).

Harary, Frank, and Edgar Palmer. 1973. *Graphical Enumeration*. New York: Academic Press.

Hazewinkel, Michiel, editor. 1988. *Encyclopedia of Mathematics*. Dordrecht: Kluwer Academic Publishers.

Hunter, J.A.H., and Joseph S. Madachy. 1975. *Mathematical Diversions*. New York: Dover Publications.

Ifrah, Georges (translated by Lowell Bair). 1985. *From One to Zero*. New York: Viking.

Ito, Kiyosi, editor. 1987. *Encyclopedic Dictionary of Mathematics*, 2nd. edition. Cambridge (Massachusetts): M.I.T. Press.

Jacob, Bill. 1990. *Linear Algebra*. New York: W.II. Freeman.

Jacobs, Harold. 1970. *Mathematics: a Human Endeavor*. San Francisco: W.H. Freeman.

James, Glenn, and Robert C. James. 1959. *Mathematics Dictionary*. Princeton: D. Van Nostrand Company.

James, Robert C. 1992. *Mathematics Dictionary*. New York: Van Nostrand Reinhold.

Karush, William. 1962. *The Crescent Dictionary of Mathematics*. New York: Macmillan.

———. 1989. *Webster's New World Dictionary of Mathematics*. New York: Webster's New World.

Kasner, Edward, and James R. Newman. 1940. *Mathematics and the Imagination*. New York: Simon and Schuster.

Korn, Granino A., and Theresa M. Korn. 1968. *Mathematical Handbook for Scientists and Engineers*, 2nd edition. New York: McGraw Hill.

Land, Frank. 1960. *The Language of Mathematics*. London: John Murray.

Lawrence, J. Dennis. 1972. *A Catalog of Special Plane Curves*. New York: Dover.

Lehmann, Charles H. 1942. *Analytic Geometry*. New York: John Wiley & Sons.

Lewis, Charlton T. 1879. *A Latin Dictionary*. Oxford: Clarendon Press.

Liddell, Henry George. 1940. *A Greek-English Lexicon*, 9th edition. Oxford: Clarendon Press.

Loeb, Arthur L. 1976. *Space Structures*. Reading (Massachusetts): Addison-Wesley.

Mackay, Alan L. 1991. *A Dictionary of Scientific Quotations*. Bristol (England): Adam Hilger.

Markowsky, George. 1992. "Misconceptions about the Golden Ratio," in *College Mathematics Journal*, 23, no. 1 (January 1992).

Marks, Robert W. 1964. *The New Mathematics Dictionary and Handbook*. New York: Bantam.

Martin, George E. 1982. *Transformation Geometry*. New York: Springer-Verlag.

McCoy, Neal H. 1968. *Introduction to Modern Algebra*. Boston: Allyn and Bacon.

McFarlan, Donald, editor. 1990. *Guinness Book of World Records 1990*. New York: Bantam Books.

Menninger, Karl (translated by Paul Broneer). 1969. *Number Words and Number Symbols*. Cambridge: M.I.T. Press.

Meserve, Bruce E., and Joseph A. Izzo. 1972. *Fundamentals of Geometry*. Reading (Massachusetts): Addison-Wesley.

Meserve, Bruce E., and Max A. Sobel. 1964. *Introduction to Mathematics*. Englewood Cliffs (New Jersey): Prentice-Hall.

Morris, William, editor. 1980. *The American Heritage Dictionary of the English Language* (New College Edition). Boston: Houghton Mifflin.

Muir, Jane. 1961. *Of Men and Numbers*. New York: Dell Publishing Company.

Mulcrone, T.F. 1957. "The Names of the Curve of Agnesi." *American Mathematical Monthly* 64, pp. 359–361.

National Council of Teachers of Mathematics. 1969. *Historical Topics in the Mathematics Classroom*. Washington: National Council of Teachers of Mathematics.

Olds, C.D. 1963. *Continued Fractions*. New York: Random House.

Onions, C.T., editor. 1982. *The Oxford Dictionary of English Etymology*. Oxford: Clarendon Press.

Owen, George E. 1971. *The Universe of the Mind*. Baltimore: Johns Hopkins Press.

Oxford English Dictionary (Compact Edition). 1971. New York: Oxford University Press.

Partridge, Eric. 1959. *A Short Etymological Dictionary of Modern English*. New York: Macmillan.

Paulos, John Allen. 1988. *Innumeracy*. New York: Hill and Wang.

———. 1991. *Beyond Numeracy*. New York: Alfred A. Knopf.

Pfeifer, Wolfgang, editor. 1989. *Etymologisches Wörterbuch des Deutschen*. Berlin: Akademie-Verlag.

Picoche, Jacqueline. 1984. *Dictionnaire étymologique du français*. Paris: Robert.

Pei, Mario. 1962. *The Families of Words*. New York: Harper and Brothers.

Richards, Stephen P. 1982. *A Number for Your Thoughts*. New Providence (N.J.): Stephen P. Richards.

———. 1987. *Numbers at Work and at Play*. New Providence (N.J.): Stephen P. Richards.

Rietz, H.L., and A.R. Crathorne. 1939. *College Algebra (4th edition)*. New York: Henry Holt and Company.

Rouse Ball, W.W. (revised by H.S.M. Coxeter). 1962. *Mathematical Recreations and Essays*. New York: Macmillan.

Saxena, S.C., and S.M. Shah. 1972. *Introduction to Real Variable Theory*. Scranton: Intext Educational Publishers.

Schaaf, William L., editor. 1963. *Our Mathematical Heritage*. New York: Collier Books.

Schuh, Fred. 1968. *The Master Book of Mathematical Recreations*. New York: Dover Books.

Shipley, Joseph T. 1984. *The Origins of English Words*. Baltimore: Johns Hopkins University Press.

Skrapek, Wayne A., et al. 1976. *Mathematical Dictionary for Economics and Business Administration*. Boston: Allyn and Bacon.

Smith, D.E. 1953. *History of Mathematics*. New York: Dover Books.

Smith, Karl J. 1984. *The Nature of Mathematics* (Fourth Edition). Monterey (California): Brooks/Cole.

Smith, William, and Theophilus D. Hall. 1889. *A Smaller Latin-English Dictionary*. London: John Murray.

Sondheimer, Ernst, and Alan Rogerson. 1981. *Numbers and Infinity*. Cambridge: Cambridge University Press.

Soukhanov, Anne H., executive editor. 1992. *The American Heritage Dictionary of the English Language (third edition)*. Boston: Houghton Mifflin.

Spiegel, Murray R. 1964. *Theory and Problems of Complex Variables*. New York: Schaum Publishing Company.

———. 1961. *Theory and Problems of Statistics*. New York: McGraw-Hill.

Stark, Harold M. 1973. *An Introduction to Number Theory*. Chicago: Markham Publishing Company.

Taylor, Barry N, editor. 1991. *The International System of Units* (Special Publication 330). Washington: National Institute of Standards and Technology, United States Department of Commerce.

Texas Memorial Museum. 1984. *Sign, Symbol & Script* [a poster]. Austin: Texas Memorial Museum.

Tower, David B. 1845. *Intellectual Algebra*. New York: Paine and Burgess.

Tucker, T.G. 1931 (1985). *Etymological Dictionary of Latin*. Chicago: Ares Publishing.

van Yzeren, Jan. 1992. "Pairs of Points: Antigonal, Isogonal, and Inverse," in *Mathematics Magazine*, 65, no. 5.

von Seggern, David. H. 1990. *CRC Handbook of Mathematical Curves and Surfaces*. Boca Raton (Florida): CRC Press.

Warner, Seth. 1971. *Classical Modern Algebra*. Englewood Cliffs (New Jersey): Prentice-Hall.

Watkins, Calvert. 1985. *The American Heritage Dictionary of Indo-European Roots*. Boston: Houghton Mifflin.

Weis, Erwin, and Erich Weis. 1965. *Taschenwörterbuch*. Stuttgart: Ernst Klett Verlag.

Wells, David. 1988. *The Penguin Dictionary of Curious and Interesting Numbers*. New York: Penguin Books.

Wentworth, George. 1888. *College Algebra*. Boston: Ginn & Company.

Wentworth, George, and David Eugene Smith. 1912. *Wentworth's Plane and Solid Geometry*. Boston: Ginn & Company.

West, Beverly Henderson, et al. 1982. *The Prentice-Hall Encyclopedia of Mathematics*. Englewood Cliffs (New Jersey): Prentice-Hall.

Williams, Elizabeth, and Hilary Shuard. 1970. *Elementary Mathematics Today*. Menlo Park (California): Addison-Wesley.

Yates, Robert C. 1974. *Curves and Their Properties*. Reston (Virginia): National Council of Teachers of Mathematics.

Yates, Samuel. 1982. *Repunits and Repetends*. Delray Beach (Florida): Samuel Yates.